International Scientific-Practical Conference: Innovations in Fuel and Energy Complex and Mechanical Engineering (FEC-2017)

IOP Conference Series: Materials Science and Engineering
Volume 253

Kemerovo, Russia
18 – 21 April 2017

ISBN: 978-1-5108-5031-6
ISSN: 1757-8981

Printed from e-media with permission by:

Curran Associates, Inc.
57 Morehouse Lane
Red Hook, NY 12571

Printed by Curran Associates, Inc. (2017)

For permission requests, please contact the Institute of Physics
at the address below.

Institute of Physics
Dirac House, Temple Back
Bristol BS1 6BE UK

Phone: 44 1 17 929 7481
Fax: 44 1 17 920 0979

techtracking@iop.org

Additional copies of this publication are available from:

Curran Associates, Inc.
57 Morehouse Lane
Red Hook, NY 12571 USA
Phone: 845-758-0400
Fax: 845-758-2633
Email: curran@proceedings.com
Web: www.proceedings.com

Table of contents

Volume 253

International Scientific-Practical Conference: Innovations in Fuel and Energy Complex and Mechanical Engineering (FEC-2017)

18–21 April 2017, Kemerovo, Russian Federation

Accepted papers received: 6 October 2017
Published online: 23 October 2017

Preface

Papers

IOP Conf. Series: Materials Science and Engineering **253** (2017) 011001 doi:10.1088/1757-899X/253/1/011001

PREFACE

The conference is devoted to the discussion of actual problems of fuel and energy complex and machine building, import substitution, operation, repair and restoration of parts and components of potentially dangerous equipment, development of welding production, technical diagnostics and industrial safety expertise. Much attention will be paid to the assessment of qualifications (No. 238-FZ "On the independent evaluation of qualifications") of personnel of enterprises of the fuel and energy complex.

Scientists and employees of universities and research institutes, technical managers, chief specialists, heads of testing laboratories of industrial enterprises, specialists of Rustekhnadzor departments, expert organizations, welding centers (certification systems for welding production), post-graduate students, doctoral students took part in the conference.

Aleksandr Smirnov
Doctor of Technical Sciences, Professor
T.F. Gorbachev Kuzbass State Technical University,
Russian Federation

IOP Conf. Series: Materials Science and Engineering **253** (2017) 011002 doi:10.1088/1757-899X/253/1/011002

Peer review statement

All papers published in this volume of *IOP Conference Series: Materials Science and Engineering* have been peer reviewed through processes administered by the proceedings Editors. Reviews were conducted by expert referees to the professional and scientific standards expected of a proceedings journal published by IOP Publishing.

Acoustic Guided Wave Testing of Pipes of Small Diameters

V V Muravev, O V Muraveva [a], V A Strizhak, Y V Myshkin

Department of quality control instruments and techniques, Kalashikov Izhevsk State Technical University, 426069, 7 Studencheskaya st., Izhevsk, Russian Federation

E-mail: [a] olgak166@mail.ru

Abstract: Acoustic path is analyzed and main parameters of guided wave testing are substantiated applied to pipes of small diameters. The method is implemented using longitudinal L(0,1) and torsional T(0,1) waves based on electromagnetic-acoustic (EMA) transducers. The method of multiple reflections (MMR) combines echo-through, amplitude-shadow and time-shadow methods. Due to the effect of coherent amplification of echo-pulses from defects the sensitivity to the defects of small sizes at the signal analysis on the far reflections is increased. An opportunity of detection of both local defects (dents, corrosion damages, rolling features, pitting, cracks) and defects extended along the pipe is shown.

Introduction

Power efficiency of any heat supply system depends on such factors as technical state of manufacturing equipment, keeping the optimal operation modes, and quick detection and elimination of the defects occurred. Nowadays more than 80% of heat energy equipment exhausted its life-time; in this connection, the problem of developing the system of control, diagnostics and monitoring of heat energy equipment becomes urgent [1].

Among the tested heat energy objects, pipes for steam boilers and pipelines are the most common. In general, pipes are made of steel blanks in accordance with the technical specification (in Russian TY 14-3P-55-2001 and TY 14-3P-190-2004). Application of pipes for units of high and super-critical steam parameters stipulates for high demands to the quality of the pipe material. As far as steam boilers and pipelines are concerned, seamless cold-deformed, warm-deformed and hot-deformed steel pipes, including hot-pressed and hot-pressed reduced pipes are applied.

By tradition, pipes for heat energy engineering are controlled at production by means of ultrasonic, acoustic-emission, magnetic and eddy-current methods of non-destructive testing (NDT); the pipe wall thickness is measured by means of ultrasonic-calibrators [2-4]. Due to a considerable length of the controlled pipes, the pointed methods have a common drawback – the necessity of scanning the pipe solid, thus requiring the corresponding mechanized equipment and large production sites. This reduces the manufacturing efficiency of the units, complicates the system of object transportation and leads to the quick wearing out of transducers. Despite of the advantage provided by contactless operation, units of magnetic and eddy current control allow for revealing only surface and near-surface defects having definite dimensions. Identification of defects becomes complicated due to the absence of the unambiguous relation between mechanical properties of the object and magnetic parameters to be measured. The essential drawbacks are also the sensitivity to variations of the clearance, magnetic properties of the object and the necessity of additional demagnetization of the object. Contact ultrasonic methods require the presence of the immersion liquid and they are hard to implement for rolled stocks having small dimensions and low quality of surface treatment (hot rolled stocks).

In recent decade the guided wave acoustic methods became popular at the NDT market [5-8], allowing for performance of the express diagnostics of pipelines at essential increase of manufacturing efficiency of control due to the absence of scanning and at local access to the object. Means of guided wave control proposed at the market are intended, as a rule, for controlling the long-distance pipelines at operation; the sensitivity of methods being rather low in this case.

The paper presents the guided wave method of controlling the pipes having the limited length, developed by the authors. It is based on the method of multiple reflections, it uses longitudinal and torsional waves; and it is implemented by means of electromagnetic acoustic (EMA) transducers.

Applied approaches

It is known that normal Pochhammer waves (longitudinal L and flexural F) and normal torsional waves T can propagate in linearly extended objects (pipes). As for the normal waves, the geometric dispersion of the velocity is specific, that is, the relation between the phase and group velocities of waves and the frequency of oscillations and transverse dimensions of the shell. Dispersion curves are the main characteristic of normal waves (Fig. 1). Application of normal waves within the essential dispersion of the speed and existence of several modes leads to distortion and attenuation of signals due to defects and to the complexity of interpretation of distorted signals appearing in the multi-mode wave guide. Distortions are caused by both variation of phase relations between signals transmitted by various types of waves with the distance increase, and by variation of distribution of oscillations along the cross-section. In this relation, when designing the apparatus for the guided wave control, the torsional wave T(0,1) of zero order without any dispersion is applied, as a rule, and also the axial symmetrical wave L(0,1) at the area of minimum dispersion of velocity. Under these circumstances, the axial symmetrical wave L(0,1) is propagating at the velocity of the rod wave

$$C_{S0} = \sqrt{E/\rho}\,, \tag{1}$$

and the torsional one T(0,1) is propagating at the velocity of the transverse wave

$$C_{T0} = \sqrt{G/\rho} \tag{2}$$

where E, G are the Young module and the shear module, correspondingly, ρ is the density of the object material.

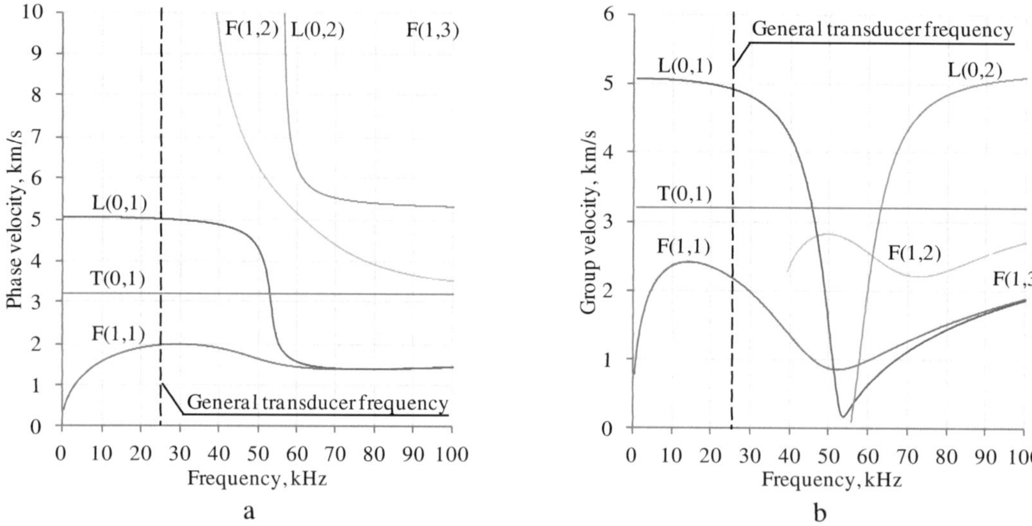

Fig. 1 Dispersion curves of phase (a) and group (b) velocities in the axial symmetrical $C_L(0,1)$ and torsional $C_T(0,1)$ waves, having the diameter 32 mm and wall thickness 4.2 mm.

IOP Conf. Series: Materials Science and Engineering **253** (2017) 012001 doi:10.1088/1757-899X/253/1/012001

There is only one angular component of displacements in torsional normal waves (Fig. 2a), it is the symmetrical motion with respect to the axis *z* and it represents the rotation of the transverse section of the pipe with respect to this axis. There are two components of displacements in longitudinal normal waves (Fig. 2b), they are the axial and radial ones; the motion takes place here symmetrically with respect to the axis of the *pipe* and the longitudinal component of displacements is prevailing.

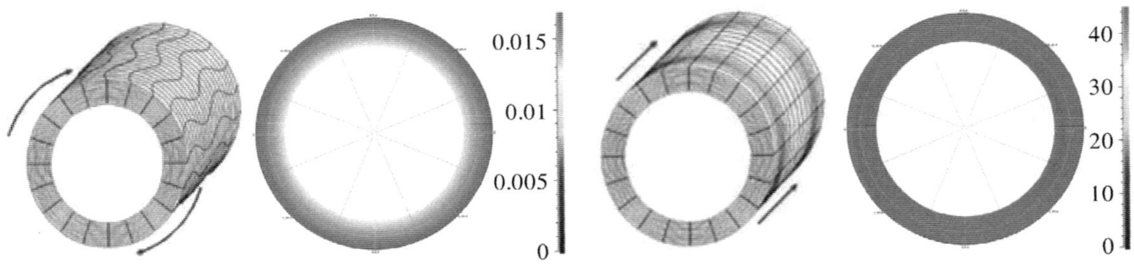

a b

Fig. 2 Normal waves in cylindrical shells and the corresponding diagrams of displacements in the torsional T(0,1) (a) and axially symmetrical L(0,1) (b) waves

Application of the fundamental mode of the torsional wave at pipe control is reasonable due to a higher sensitivity to defects and low attenuation [9]. The joint application of the symmetrical and torsional waves allows for providing the best identification ability for defect of different orientation (transverse and longitudinal cracks) [10-11].

In order to assess the possibilities of the guided method of control as applied to pipelines of different dimension types, the preliminary analysis of the acoustic path of the method has been performed to substantiate the parameters of control. In particular, basing on the generalized integral solutions of the problem for excitation and propagation of torsional waves by EMA transducers in free pipes, the efficiency of excitation of acoustic waves in them has been investigated with regard to geometrical dimensions of a pipe and parameters of excitation [9, 12].

The acoustic path of the method for multiple reflections at the pipe control is described by the formula [13]

$$U_{dn} = U_0 \cdot n \cdot D_d^{2(n-1)} \cdot R_d \cdot R_e^{n-1} \cdot e^{-2\delta L(n-1)} \cdot e^{-2\delta l}, \tag{3}$$

where U_{dn} is the amplitude of the echo-pulse due to the defect at the *n*-th reflection, R_e is the factor of reflection from the boundary "object surface – transducer", δ is the attenuation factor of the object, *L* is the length of the object, *l* is the distance to the defect, R_d and D_d are the factors of reflection and transparency of acoustic waves from the defect respectively.

In terms of multiple re-reflections from defects and face ends of the pipe, the effect of a coherent amplification of echo-signals from defects for the consequent reflections (proportionally to n) is observed, thus allowing to improve the sensitivity to low-sized defects.

In accordance with the laws of distribution of plane acoustic waves in wave guides, reflection and passing through the defective area of the wave guide is determined by its mechanical impedance Z = ρ*CS (s is the area of the wave guide cross-section)*. When the properties of the wave guide material do not vary (ρ*C* = *const*) and only its cross-section is changed, the factors of reflection R_d and transparency D_d through the defective area are determined by the following formulas [14]:

$$R_d = \frac{S_2^2 - S_1^2}{\left(S_1^2 + S_2^2\right)^2 + 2iS_1S_2\cot(\omega l_d / C_0)}, \tag{4}$$

$$D_d = \frac{2S_1 e^{i\omega l_d / C_0}}{S_1 + S_2}, \tag{5}$$

where S_1 up to S_2 is the transverse section of the defect-free area of the wave guide and the area with the defect, l_d is the length of the defect along the axis, ω is the cyclic frequency of the wave, C_0 is the velocity of the rod wave.

Formulas (3) – (5) describe the acoustic path at implementation of the method for multiple reflections of long-distance objects with application of rod and torsional waves at the absence of velocity dispersion; and they allow for analyzing the influence of all the pointed factors on the efficiency of control and the accuracy of defining its information parameters.

It should be noted that when applying the torsional wave, the reflection from the defect accompanied by variation of the transverse section of the rod from S_1 to S_2 takes place due to both the decrease in the transverse section of the rod and reduction of the velocity of the wave propagation.

As for the defects which are the concentrators of mechanical stresses (cracks with low disclosure) and which do not change significantly the mechanical impedance of the pipe, we deal with the mechanism of radiation for acoustic emission waves from areas where the stresses exceed the average value along the section of the object of control when passing through the sounding pulse (tension-compression or torsion) [12]. Note, that these defects can be developed at operation of the heat energy equipment.

Equipment and results of control

The developed acoustic guided wave flaw detector is intended to control the pipes having the limited length 3-12 meters and small diameters ∅ 5 – 90 mm. The generalized structural scheme of the flaw detector is shown in Fig. 3. The principle of operation of the flaw detector implies the methods for multiple reflections combining the echo-through, amplitude-shadow and time-shadow methods. The method for multiple reflections is based on registration of echo pulses that are multiply (5-10 times) reflected from external and internal defects of the pipe wall and from the opposite face ends of the pipe. In terms of multiple re-reflections from defects and from the face ends of the pipe, the effect of coherent amplification of echo signals from defects at the consequent reflections is observed, thus allowing to improve the sensitivity to small-sized defects when analyzing the signal at long-range reflections.

The generator of the unit of the generator and preamplifier block (GPA) produces the electrical pulse supplied to the radiator of the unit of electro-acoustic transducers (EAT) thus leading to the appearance of the acoustic pulse propagated in the pipe at the velocity of the torsional or rod wave depending on orientation of the EAT with regard to the generating line of the pipe. The acoustic pulse, multiply reflected from defects of the pipe like continuity violation and from its face end, it received at the same face end by the EAT unit and it is transmitted to the preamplifier of the GPA unit as an electric signal. The amplifier performs further amplification of signals with the controlled coefficient; its amplification coefficient is chosen automatically, so that the measured signal could not face the limitation. The electric signal from the amplifier is transmitted to the input of the analog digital converter and then to the memory of the personal computer. Operation of the flaw detector is performed by means of a specialized software program.

Fig. 4 shows the registered oscillogram at multiple sonic test of the flawless pipe and the pipe with the local defect. The presence of defects in the pipe leads to appearance of echo pulses from defects at the area between the probing pulse and echo pulse from the face end, and also at consequent reflections as echo pulses that are symmetrically arranged between two neighboring reflections. In order to eliminate the influence of the quality of an acoustic contact and efficiency of excitation receipt of waves for various grades of steels on the results of control, the amplitude of the echo pulse from the defect is regulated in accordance with the amplitude of the first echo pulse from the face end of the object. The regulated amplitude of the echo pulse from the defect is the basis for the natural assumption of the validity or defectiveness of the pipe according to criteria of sorting out which can be determined with regard to the limiting deviation of the wall thickness from the nominal one. Criteria of sorting out of pipes with regard to the velocity and attenuation are chosen basing on the accumulated statistical information on types and degree of the hazard for the defects revealed at inspection.

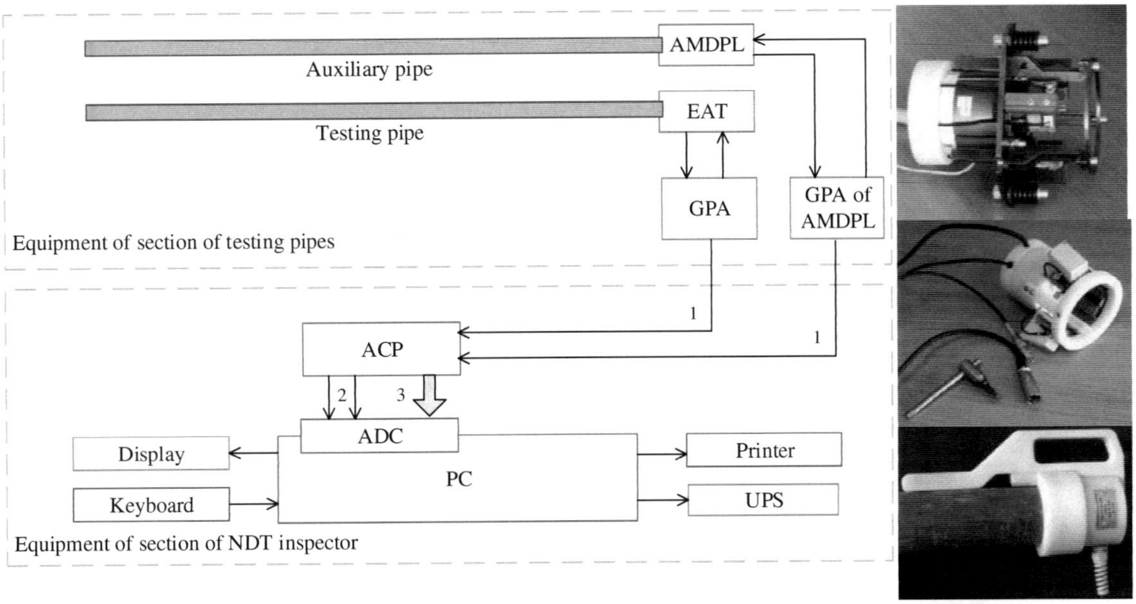

Fig. 3 The generalized flow diagram of the flaw detector and the type of EAT: AMDPL – acoustic measuring device of pipe length, EAT – electro acoustic transducer, GPA – generator and preamplifier block, ACP – programming amplifier, commutation and power supply, ADC – analog digital converter, PC – personal computer, 1-3 – data cables

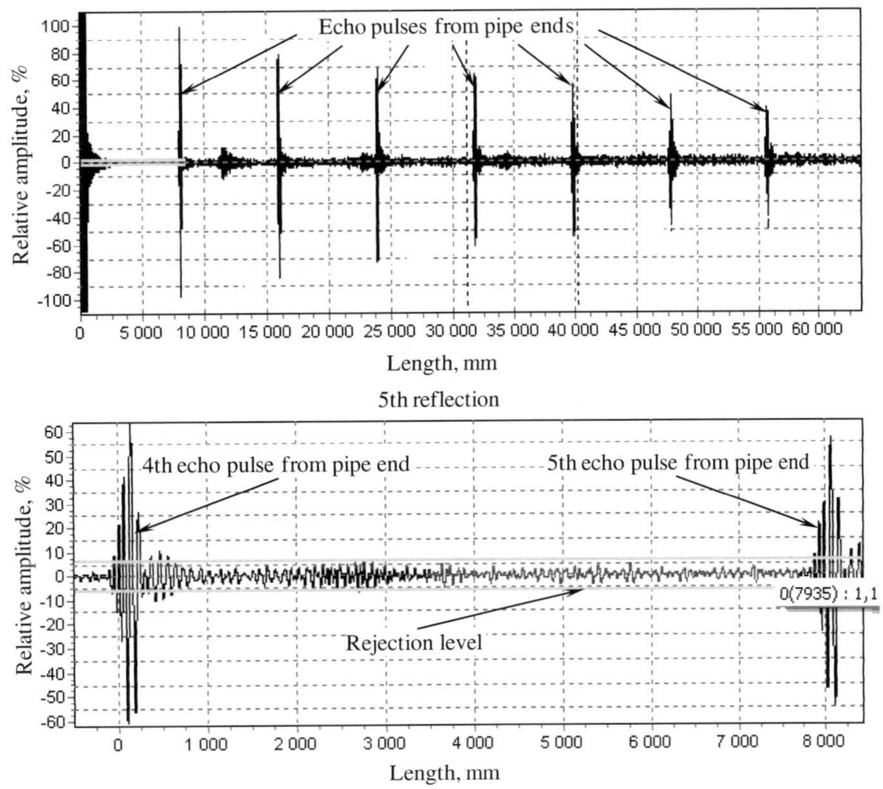

a

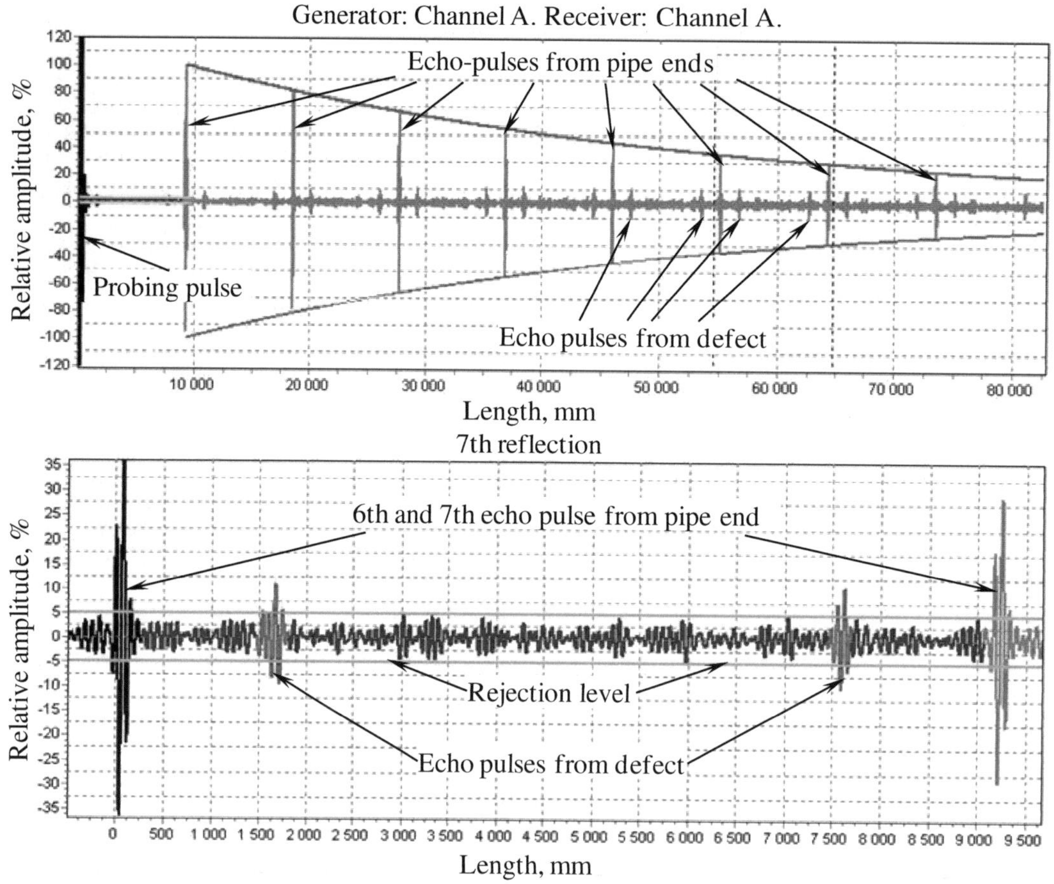

b

Fig. 4 Flaw patterns of the valid pipe (a) and the pipe with the local defect (b)

The basic measured characteristics of defects are:

- the amplitude of the echo pulse from the defect expressed in percent of the echo pulse from the opposite face end of the pipe (the factor of reflection for the first reflection R_1, %);
- the amplitude of the echo pulse from the defect at the N-th reflection expressed in percent of the first echo pulse from the opposite face end (the factor of reflection for the N-th reflection R_N, %);
- coordinates of the revealed defects along the pipe length (the distance from the transducer to the defect).
- the attenuation factor for echo pulses reflected from face ends of the pipe. This factor can be significantly increased at the presence of long-distance defects like corrosion on the internal or external surface of the pipe;
- the velocity of propagation of the torsional wave which can be significantly decreased at the presence of long-distance defects like corrosion, wall thickness variation and cracks.

For precise definition of the velocity and attenuation of waves, the flaw detector is equipped with the acoustic measuring device of the pipe length (AMDPL), its principle of operation being based on measuring the time of acoustic wave passing through the air space inside the pipe. The method provides the accuracy of the pipe length measuring within ±10 mm; and it can be applied, among other, as the measuring device for pipe lengths.

The echo pulse method allows for detecting the local defects: dimples, rolling laps, shells, corrosion damages (Fig. 5). The method for multiple reflections allows for revealing both localized and long-distance defects along the pipe length; it improves the sensitivity to small-sized defects and reduces the uncontrollable dead zone from the side of the acoustic signal input which is inherent to the echo pulse method. This reduction is due to the way of controlling from one of the face ends of the pipe.

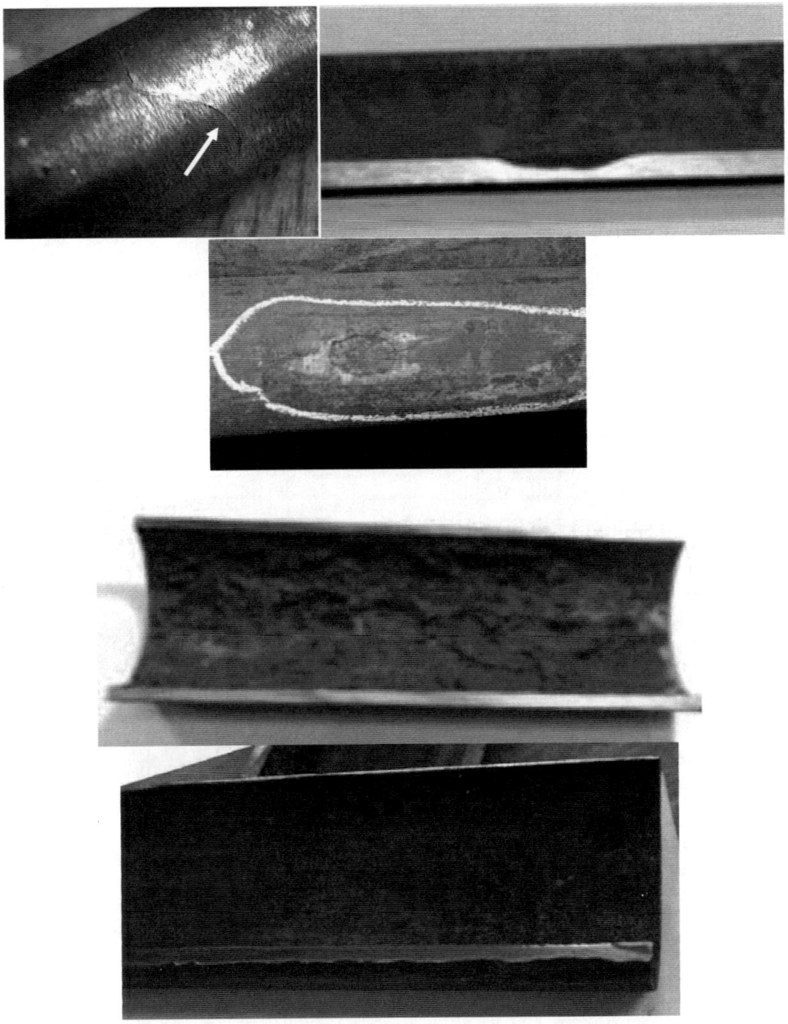

Fig. 5 The revealed defects: a crack, internal and external local corrosion, internal and external multiple corrosion

Conclusions

The developed technique of the guided wave acoustic control of pipes possesses the following advantages: it does not require scanning and application of contact and immersion liquids and any kind of preparation of the surface of the controlled object; it possesses high efficiency; it has a rather high sensitivity to defects independently on the depth of their location and the distance to the transducer; it reveals the most risky defects which influence on the pipe life time.

There is a possibility to apply the developed techniques when controlling the pipes at their operation; the sensitivity being reduced a little in this case due to impossibility of obtaining the series of multiple reflections.

This work was supported by the Russian Science Foundation, project no. 15-19-00051.

IOP Conf. Series: Materials Science and Engineering **253** (2017) 012001 doi:10.1088/1757-899X/253/1/012001

References

[1]. A. N. Smirnov, et al., Criteria for the evaluation of the technical state of the long-lived metal of HPP equipment based on acoustic structuroscopy, Russian Journal of Nondestructive Testing 51.2 (2015) 94-100.
[2]. O. A. Bulychev, et al., Revealing surface microcracks in metal articles by excitation of high-frequency eddy currents with subsequent infrared-camera imaging, Russian Journal of Nondestructive Testing 52.8 (2016) 457-462.
[3]. A. F. Matvienko, et al., The quality control of underground gas pipelines via the electromagnetic and acoustic method, Russian Journal of Nondestructive Testing 51.9 (2015) 546-553.
[4]. D. A. Boreiko, I. Yu. Bykov, A. L. Smirnov, The sensitivity of the acoustic-emission method during the detection of flaws in pipes, Russian Journal of Nondestructive Testing 51.8 (2015) 476-485.
[5]. Riichi Murayama, A New Guide Wave Inspection System Using Three Polarized Transverse Wave EMATs without Any Couplant, Journal of Sensor Technology 6.04 (2016) 110.
[6]. P. S. Lowe, et al., Inspection of Pipelines Using the First Longitudinal Guided Wave Mode, Physics Procedia 70 (2015) 338-342.
[7]. O. V. Muravieva, et al., Acoustic guided wave testing of downhole pumping equipment elements (Russian), Oil Industry Journal 2016.09 (2016) 110-115.
[8]. Kehai Liu, et al., Guided waves based diagnostic imaging of circumferential cracks in small-diameter pipe, Ultrasonics 65 (2016) 34-42.
[9]. O. V. Murav'eva, et al., Factors that affect the excitation effectiveness of torsional waves during waveguide inspection of pipes, Russian Journal of Nondestructive Testing 52.2 (2016) 78-84.
[10]. A. A. Nasedkina, A. Alexiev, J. Malachowski, Numerical Simulation of Ultrasonic Torsional Guided Wave Propagation for Pipes with Defects, Advanced Materials, Springer International Publishing, 2016. pp. 475-488.
[11]. D. N. Alleyne, T. Vogt, P. Cawley, The choice of torsional or longitudinal excitation in guided wave pipe inspection, Insight-Non-Destructive Testing and Condition Monitoring 51.7 (2009) 373-377.
[12]. O. V. Murav'eva, S. V. Len'kov, S. A. Murashov, Torsional waves excited by electromagnetic–acoustic transducers during guided-wave acoustic inspection of pipelines, Acoustical Physics 62.1 (2016) 117-124.
[13]. O. V. Murav'eva, D. V. Zlobin, The acoustic path in the method of multiple reflections during nondestructive testing of linearly extended objects, Russian Journal of Nondestructive Testing 49.2 (2013) 93-99.
[14]. G. A. Budenkov, O. V. Nedzvetskaya, Principal regularities of Pochhammer-wave interaction with defects, Russian Journal of Nondestructive Testing 40.2 (2004) 99-108.

Modern Methods of Rail Welding

Nikolay A Kozyrev [1, a], Olga A Kozyreva [2], Aleksander A Usoltsev [1], Roman E Kryukov [1], Roman A Shevchenko [1]

[1] Siberian Industrial State University
654007, Russia, Kemerovo region, Novokuznetsk, Kirova street, 42
[2] Kemerovo State University, 6, Krasnay street, Kemerovo, 650000

E-mail: [a] Kozyrev_na@mtsp.sibsiu.ru

Abstract. Existing methods of rail welding, which are enable to get continuous welded rail track, are observed in this article. Analysis of existing welding methods allows considering an issue of continuous rail track in detail. Metallurgical and welding technologies of rail welding and also process technologies reducing aftereffects of temperature exposure are important factors determining the quality and reliability of the continuous rail track. Analysis of the existing methods of rail welding enable to find the research line for solving this problem.

Introduction

Railroad communication in Russian Federation and its condition have a high effect on the economical development in the country. Length of railroad and intensity of freight traffic put in new claims on truck superstructure, especially on its main element – rails. Nowadays on Russian railways as well as abroad jointed track construction is rejected. One of the main disadvantages of the jointed track is a rail joint. Development of the technologies enabling to ensure the possibility of continuous rail track realization is a technological direction of a current interest [1-6]. It should be taken into account that maintenance of rail tracks in Russia takes place under difficult climatic and operational conditions (in Russia, unlike in European countries, the same tracks for passenger and industrial traffic are used).

Advantages of continuous rail track are the following [7-8]:

- 30-40 % decrease of track maintenance costs, higher safety of traffic, reliability of construction;
- 8-10 % reduced specific train resistance and, therefore, saving of fuel and electrical energy for the haulage, which is relevant in the context of continuous growth of energy sources' prices;
- increase of service life of the truck superstructure due to lower rails' damageability than in jointed track (cracks in bolt holes' edges, indent of a railhead, crushing and saddles). So, defect failures (contact and in joints) of continuous string occur 1.8-2.0 times more rarely than for jointed rail track, and without including equalizing span – 3-4 times more;
- reduced volume of work on track surfacing (up to 25-30 %), connected to low spots in joints, especially liquidation of pumping, which becomes a big problem with increase of axle loading;
- reduced intensity of external rail's side wear in curves and, therefore, rail damage is reduced in 1.5-1.6 times;
- 1.5-2.0 times reduced demand for ballast stone purification on coal-mineral routes;
- reduced take of metal for joint assemblies (up to 4.5 t per km);
- reduce of costs for trucks' repair;
- increase of travelling comfort for passengers;
- increase of working reliability of electrical track circuits of automatic block.

One of the important advantages of continuous track construction is also that there is a possibility to use reinforced-concrete slab, which increases safety against buckling, longitudinal and lateral motion resistance and ensure equal rigidity of track lengthwise. Also use of reinforced-concrete sleepers reduces the take of industrial wood.

The most damageable place of the rail is a welding joint. Withdrawing from service due to welding defects may reach 30% of total withdrawing, however total length of joints is less than 2% of rail length. In order to fit purpose and for consistent performance welded joints have to have the same properties as the rails itself. They have to be firm (moment of inertia and resisting moment have to be big enough for bending and torsional stresses not to exceed allowed values); long-lasting (to have high hardness, wear resistance and toughness, including at low temperatures); to have high contact-fatigue endurance. Also, vertical and horizontal straightness have to be ensured, major defects such as reed marks, microcracks, nonmetallic inclusions have to be eliminated.

One of the key technologies for construction, repair and maintenance of continuous rail track is rail welding, for which the following methods are developed and patented:

- Pressure welding: electro-contact, gas-pressure, induction, laser, friction, etc.;
- aluminothermic welding;
- electric arc welding: covered-electrode, submerged, gas-shielded, electro-slag, flux-cored, etc.

Method of pressure welding is based on heating the rails' ends up to plastic state temperature (more than 1000 °C) and squeezing them with a certain force (depending on cross-section area and physical and mechanical properties of metal of the rails)[9]. Rails' end can be heated up with electric current - electro-contact method, with gas welding torch – gas pressure method, with high-frequency current (inductor), with lasers, plasma, friction heat, etc. In pressure welding filler metals are not used, i.e. two rails' ends are welded to each other directly.

Aluminothermic welding method is based on desoxydation of metals and alloys in the exothermic reaction with aluminum. Great amount of heat is generated while the reaction proceeds. Liquid metal is poured between welded rails[10].

Electric arc welding method is based on the fusing of electrode metal (bar or wire) with an electric arc and filling the gap between rails with it.

Aluminothermic and electric arc welding methods differ from welding in pressure in the fact that weld of width 15-25 mm and more consists of filler metal which has cast structure.

In all welding methods there is heat-affected zone (HAZ) – area of changed structure of base rails' metal bordering on the weld. It leads to residual stresses and, as following, to reduced strength properties of the weld. The width of HAZ depends on the high temperatures exposure time on the base metal, on the mass of filler metal, welding method and parameters.

For the welding joints made by pressure welding it is possible to use thermal treatment to increase their mechanical properties (including differential treatment with rail-end hardening in HAZ)[11].

In the case of aluminothermic and electric arc welding there is a filler metal in the weld, which highly differs from rails' metal in constitution and structure, so thermal treatment of these weld joints doesn't increase their mechanical properties very much.

From all the methods of welding with pressure mentioned above the most commonly used one is electro-contact method (EC) which is applied in more than 95% of cases. On Russian railways only EC is applied and about 600 thousands of weld joints are made annually by rail-welding train and about 50 thousands joints by moving rail-welding machines [12]. Let's consider advantages and disadvantages of existing EC technologies.

Advantages of EC:

- high quality of weld;
- control system in rail-welding machine enable to control deviation of welding variables;
- high level of mechanization and automation of operations (under stationary conditions);
- high processing rate.

Disadvantages of EC:

- impossibility of joint welding in switch zones;

- expensive equipment;
- need for long occupations during welding works.

Nowadays the most widely applied method of electro-contact welding is pulse flash welding. This method of EC welding is more economical and processible comparing to continuous flash. Under EC welding as well as under other welding methods continuous heating and cooling of metal occurs in HAZ. Depending on steel constitution welding procedure is chosen: either continuous or pulse flash welding. The flash type determines linear quantity and temperature profiles in weld's HAZ [13, 14]. The choice is based on the elimination of hardening structures formation (martensite and bainite), which cause additional stresses and cracks leading to destruction of the rails [8]. As a result [15], the development of such procedures for rails of high-speed networks made of chord steel is coming into sharp focus.

However, when applying continuous rail track structure some questions are still unsettled, in particular questions concerning the increase of strength of weld and HAZ, because the amount of defects in this area is about 13-15% of total amount of defects on a rail bar.

This problem is solved during the fabrication of continuous rail track by compulsory heat treatment of welding joint. Heat treatment is done with induction-heating installation which increases costs. This problem in practice is advised to be solved by combining continuous and pulse flash welding methods, changing the heating rate during welding process and regulating the cooling speed [15]. At that, while partly using continuous flash method, there is a possibility of welding defects.

The paper [16] introduces after the welding during the cooling to carry out quazi-isothermal holding in the interval of highly disperse structure formation temperatures by flashing alternate current through the welding joint and maintaining the temperature till the end of transformation. Using the quazi-isothermal holding at a temperature 600 – 650 ^{0}C allows getting highly disperse structure of the weld without any additional heat treatment.

Because of the impossibility to weld rails in switch zones by welding machines the aluminothermic rails' welding (ATRW) have come into operation on Russian railways since 1995 year. Questions concerning application of ATRW not only in switch zones, but also on the runs are being discussed currently.

Advantages of ATRW:

- mobility;
- short occupations;
- welding process is carried out without electrical energy.

Disadvantages of ATRW:

- it is not possible to supervise and directly control the welding process;
- wide and uncontrolled heat-affected zone;
- quality of welding thermite mixture affects the quality of weld;
- weld quality highly depends on the welder's experience.

The technology in general is described in works [17-21]. The thermite used for rails' welding is made from iron dross – steel-rolling industry wastes and metallic aluminum. It is crushed to get grains of 0.1-2.5 mm size. Primary aluminum has to contain no less than 98–99 % of pure aluminum. Before being crushed the iron dross has to be well burnt in order to remove moisture and oil from it. The dust from crushed dross and aluminum is removed by separator. Chemical constitute of the dross differs. The less silicon dross contains, the better it fits for producing of welding thermite. Oxygen content in the dross shouldn't be less than 25%. Iron and oxygen can form three oxides: FeO, Fe_2O_3 and Fe_3O_4. Under the normal conditions components of thermic mixture do not interreact, but if the mixture is heated up to 1100– 1200 °C the reaction starts between its components. As the result of this reaction metallic aluminum oxidizes and turns into aluminum oxide and iron dross reduces to iron. The reaction is spontaneous and uncontrollable and it proceeds with the release of big amount of heat. Thermite mixture contains by weight 23.7 % of aluminum and 76.3% of dross. 1 kg of thermite mixture releases 3188.22 kJ of heat during its combustion, which produces the temperature of heated metal about (2700–3000) °C.

IOP Conf. Series: Materials Science and Engineering **253** (2017) 012002 doi:10.1088/1757-899X/253/1/012002

The dross and aluminum might contain different impurities. That's why the percent of aluminum and dross content in thermite mixture is calculated taking into account the purity of aluminum and content of oxygen in the dross. Practice of thermite welding has shown that the dross content in thermite mixture has to be 7-8% higher than calculated content value. In that case thermite metal is denser and is welding with rail's metal much better.

For the more effective use of the heat released in reaction and for increase of metal fall, little pieces of steel, which are nail industry waste, are added to the thermite mixture. The steel, when it is melted, increases metal fall and decreases the initial temperature of thermite reaction products. This waste (called cuttings) is added to thermite mixture at the rate of 12 to 20% of the weight of assumed thermite metal fall depending on the weight of thermite portion. Ferroalloys, mostly ferromanganese, ferrosilicon, ferromolybdenum, ferrotitanium, ferrovanadium, are added to the thermite mixture to improve mechanical properties of the metal. Ferroalloys enable to get the thermite metal with mechanical properties close to mechanical properties of the metal of welded rails. Therefore, chemical constitute of the metal occurred from thermite mixture is very far from the constitute of welded rails' steel with all that it implies. Besides, the thermite itself is a source of oxides, both – exogenous (unreacted iron oxide) and endogenous (generated in oxidation-reduction reactions). Also it should be pointed that slug inclusions and gases generated in reactions are not always have time to float from the reaction zone. As the result, welding zone is contaminated with non-metallic oxide inclusions and micropores which are the centers of crack initiation.

At the present time thermite fusion welding method might be applied for rails welding. In this method from all the thermite reaction products only liquid metal with no slag is used which increases the quality. However, ATRW produces joints with casting processes during which there is a possibility of formation of non-metallic inclusions, pores, caverns, bleeds, slag inclusions, internals cracks and micropores.

Huge problems might occur when using ATRW in winter periods. At low temperatures the cooling speed increases, for which cause crystallization of weld pool molten metal also increases. As a result, evolved gases and slag particles don't have enough time to float and fill the metal with pores and slag inclusions. Increased level of heat removal from heated metal and increase of gas content in it enable crack formation in the weld and weld-affected zone. If storage regulations are not observed there might be some moisture in materials leading to saturation of weld with hydrogen.

The paper [22] show the results of operational tests for weld joints welded with the use of different technologies. It is shown that the main operation influencing the generating of needed structure and ensuring required quality of weld is a preheat cycle. Experiments proved that if the preheat is not enough or if the temperature of environment is low then temperature of rails' ends decreases rapidly leading to increase of cooling speed of metal in weld and weld-affected zone. It gives rise to hardening structures, embrittlement and formation of microcracks. So, when rails are welded at the ambient temperature 15 ° C cooling speeds on peripheral area of rail base are almost twice higher than in the top of rail.

The paper [23] introduces experiments in ATRW at the temperatures below zero. Ambient temperature adversely affects the quality of welds. Experimental results obtained at different welding temperatures don't meet the regulatory requirements.

It should be pointed that as the experience of operating of ATRW weld joints on the railway network and tests on All Russian Scientific Research Institute of Railway Transport Test Loop show, the quality of these joints is lower than EC joint's quality due to the welding method itself. That's why main companies, performing ATRW on the railway network ("Snaga", "Welding Filler Company", "Aluminohermic welding", "Railtech", "Welding technologies") claim this method to be fitted only for rail welding within switches, on the bridges, elevated structures, in the tunnels when it is difficult to organize occupations for using welding machines.

The paper [24] introduces calculation showing that electro-contact welding is 41.8 % more expensive than aluminothermic welding. However, if we consider warranty period of joints welded by EC method and ATRW depending on the freight traffic (for EC accordingly to RR Company Standard 1.08.002 – 2009 for rails' type R75 and R65 – 150 million tons gross; for rails' type R50 – 120 million tons gross; for ATRW accordingly to TR 0921 – 127 –01124323 – 2005 for rails' type R75 and R65 – 120 150 million tons gross; for rails' type R50 – 100 million tons gross) the cost of joint for each million tons gross is changing. For EC: for rails' type R75 and R65 – 81.33 rub., for rails' type R50–101.67 rub. For ATRW: for rails' type R75 and R65– 59.17 rub., for rails' type R50– 71 rub. Therefore, present value of EC welding is 27.25 % higher than ATRW for R75, R65 rails' welding and 30.17% higher for R50 rails' welding. It also should be pointed that analysis of data of 897 defected ATRW welded joints withdraw during 2009 and 9 months of 2010 year done in the work [25] has shown that 89% of ATRW joints (710 pcs) haven't worked the whole warranty period. It also reduces the difference between EC and ATRW costs. ATRW cost increases with the continued operation of the joints. Inspection period of ATRW joints (no longer than 6 months after welding, then at least once a year) most of which are operated with safety strips is longer than inspection period of joints welded by EC method and operated without safety strips (at least once a year during first two years after welding, than at least once in two years) [25].

Conclusions

Consequently, ATRW welding haven't proved itself to be reliable and quality method for connecting rails, economical point is connected with quality characteristics and traffic safety.

We should point that both welding methods are developing continuously, for instance, German company «Electro-Thermite GmbH & Co. KG» claims about succeeded increase of ATRW joint quality. Despite the good advertising, before this rail welding technology will become widely applied, there must be conducted large-scale researches, laboratory, field and operating tests on track sections of different geometry, traffic flow and climate conditions.

References

[1] Meade, B., Railroad welding demands specialized processes // Welding journal, Volume 76 - Issue 9, 1997, P. 47-52.

[2] Kargin, V.A., Tikhomirova, L.B., Abramov, A.D., Galai, M.S., Effect of vibroimpact treatment on the physical and mechanical properties of the surface layer of welded joints in rails // Welding International. Volume 28 - Issue 3, 2014, Pages 245-247.

[3] Yamamoto, R., Komizu, Y., Fukada Y., Experimental examination for understanding of transition behaviour of oxide inclusions on gas pressure weld interface: joining phenomena of gas pressure welding // Welding International. Volume 28 - Issue 7, 2014, Pages 510-520.

[4] Karimine, K., Uchino, K., Okamura, M., Susceptibility to and occurrence of HAZ liquation cracking in rail steels: Study of rail welding with high-C welding materials (4th Report) // Welding International. Volume 11 - Issue 6, 1997, Pages 452-461.

[5] Kuchuk-Yatsenko, S.I., Shvets, Yu.V., Didkovskii, A.V., Chvertko, P.N., Shvets, V.I., Mikitin, Ya. I., Technology and equipment for resistance flash welding of railway crossings with rail ends through an austenitic insert // Welding International. Volume 22 - Issue 5, 2008, Pages 338-341.

[6] Irving, B., Long Island Rail Road orders an all-welded fleet // Welding journal, Volume 22 - Issue 9, 1997, Pages 33-37.

[7] Continuous rail track - progressive construction of railway track [Text] / L.B. Klimenko // Annex to the magazine "World of Transport" MKJT MPS RF, 2004. № 1. 88-93.

[8] Kozyrev N.A. Railway rails of electric steel [Text] / N.A. Kozyrev, V.V. Pavlov, L.A. Godik, V.P. Dementiev. - Novokuznetsk: Evraz, Novokuznetsk Metallurgical Combine, 2006. 388 p.

[9] Kuchuk-Yatsenko, S.I., Shvets, V.I., Didkovskii, A.V., Taranova, T.G., Investigation of the weldability by resistance welding of rails strengthened by heat treatment // Welding International. Volume 24 Issue 6, 2010, Pages 455-461.

[10] Wegrzyn, J., Maxur, M., A thermit mixture for rail track welding // Welding International. Volume 6 - Issue 1, 1992, Pages 5-8.

[11] Genkin, Z., Welding and heat treatment of joints in railway rails in induction equipment // Welding International. Volume 19. Issue 2, 2005, Pages 160-164.

[12] Kalashnikov E.A. Techniques of welding rails: trends in Russia and abroad [Text] / E.A. Kalashnikov, Yu.A. Korolev // Path and track economy. 2015, No. 8. P. 2-6.
[13] Mitsuru F. Rail flash-butt welding technology/ F. Mitsuru, N. Hiroaki, N. Kiyoshi// JFE Technical Report. 2015. 20. P. 159 – 163.
[14] Saita K. Trends in rail welding technologies and our future approach/ K. Saita, K. Karimine, M. Ueda, K. Iwano, T. Yamamoto, K. Hiroguchi // Nippon steel & sumitomo metal technical report. – 2013. – № 105. – P. 84 – 92.
[15] Shur EA Complex method for resistance welding of rails: [Text] / EA Shur, VA Rezanov // Herald VNIIZhT. 2012. 3. pp 20 - 22.
[16] Shevchenko R.A. Improving the technology of welding of rail steel with high chromium content [Text]/ R.A. Shevchenko // Science and youth: problems, searches, solutions: Proceedings of the Scientific Conference of students and young scientists. Issue 20 - P.III - Novokuznetsk: SibSIU, 2016. P. 259-261.
[17] Voronin N.N. Technology of aluminothermic welding of rails: [Text]: studies. benefit / N.N. Voronin V.V. Zasipkin, V.I. Konenkov, E.V. Sparrows, O.N. Trynkova; ed. N.N. Voronin. M.: MIIT, 2008. 117 p.
[18] Voronin N.N. Aluminothermic welding of rails: [Text]: studies. benefit: / N.N. Voronin, V.V. Zasipkin, V.I. Konenkov, E.V. Sparrows, O.N. Trynkova; ed. N.N. Voronin. M.: "Training Center for Education in railway transport", 2013. 195 p.
[19] Sergejevs, D., Mikhaylovs, S., Analysis of factors affecting fractures of rails welded by alumino-thermic welding // Transport problems. Part 3, 2008, Pages 33-37.
[20] Karguin, V.A., Tikhomirova, L.B., Galay, M.S., Kuznetsova, Ye.S., Improving service properties of welded joints produced by aluminothermic welding // Welding International. Volume 29 Issue 2, 2015, Pages 155-157.
[21] Lee, F.T., Managing thermite weld quality for railroads // Welding journal, Volume 85. Issue 1, 2006, Pages 24.
[22] Voronina O.N. The development of construction of rails, their joints and processing technology [Text] dis. Phd. tehn. Sciences: 07.00.10 / O.N. Voronina. Moscow, 2014. 228 p.
[23] Voronin N.N. Aluminothermic welding of rails in winter. [Text] / N.N. Voronin, O.N. Trynkova, O.V. Fomichev // The world of transport. 2012. 4. p. 56-59.
[24] Velichko D.V. Economic evaluation of the contact and the aluminothermic welding of rails [Text] / D.V. Velichko. // Actual problems of modern science: a collection of articles of the International scientific-practical conference: In 4 parts, 2013. pp 93-96.
[25] Rukavchuk Y.P. Defectiveness joints aluminothermic welding of rails [Text] / Y.P. Rukavchuk, S.A. Christmas, I.Z. Etingen // "Path and track facilities» 2011. 4. p.26-27.

IOP Conf. Series: Materials Science and Engineering **253** (2017) 012003 doi:10.1088/1757-899X/253/1/012003

Fatigue Strength of Blade with Disk Joins, Obtained by Linear Friction Welding

Alexander Y Medvedev [1, a], Vladimir M Bychkov [1], Valeriy V Atroshenko [1], Naila F Ismailova [2], Andrey V Supov [2]

[1] Ufa State Aviation Technical University (USATU), K. Marx Street 12, Ufa, The Republic of Bashkortostan, Russian Federation, 450008
[2] Ufa Engine Industrial Association (UMPO), Ferin Street 2, Ufa, The Republic of Bashkortostan, Russian Federation, 450039

E-mail: [a] medvedevalexandr@inbox.ru

Abstract. Fatigue tests of blisk welded sections is described. The bimetal blisk contains VT6 alloy (Ti6Al4V) blades and VT8-1 alloy (Ti6.5Al3.5Mo1.2Sn1.2Zr0.2Si) disc, joined by linear friction welding. The test results indicate that the combination of materials VT6 + VT8-1 can be used for bladed disks manufacturing, and technological factors, including availability of weld does not have a decisive influence on the fatigue strength of the investigated structure.

Introduction

Modern gas turbine engines have a wide application of blisks instead of constructions with mechanical fastening blades to discs. Now,Russian industry has mastered technology of manufacturing bladed disks from solid billets. However, with increasing height blades increases processing time and falls material utilization rate. In addition, this technology prevents the receipt of structures in which the disk and blades are made of different materials. Listed shortcomings can be remedied by the introduction of welds in the blisks design. Currently, the only method used industrially manufacture of welded blisks is linear friction welding (LFW).Linear friction welding is a type of friction welding with reciprocating movement of the parts used in foreign aircraft building industry since the early 2000 's, and mastering the domestic industry currently.

In existing modular compressor stages a combination of alloys VT8-1 (as the disk material) and VT6 (the blades) is quite often used. This combination is promising also such joint has high rates of static strength and toughness [1, 2], however, for application of welding technology in aviation engines the welded parts fatigue strength need to be evaluated [4-11].

The methodology of experimental research

As the object of study engine PD-14 II stage high pressure compressor blisk was chosen. The original version of blisk made from alloy VT8-1, was amended on welded structure. Disc material was kept unchanged, blades material was changed to VT6 alloy. Welded blades have height 48 mm, length of the chord in the plane of the weld interface was 53 mm, radius transition from shoulder blade to disk was 2 mm.

Samples (blisk sections) are represented at figure 1: left photo - after welding, right photo - after machining.For tests 21 samples in 2 batches were made. In the first batch, the blades manufacturing alloy VT6 in structural state has been used, it's traditionally used for making GTE blades. Second blades batch was made from VT6 alloy in ultra fine-grained condition.

Fig. 1. Blisk section:
left–after the welding; right–after the final machining.

Bladed disk sections from the first batch were annealed, and then hammered by 2 mm diameter steel balls. Fatigue tests were conducted with transverse bend under the influence of inertial loads using electrodynamic vibro-stends. The study of the samples included the search for cracks by the method of capillary flaw detection and their fractographic investigation.

Results and discussion

It was determined, that the fatigue crack in all cases is formed on the transition radius from blades to the disk. Analysis of macro- and microstructure, microhardness measurement revealed that metal structure in cracked areas corresponds to the original structure of the alloy VT8-1. Here is no thermal or mechanical affect of welding on metal structure in the fracture zone. Cracks in the first installment were atinput and output edges (fig. 2). Fractographic study revealed inferior quality of machining the transition radius from blades to the disk (fig. 3), the presence of scratches and facets.

Based on the results obtained, before the fatigue test the second installment, was held an additional polishing of input and output edges, as well implantation of sections by nitrogen ions. The measures taken have made it possible to exclude the destruction of samples from the second installment on edges – the fracture was displaced on the back of the blade (fig. 4). Despite this, it is easy to see that the endurance curve for the samples from the second batch is lower than for the first batch. Available number of samples was insufficient to determine σ_{-1}, as for the first installment and for the second. However, from the information provided on the fig.5, you can see that we should expect the value of fatigue limit of sections of the first batch of about 255 MPa, and slightly lower for sections of the second installment.

Fig. 2. The location of the weld seam relative to the crack, identified by etching (for 1[st]batch).

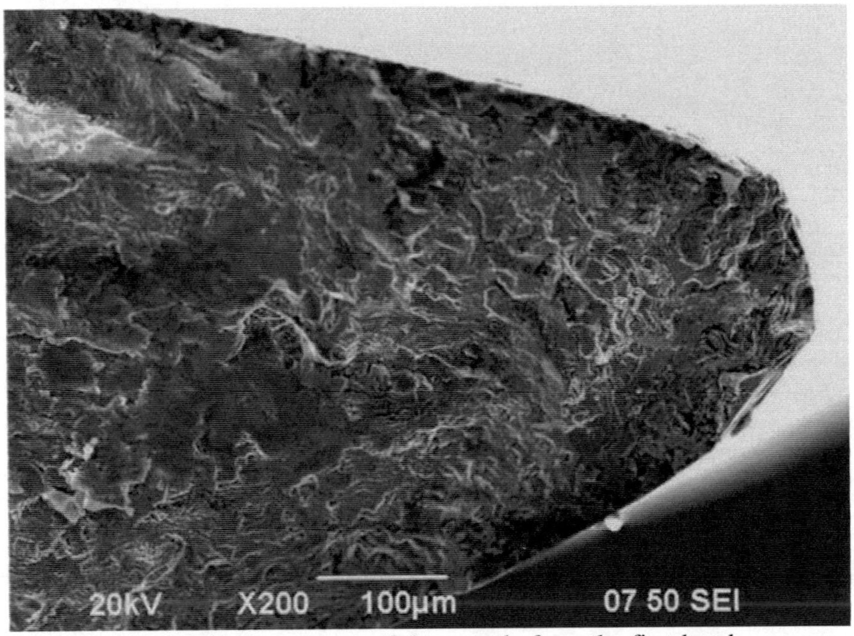

Fig. 3. The fractogram of the sample from the first batch.

IOP Conf. Series: Materials Science and Engineering **253** (2017) 012003 doi:10.1088/1757-899X/253/1/012003

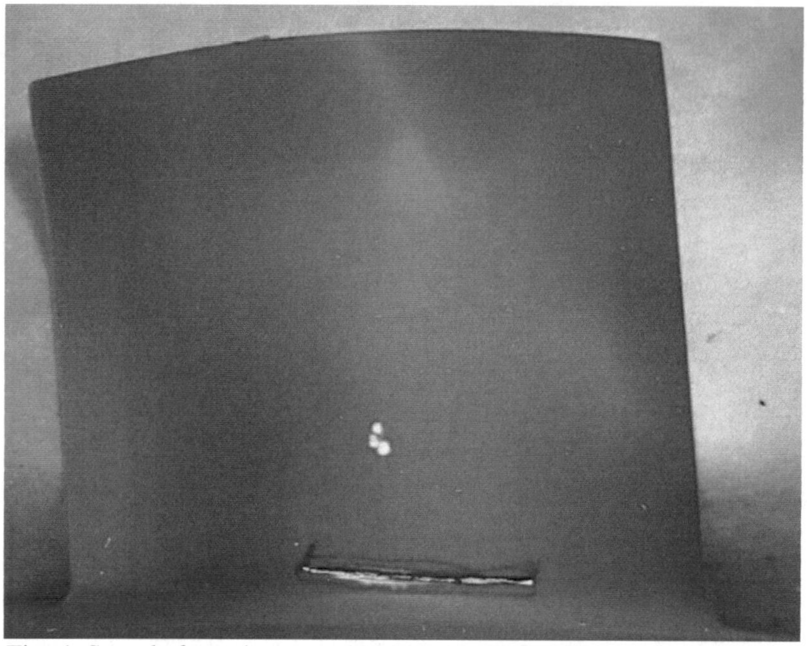

Fig. 4. Sample from the second batch after the capillary flaw detection.

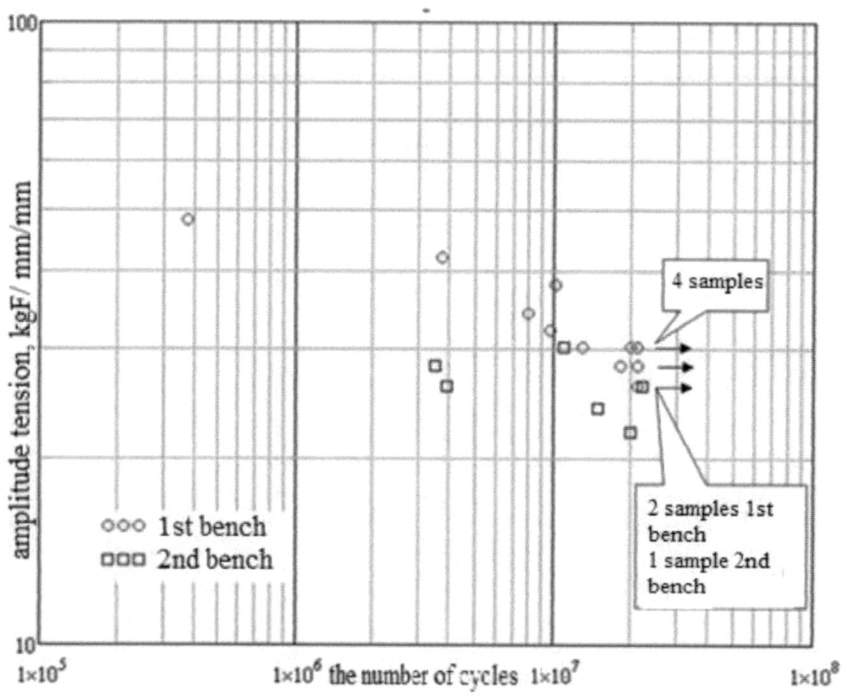

Fig. 5. Blisk section endurance diagram.

Reducing of fatigue strength, despite the improvement of the quality of the surface, may be triggered by few causes.

Heat treatment was excluded from technological process of second installment manufacturing because of heightened sensitivity to heating of ultrafine-grained microstructure. This could reduce endurance, since the cracking region is in the zone of action of tensile welding stress [3].

IOP Conf. Series: Materials Science and Engineering **253** (2017) 012003 doi:10.1088/1757-899X/253/1/012003

The obtained results can be related to difference between elastic properties of blades with ultra fine-grained and traditional condition – for the same deformation amplitude, the frequency of loading for a second batch were on average 6% higher.

Conclusions

Clear localization of fractures on samples with differing technology mechanical, heat treatment and surface hardening, indicates that from a list of factors (constructional features, microstructural condition of the blades, the thermo-mechanical effects of welding, machining quality, technology of heat treatment and surface hardening), the most significant factor is the concentration of stresses in the transitional radius from blades to disk.

Location of fatigue cracks showed that the welds are less dangerous than the stress concentrators already exists in the blisk structure, and so linear friction welding can be used for blisk manufacturing in gas turbine engines.

References

[1] A.Y. Medvedev, V.M. Bychkov, A.S. Selivanov, S.P. Pavlinich, S.K. Dautov, A.V. Supov, Linear friction welding of two-phase titanium alloys VT6 and VT8-1, Welding International 29 (2015) 66-69.

[2] A.Y. Medvedev, V.M. Bychkov, E.E. Tarasenko, M.P. Savichev, A.V. Supov, ISPYTANIYA SOYEDINENIY, POLUCHENNYKH LINEYNOY SVARKOY TRENIYEM, NA UDARNYY IZGIB. scientific journal of USATU 16 (2012) 94-97 (in Russian).

[3] R. Nikiforov, A. Medvedev, E. Tarasenko, A.Vairis, Numerical simulation of residual stresses in linear friction welded joints, JOURNAL OF ENGINEERING SCIENCE AND TECHNOLOGY REVIEW, 8 (2015) 50-53.

[4] Zaruba I.I., LebedevV.K., ShejkoP.P., BasanskijV.V.. Svarka modulirovannym tokom // Avtomaticheskaya svarka. 1968. №11. p. 35-40.

[5] KuznetsovV.D., StepanovD.V. Struktura i svojstv ametalla svarnogo shva, modifitsirovannogo nanooksidami. // Avtomaticheskaja svarka. 2015. № 11. p.19 - 24.

[6] Sidoruk V.S., Dudko D.A., Zatserkovnyj S.A., Taraborkin L.A. Vlijanieparametrovrezhimaruchnojdugovojsvarkimodulirovannymtokomnapervichnujustrukturu metallashva. // Svarochnoeproizvodstvo. №5. 1988. p. 4 - 6.

[7] Dudko D.A., Zatserkovnyj S.A., Sidoruk V.S., Taraborkin L.A., Mahlin N.M. Zavisimost' himicheskogo sostava metalla shva ot parametrov rezhima dugovoj svarki modulirovannym tokom. // Avtomaticheskaja svarka. 1989. 2. p.27 - 29.

[8] Larionov V.P., Sleptsov O.I., Mihajlov V.E. Osobennosti tehnologii svarki nizkolegirovannyh stalej povyshennoj i vysokoj prochnosti pri nizkih temperaturah // Dokl. na XIX komissiistran-chlenov S`EV poprobleme "Svarki". PNR, 1986, 44 p.

[9] N.V Kobernik, R.S. Miheev, A.S. Pankratov, A.A. Linnik. Sovremennye predstavlenija o modifikatsii naplavlennogo metalla i metalla shva nanorazmernymi chastitsami. // Svarka i diagnostika. 2015. 5. p. 13 – 18.

[10] Naplavochnyj material s `effektom vtorichnogo tverdenija dlja povyshenija stojkosti metalloobrabatyvajuschego instrumenta / E.N. Eremin, A.S. Losev, A.E. Eremin, G.P. Rumjantsev // Porodorazrushajuschij i metalloobrabatyvajuschij instrument - tehnika i tehnologija ego izgotovlenija i primenenija: Sb. nauch. tr. - K.: INM im. V.M. Bakulja NAN Ukraïni, 2011. Vip. 14. p. 453 - 455.

[11] Getskin O.B., Poloskov S.I., Erofeev V.A., Vit'ko O.P. Phiziko-matematicheskaya model' sistemy "istochnik pitaniya-duga" dlya svarki plavyash'imsya elektrodom v zash'itnyh gazah [Physical and mathematical model of system "power source - arc" for consumable electrode shielded welding]. Tyazheloe mashinostroenie, 2008, no.6, pp. 18-20.

Forecasting Resource as a Method of Increasing the Security of Technical Devices

Anatoly P Cherepanov, Pavel K Lyapustin

Federal State Funded Educational Establishment of Higher Education
"Angarsk State Technical University"(ASTU)
665835, Irkutsk Oblast, Russia, Angarsk, Tchaikovsky str., 60.

E-mail: boning89@mail.ru

Abstract. The article shows a method of increasing the safe operation of technical devices at various stages of the life cycle according to the proposed classification parameters of the resource by applying the model of resource prediction. The model takes into account the presence of defects, the rate of corrosion and corrosion resistance of the material, the volume of technical diagnosis, the degree of risk in case of failure or damage of technical devices. The article shows the application of the model resource of the technical device from the manufacture to the end of its service life.

Introduction

In accordance with the Federal norms and rules in the field of industrial safety [1, 6-8] technical device should be safe during the whole service life. The generalized factor of safety of technical devices is risk. The risk increases when a critical technical condition of the device. Technical condition (TC) is determined by the strength and reliability of bearing elements for a nominal (normal) and abnormal (emergency) situations [2].

In [3] proposed a deterministic risk analysis of chemically hazardous facilities with the introduction of the following main assumptions: the allowable resource [τ], which is defined in the design, takes into account the degree of damage accumulation during the operation of technical devices. Over time the resource τ is reduced according to a linear law:

$$D_T = F_D \left\{\frac{\tau}{[\tau]}\right\} = C_D \frac{\tau}{[\tau]}, \quad (1)$$

whereC_D —severity rate given the degree of damage.

At each time interval τ measured in years, the conditional probability of reaching the limiting state will be equal to:

$$P = \frac{1}{n_\tau \cdot [\tau]}\tau, \quad (2)$$

Where value: $\frac{1}{n_\tau \cdot [\tau]}$ this increment in the conditional probability of reaching a dangerous state in a year. Resource $n_\tau$taken in the range 2÷10.

Thus the allowable design resource [τ] is determined by the time n_τ achieve the limiting state of formula:

$$[\tau]=\frac{\tau_{op}}{n_\tau}. \tag{3}$$

The life cycle in [2] is determined by the regularity of reducing the technical condition in the process of operation to the limit when reaching the resource, is equal to zero, at which the operation becomes dangerous.

In the section shown an explanation of how the reduction of the period of safe operation depending on the reduction of stocks of thicknesses and strength reserve of the walls, considered in work [2].

The thickness of the vessel wall with wear and tear due to corrosion (chemical or electrochemical) or mechanical effects (erosion) reduced time intervals between the first, second and subsequent time interval and is determined by calculation.

The time intervals between the first (t_1), – second (t_2),...,–*i*–m (t_i), if k=1...*i*.

Data for calculation of reserves of strength are taken from measuring the actual thickness of the vessel walls during technical diagnosis. To determine the time intervals between overhauls designed the classification parameters resource at all stages of the life cycle of TU [2], which is given in the table.

Table. The classification parameters of the resource at various stages of the life cycle technical device

Stage of the life cycle	Option technical state	Design parameterswall			The estimated parameters resource	
		Thickness	Reserve of wall thickness	Reservesofst rength		
Design	Design	S_p	n_p^s	n_p	Design	T_p
Manufacture	Initial	S_u	n_u^s	n_u	Initial	T_u
Service life	Actual	$S_к$	n_k^s	$n_к$	Remaining	$T_к$
Service life termination	Limit	S_p	n_n^s	n_n	Residual	T_o
Damage probability	Critical	S_{min}	n_{lim}^s	n_{lim}	Limiting	T_{lim}

In accordance with the table in the time intervals from the start of operation until the end of the current resource, then before the end of residual life and on up to the end of the full resource, determine the stock sizes of wall thickness, and reserves of strength. Conversely, knowing the intervals between the value of the stocks of wall thickness and reserves of strength, determine the time intervals from the start of operation until the end of the current resource, then before the end of residual life and on up to the end of the full resource.

In accordance with the table in the time intervals from the start of operation until the end of the current resource, then before the end of residual life and on up to the end of the full resource, determine the stock sizes of wall thickness, and reserves of strength. Conversely, knowing the intervals between the value of the stocks of wall thickness and reserves of strength, determine the time intervals from the start of operation until the end of the current resource, then before the end of residual life and on up to the end of the term safe operation. Because in the process of operation in accordance with these time intervals are reduced, the stocks of wall thickness, and reserves of strength, the time intervals from design to achievement of the ultimate limit state, it is possible to determine by the formulas:

– reserve of wall thickness:

$$T_p^s \geq T_u^s \geq T_k^s \geq T_o^s \geq T_{lim}^s = f\left(n_p^s \geq n_u^s \geq n_k^s \geq n_n^s \geq n_{lim}^s\right); \tag{4}$$

– reserves of strength:

$$T_p \geq T_u \geq T_k \geq T_o \geq T_{lim} = f(n_p \geq n_u \geq n_k \geq n_n \geq n_{lim}). \tag{5}$$

Under the terms of the (2, 3) proposed the prediction model of the resource [2]. Figure 1 shows the model diagram of the service life forecasting based on transition regularities from the initial and actual technical state to the limit technical state.

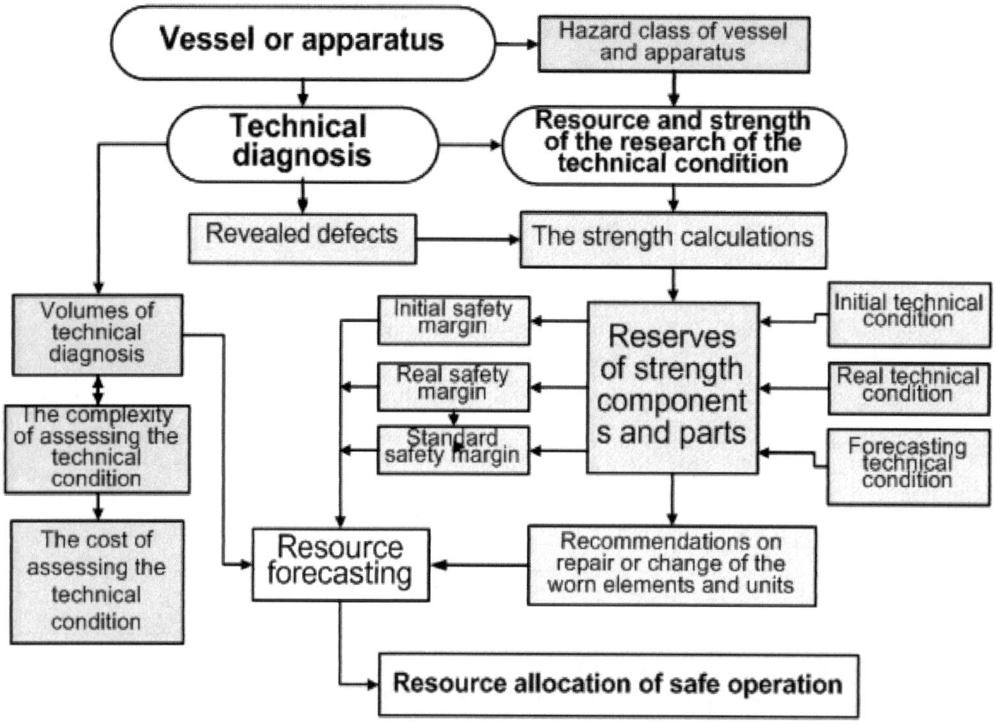

Fig. 1 - The model diagram of the service life forecasting based on transition regularities from the initial and actual technical state to the limit technical state

To implement the model is applied to the transition function from the initial (source) and the actual (real) the technical condition. The model shows the relationship between the past the technical condition (before commissioning) and the present technical state (moment of technical diagnosis), to predict the future of the technical condition under which a high probability of failure. The model allows to calculate the residual life and to determine the time of further safe operation of technical devices.

As you know, reserves of strength as wear of the walls decrease with time and are fixed parameters determined at specific points in time, the technical diagnosis, but for predicting the resource they are not enough. Therefore, for the generalized characteristic for the resource definition in [4] first applied the rate of reduction in reserves of safety and reserves of the wall thickness, which ensure the safety of the technical device for the subsequent period of operation, that is, from the current technical inspection (technical diagnostics) to achieve the future technical condition, in which a high probability of failure.

The degree of reduction in reserves of strengthand the degree of reduction of the stock of the wall thickness can be determined by the formulas:

$$K_k = \frac{n_k}{n_u - n_k} \text{and} K_k^S = \frac{n_k^S}{n_u^S - n_k^S}. \qquad (6)$$

where k - the ordinal number of technical diagnostics and the resource of stress research, defined as k=1...i;

n_u and n_u^S- initial reserves of strength and the reserves thickness of the wall at the time of manufacture of the vessel;

n_k and n_k^S-the current reserves of strength and reserves of the wall thickness;

n_n and n_n^S -regulatory reserves of strength and reserves of the wall thickness.

The degree of reduction in reserves of strength K_k and the degree reduction of stock of the wall thickness K_k^S show the wear for the period from the initial or current state, and the limiting state, and the adopted as a safety factor, as proposed in [3].

The degree of reduction in reserves of strength $K_k$and the degree reduction of stock of the wall thickness K_k^S show the wear and tear of the vessel wall, which is possible over a period of time from the initial or the current state until reaching a limit state. These values are taken as the safety factor that has been proposed in [3].

Applying the formula (4) as a radical expression, obtain a quadratic dependence for calculation of the resource of the technical device first shown in [4], which is represented by the formula:

$$T_k = t_k\sqrt{\frac{n_k}{n_u - n_k}}. \quad (7)$$

The results and discussion

According to the formula (5) is designed and constructed by the exponential dependence of the resource from the reserves of wall thickness. A graph of the resource from the current stock of wall thickness is shown in figure 2.

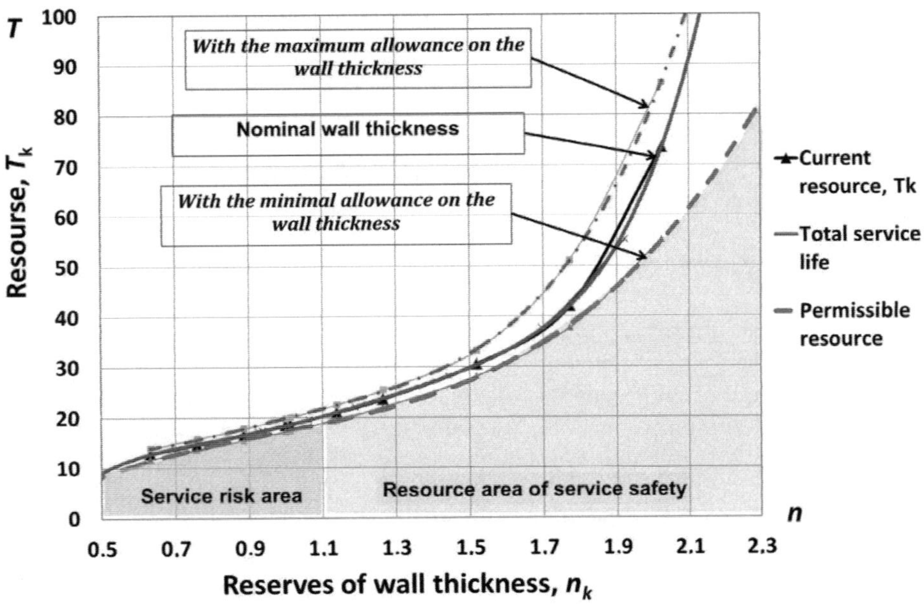

Fig. 2 - A graph of the resource from the current stock of wall thickness

However, the formula (5) does not reflect the impact on the resource volume control in the technical diagnostics and does not reflect the impact of defects and the degree of danger of destruction of technical devices, as shown in the diagram (figure 1) and is given by formulas (2, 3). To improve the reliability and accuracy of the calculation of the resource [4], shows the formula, which takes into account the volume control in the technical diagnostics, the presence of defects and the degree of danger of destruction of technical devices:

– original (source) of the resource at time of manufacture technical devices:

$$T_u = f\,(n_n;\, n_u;\, W_f;\, \xi;\, \beta); \quad (8)$$

– current (settlement) of the resource for the period of technical diagnosing at time k:

$$T_k=f\,(n_n;\ n_k;\ W_{f\,k};\ \xi_k;\ \beta_k,); \tag{9}$$

– residual life determined by the results of the initial or subsequent technical diagnostics for a period of time (k+1):

$$T_o=f\,(n_n;\ n_{(k+1)};\ W_{f\,(k+1)};\ \xi_{(k+1)};\ \beta_{(k+1)}), \tag{10}$$

wheren_k– the stock (strength, stability, etc.) specified for the period of technical diagnosing at time k;

$n_{(k+1)}$ – stock (strength, stability, etc.) in specified for a period of time (k+1), which will come over the next technical diagnosis;

W_f– effective technical diagnostics;

ξ– coefficient, of responsibility of the technical device;

β– coefficient taking into account the presence of valid or invalid of defects identified during the technical diagnostics that will be addressed during the repair.

For example, the vessels depending on the capacity or nominal diameter, and also depending on the maximum allowable working pressure is determined by the category or categories of risk (1st, 2nd, 3rd and 4th), which in [2] are characterized by the coefficient of responsibility. Because the definition (refinement) of the corrosion rate can be made at various stages of the life cycle of a technical device, respectively, and the allowable resource can be defined (to be refinement) at different stages of the life cycle.

Substituting in the formula (5) reserves of strength or thickness of the housing wall and that the reduction factors (6, 7, 8), we obtain formulas for the calculation:

– original (source) of the resource at time of manufacture technical devices:

$$T_u^S = [\tau_u]\cdot\sqrt{W_f\cdot\beta\cdot\left(q\frac{S_p}{S_u-S_p}\right)^{\xi}}. \tag{11}$$

– current (settlement) of the resource for the period of technical diagnosing at time k:

$$T_k^S = [\tau_k]\cdot\sqrt{W_k\cdot\beta_k\cdot\left(q_k\cdot\frac{S_k}{S_u-S_k}\right)^{\xi}} \tag{12}$$

– residual life determined by the results of the initial or subsequent technical diagnostics for a period of time (k+1):

$$T_o^S = [\tau_o]\cdot\sqrt{W_o\cdot\beta_o\cdot\left(q_o\frac{S_o}{S_u-S_o}\right)^{\xi}}. \tag{13}$$

Thus, the prediction model of the resource of the technical device can be implemented at any stage from design to achievement of a limit state.

On the given formulas it is possible to estimate the resource with the replacement or strengthening of worn-out elements of the technical device in the process of repair.

According to the proposed method from 2002 to 2011 was carried out to estimate the technical condition of the devices in number 3677 pieces, which are operated at hazardous production facilities of petrochemical industry. Evaluation of the resource was carried out by results of technical diagnostics and strength calculations of elements with respect to the amount of technical diagnostics, groups and classes of risk, indicators of corrosion and corrosion resistance of the metal, the presence of defects and other factors that caused the deterioration of elements of technical devices.

Conclusions

1. It is shown that the knowledge of the actual reserves of strength and reserves of the thickness of the walls, increases the accuracy and reliability of the assessment of the resource, and assignment of term of safe operation throughout the entire life cycle from manufacture to disposal.
2. The proposed method of predicting resource makes it possible to reduce the risk in operation of technical devices of hazardous production facilities throughout the entire life cycle from manufacture to disposal.

IOP Conf. Series: Materials Science and Engineering **253** (2017) 012004 doi:10.1088/1757-899X/253/1/012004

3. It is proposed the improvement of existing and creation of new guidelines for predicting resource of technical devices of hazardous production facilities, which is shown in the materials of all-Russian scientific-practical conference "Actual issues of industrial safety" [5].

References

[1] The Federal norms and rules in the field of industrial safety "Rules of carrying out of examination of industrial safety". Approved by Order of Rostechnadzor dated 14.11.2013 № 538 (Registered in Ministry of justice of Russia 26.12.2013 № 30855).
[2] Russia's Security. Legal, socio - economic and scientific - technical aspects. Manage resource exploitation of high-risk objects / Under total Ed. MahutovN. M.: MGF "Knowledge", 2015, 600 p.
[3] Makhutov N.A. Structural strength, resource and technogenic safety: 2 h / N..Makhutov N.A. – Novosibirsk: Nauka, 2005. Part 1: Criteria of strength and life. 494 p. Part 2: Justification of resource and security. 610 p.
[4] Cherepanov A.P. Predicting resource of technical devices petrochemical plants. // Monograph. FSFEE of HE "Angarsk state technical University". Angarsk. 2017. 224.
[5] Cherepanov A.P. Adaptation of standards to the conditions of the examination prior to the application of technical devices at hazardous production facilities. Safety and survivability of technical systems: papers and reports / V all-Russian conference (Krasnoyarsk, October 12-15, 2015): 3; scientific. the editorship of V.V. Moskvichev. Krasnoyarsk: Sib. Feder. University, 2015. T. 2. 272.
[6] FNP v oblasti promyshlennoy bezopasnosti Pravila provedeniya promyshlennoy bezopasnosti p.28 Prikaz Rostekhnadzora ot 14.11.2013 №538 «Ob utverzhdenii federal'nykh norm i pravil v oblasti promyshlennoy bezopasnosti» (Zaregistrirovano v Minyuste Rossii 26.12.2013 №30855); In Russian.
[7] FNP v oblasti promyshlennoy bezopasnosti Pravila promyshlennoy bezopasnosti opasnykh proizvodstvennykh ob"yektov, na kotorykh ispol'zuyetsya oborudovaniye, rabotayushcheye pod izbytochnym davleniyem p. 415 pp. g. Seriya 20.Vypusk16.-M.: Zakrytoye aktsionernoye obshchestvo «Nauchno-tekhnicheskiy tsentr issledovaniy problem promyshlennoy bezopasnosti», 2014. 254 p. ISBN 978-5-9687-0609-6.; In Russian.
[8] Makhutov N.A., Gadenin M.M. Tekhnicheskaya diagnostika ostatochnogo resursa i bezopasnosti. – M.: Spektr, 2011.– 185 p.; In Russian.

IOP Conf. Series: Materials Science and Engineering **253** (2017) 012005 doi:10.1088/1757-899X/253/1/012005

Remote Acoustic Emission Monitoring of Metal Ware and Welded Joints

Boris I Kapranov, Vladimir A Sutorikhin

National Research Tomsk Polytechnic University
634028, Tomsk, Savinykh str. 7, Russia

E-mail: introbob@mail.ru

Abstract. An unusual phenomenon was revealed in the metal-ultrasound interaction. Microwave sensor generates surface electric conductivity oscillations from exposure to elastic ultrasonic vibrations on regions of defects embracing micro-defects termed as "crack mouth." They are known as the region of "acoustic activity,"method of Acoustic Emission (AE) method. It was established that the high phase-modulation coefficient of reflected field generates intentional Doppler radar signal with the following parameters: amplitude-1-5 nm, 6-30 dB adjusted to 70-180 mm. This phenomenon is termed as "Gorbunov effect," which is applied as a remote non-destructive testing method replacing ultrasonic flaw detection and acoustic emission methods.

Introduction

Until recently, it was supposed that acoustic emission becomes apparent as temporary short acoustic signal sequences, generated by electric coupling failures [1]. These failures result in emerging density variations, acoustic waves spreading throughout the metal in all directions at acoustic velocity. Registering these 0.1-25nm mechanical oscillations by contact-type piezoelectric sensors, oscillation onset location and hazard level of emerging defect were determined. The disadvantages of this method included: essential mechanical loading to obtain desired signals and required application of contact highly-sensitive piezoelectric sensors, involving the application of immersion liquid. Both disadvantages could be eliminated by applying ultra-high frequency radar. However, practical testing demonstrated that this radar has low-level degree of sensitiveness (1-5 micrometer). i.e. 3-4 times more than required.

Theory (Theoretical justification)

In numerous experimental investigations [2-4] the so-called "Gorbunov effect" which emerges with acoustic signals has been described. Pertinently, this effect is when emerging source of acoustic emission, related to as" active" (i.e. counting rate of acoustic impulses increase proportionally to i degree of breakage force), furthers an unusual change within the metal around the source of AE. In fact, even after the mechanical loading back period this region retains its unusual properties. If this region is subjected to elastic oscillations (low-power ultrasound waves of up to 5-10W/cm^2) one could observe synchronous changes of surface electric conductivity. Lateral spectral component of phase-modulated signal, reflected from investigated surface, exceeds the noise spectral power to more than 6-20 dB. In the case of no "active" region, lateral spectral component of phase-modulated signal is not higher than the noise level. Actual microwave sensor frequency is 33 ghz. This interaction effect, discovered in

1985, is compatible with such well-known theories as failure process within metal objects[4] and metal ionization being of X-ray radiation source.

A characteristic feature of the "Gorbunov effect" is its time of occurrence. For example, if the "active" AE source time proceeds for a fraction of a millisecond and coincides with mechanical loading time, then the detection time for "Gorbunov effect" is from several minutes to ten hours after load release. Registering the "Gorbunov effect" by microwave sensor is possible at a significant distance (70-180 mm)from the object surface. Exposure zone is not associated with the "activity" zone but embraces the metal surface completely. In this case, no scanning is necessary as in previous AE diagnostic methods.

Experimental technique

Mechanistic potential of this new diagnostic method has not been completely described. Nevertheless, designed Doppler radar-based device (remote active defect indicator) has a wide application spectrum-investigating hazard metal defects, monitoring motor block and turning strength and expanding nondestructive testing method application. The steady loading could be excluded in the case of applying the acoustic emission (AE) method for testing. It was revealed that previously detected defects by AE method could be also identified by the "Gorbunov method," without applying additional mechanical loading. Super-high frequency crack (flaw) detector (DIAD-1) operates in combination with ultrasonic generator of the following parameters: frequency – 50 Kh; power outlet- $P= U_{out}^2/R_{tr}=(80^2/300)=21$ *W,* where U_{out} is generator ultrasound output voltage; R_{tr} – active resistance of piezoelectric transmitter of MA40E9-1 type. Ultrasonic transmitting unit area is 1см^2. Operating distance from super-high frequency sensor to examined sample is 90-120 mm with step 3.3 mm. Operating parameters of new device were compared to those investigated samples by ultrasonic crack detector. Two additional samples were tested on the X-ray unit (fig. 1 a, b). Photos of samples are presented in fig. 2 a-d. Desired signal level obtained through DIAD and fast Fourier transformation (FFT) program are depicted in fig. 3 a-d.

Figure 1a X-ray of object joint № 1-20. Defect of weld joint as a result of procedure violation (rapid water cooling); desired signal 14 dB. Crack under weld.

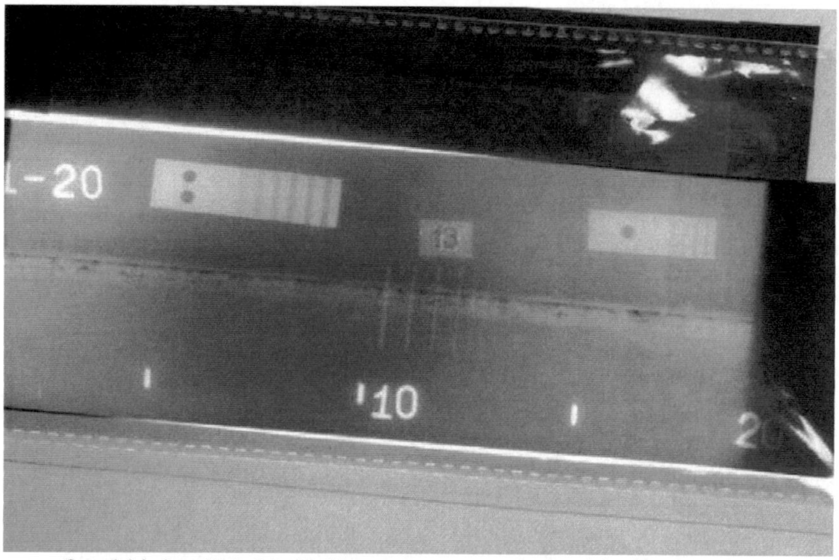

Figure 1b X-ray of weld joint №1-24. Hollow aperture under weld. Desired signal 12dB.

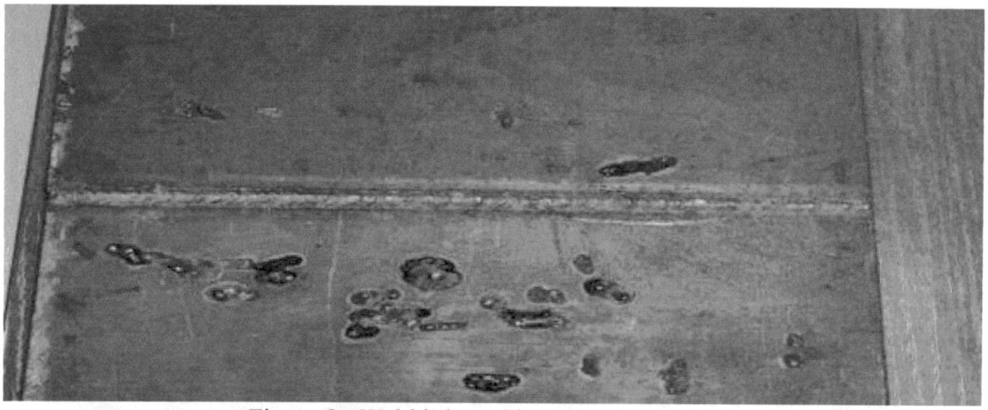

Figure 2a.Weld joint, object №1-20 Desired

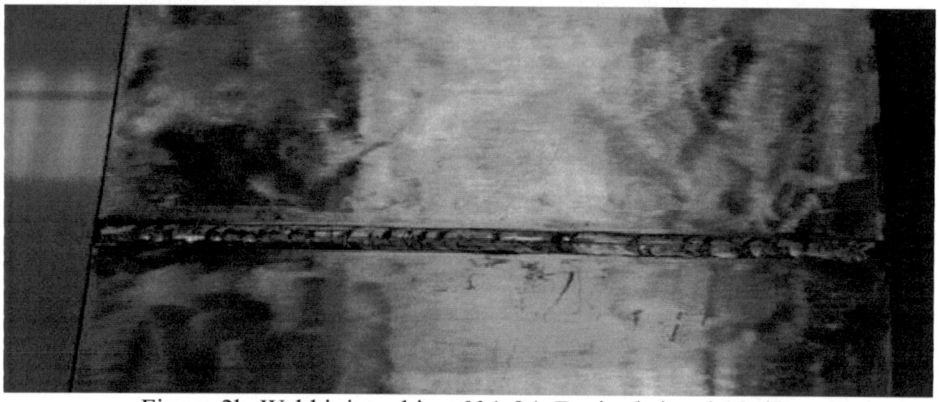

Figure 2b. Weld joint, object №1-24 Desired signal 12 dB.

IOP Conf. Series: Materials Science and Engineering **253** (2017) 012005 doi:10.1088/1757-899X/253/1/012005

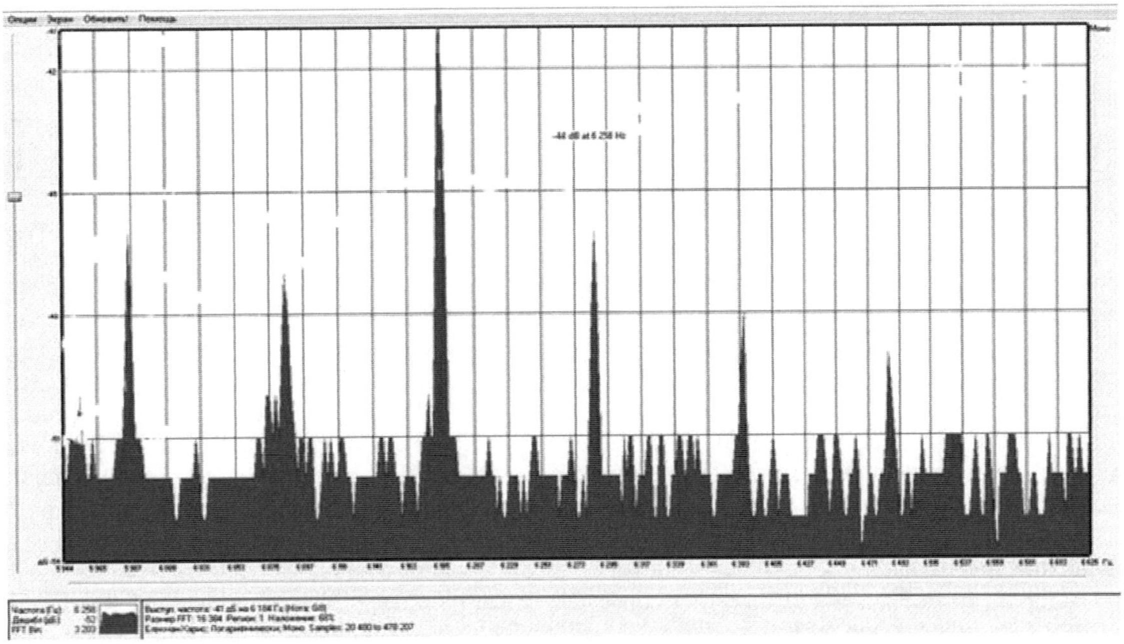

Figure 3a. Desired signal 14 dB. Object №1-20

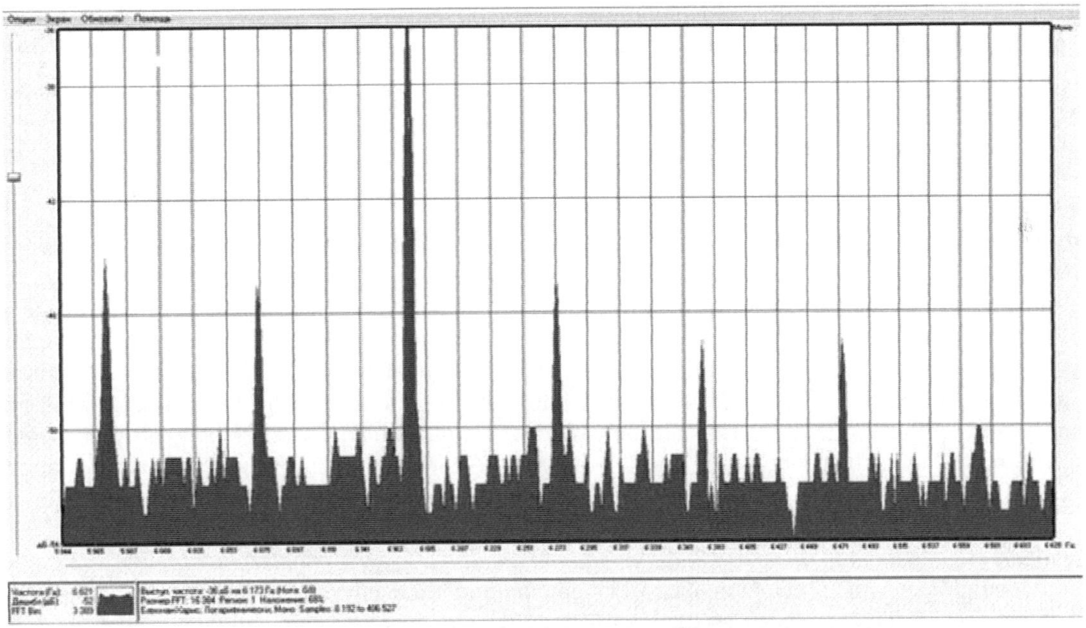

Figure 3b. Desired signal12 dB. Object №1-24

Brief explanation of acoustic-electronic interaction theory

Surface impedance is written on the basis of reflection coefficient ρ [5]

$$\rho = H_t(x)/H_t(0) = (1 - cZ/4\pi)/(1 + cZ/4\pi) \tag{1}$$

where, H_t - magnetic field vector; c- light velocity (3 10^8m/s); $Z(\omega,\sigma)$ –surface impedance (s/m);
$Z(\omega,\sigma) = (4\pi / c)\sqrt{\varepsilon}(\omega,\sigma), \varepsilon = \varepsilon_l + j4\pi\sigma / \omega$

where, $\varepsilon(\omega,\sigma)$ - complex dielectric constant; ε_l - real part; ω – cyclic frequency (rad/s); and σ - electrical conductance of metal (1/s)- arguments of surface impedance. The calculations showed surface conductance change (10%) resulting in significant reflection coefficient phase change (5%) and insignificant module change (0.025%). Likewise, the derived phase of reflection coefficient to conductivity(dφ/dσ) is less than zero. Increasing conductivity reveals decreasing reflection coefficient phase. At the same time calculations showed that mechanical surface oscillations result in inverse phase shift relationship. Reflection coefficient phase for surface is opposite to reflection phase of conductivity. Emerging elastic-surface-wave results in changing crystalline lattice density. Crystalline lattice density decreases maximum and increases minimum in S-waves. Thus, detecting surface elastic oscillations is characteristic of AE effect, and possibly, Doppler radar; in the case of emerging phase shift. Conditioned change of surface conductivity associated with mechanical oscillations is evidently observed and could be described as "Gorbunov effect."

Results and discussion

Described practical results of detecting concealed defects in metals by dynamic effect detector on microwave sounding-based proved the practical utility of this new non-destructive testing method. This new method embraced not only the ultrasonic diagnostic method and acoustic emission method, but also microwave sensor method for defect control. AE method requires significant mechanical loading and is incompatible with ultra- sound. Ultra-sonic zoning method does not interact with the microwave sounding method. Practically, all three methods successfully interact with each other demonstrating newly up-dated results. It should be noted that ultrasonic contact transmitters being substituted by ultrasonic laser generator of elastic oscillations are applied in diagnostics, which, in its turn, enhances this as a future non-contact method [6-10].

Conclusion

Above-described results in studying the new non-destructive testing method by applying Doppler radar showed that based on comparative control analysis results the welded joints are different from those in previous ultrasonic control methods (in this case, it is not necessary to compare experimental samples to reference samples without defects, but having complex geometrical shapes). AE method has numerous positive advantages – no additional mechanical loading and applicable with mobile units and / or units. The implementation of this method would decrease the diagnostic time, improve research reliability and simplify operations in auto-detection of fatigue effect in metal objects in-process. It should be highlighted that continuous control of welded joints after production is possible without complex ultrasonic equipment "X-raying."

References

[1] Kostoglotov A.M., Prist A.V., Acoustics-emission method of modeling of the processes of the destruction material. Defektoskopiya, 2002, p 36.

[2] Gorbunov V, Sutorikhin V. Possible control of elastic deformation limits by the microwave (MW) method. Defectoscopy, 1999,7:75-80.

[3] Gorbunov VI, Sutorihin VA. Possibility defectoscope of the metal details of the microwave oven afield. Electronic magazine "Technical acoustics", 2010,60-66.

[4] Kapranov B., Sutorikhin V. The Doppler Detection Fault, Physical Science International Journal, SCIENCEDOMAIN international www.sciencedomain.org., 13(1): 1-5, 2017; Article no. PSIJ.30562, 2017, p 5.

[5] Leontovich MA. Approximate boundary conditions for electromagnetic fields on surfaces of well-conductive bodies. Research in Radio-wave Propagation, 1948.

[6] Makhutov N.A. Structural strength, resource and technogenic safety: 2 h / N..Makhutov N.A. Novosibirsk: Nauka, 2005. Part 1: Criteria of strength and life. 494 p. Part 2: Justification of resource and security. 610 p.

[7] Makhutov N.A., Gadenin M.M. Tekhnicheskaya diagnostika ostatochnogo resursa i bezopasnosti. M.: Spektr, 2011. 185 p.; In Russian.

[8] FNP v oblasti promyshlennoy bezopasnosti Pravila promyshlennoy bezopasnosti opasnykh proizvodstvennykh ob"yektov, na kotorykh ispol'zuyetsya oborudovaniye, rabotayushcheye pod izbytochnym davleniyem p. 415 pp. g. Seriya 20.Vypusk16.-M.: Zakrytoye aktsionernoye obshchestvo «Nauchno-tekhnicheskiy tsentr issledovaniy problem promyshlennoy bezopasnosti», 2014. 254 p. ISBN 978-5-9687-0609-6.; In Russian.

[9] FNP v oblasti promyshlennoy bezopasnosti Pravila provedeniya promyshlennoy bezopasnosti p.28 Prikaz Rostekhnadzora ot 14.11.2013 №538 «Ob utverzhdenii federal'nykh norm i pravil v oblasti promyshlennoy bezopasnosti» (Zaregistrirovano v Minyuste Rossii 26.12.2013 №30855); In Russian.

[10] The Federal norms and rules in the field of industrial safety "Rules of carrying out of examination of industrial safety". Approved by Order of Rostechnadzor dated 14.11.2013 № 538 (Registered in Ministry of justice of Russia 26.12.2013 № 30855).

IOP Conf. Series: Materials Science and Engineering **253** (2017) 012006 doi:10.1088/1757-899X/253/1/012006

Increase in the Technical Level of Mine Haul Trucks

Yuri Voronov [a], Aleksey Khoreshok, Artyom Voronov, Sergey Grishin, Aleksey Bujankin

T.F. Gorbachev Kuzbass State Technical University, Kemerovo, Russian Federation

E-mail: [a] ya.voronov454@yandex.ru

Abstract. The study presents theoretical and methodological fundamentals of minehaul trucks optimal design. It provides methods based on the systematic approach to integrated assessment of truck technical level and methods for optimization of truck parameters depending on performance standards. The results of these methods application are given.

Introduction

Technical re-equipment of open-pit mining facilities and further development and upgrades of automotive mining machinery are inextricably linked. Efficiency in execution of hauling operations determines the technical and economic performance indicators of coal production on the whole, as transportation costs make up more than 50% of the overall mining costs.

So far skyrocketing payload capacity as the main tendency for mining automotive machinery development has not been accompanied by a corresponding gain in performance and increase in other figures of merit. However, it should be noted that the engineering policy meant to boost the payload and aimed at further mining truck redesign without major quality changes is outdated. The evolution of mining haul trucks and empirical approach to their design will never result in any quality breakthroughs in the future. At the same time power-to-weight ratio of the new wheeled mine transport is to be increased together with its reliability, service life and performance; empty mine haul truck weight is to be lowered; all truck parameters are to be balanced. Open-pit mining transport enhancement can be achieved, in the first place, by optimal tuning of all the structural components, units and systems of a truck not only among themselves, but with the facility environment as well. So, when designed, a mine haul truck must be regarded as a complex technical aggregate being in its turn a part of the total surface mining transport system.

These areas are closely connected with the problem of improvement in the industrial output quality, which, being a part of the widely spread in the Western world "total quality control" ideology, is one of the major issues for the Russian economy.

Vast amounts of mined bulk to be transported by haul trucks, influence of hauling operations upon the speed of open-pit mining and their high labour intensity stimulate the constant search for large-scale improvements to the mining transport.

M.V. Vassiliev, A.A. Kuleshov, M.G. Potapov, K.N. Trubetskoy, S. Alarie, M. Gamache [1], C.N. Burt, L. Caccetta [2, 3], R.A. Carter [4, 5], V.A. Temeng [6, 7] and others contributed greatly to the theory and practice of automotive mine machinery usage at open-pit facilities. In their papers, they determined the sphere of haul truck efficient use in open-pit mining [1, 3, 5, 8-10], summed up the experience of its operation [2, 4, 7, 11-14], highlighted the issues of techno-economic studies devoted to automotive mine machinery usage efficiency [1-7, 15-17].

Belarusian Automobile Plant(OJSC «BELAZ», Republic of Belarus)is a major world manufacturer of heavy dump trucks used in mining and construction. Mainly its produce consists of mine dump trucks with a payload capacity of 30 to 360 t. During all the years (since 1948 when the company was founded) BelAZ specialists have been doing their best to improve the truck designs. These designs implementations contributed to providing mining companies with high-performance equipment.

However, in recent years BelAZ designing engineers have had to face the following problem: increased payload did not lead to the proportional rise in truck performance. The power-to-weight ratio is still insufficient, the payload ratio and life mileage are too low as well as the truck technical level in general and only production and maintenance costs go up with every passing year.

This situation can be accounted for by the fact that the theory of mining machinery optimal design (including mine dump trucks) has not been developed sufficiently yet. Mine haul truck design framework still needs the procedures for ultimate solutions synthesis. The systematic approach as a basis for this synthesis has actually not been used for BelAZ dump trucks design and their parameters optimization until recently.

Results and discussion

In order to estimate the technical level and quality of the mining machinery (including automotive machinery) adequately, the most agreeable alternative is the non-expert assessment method, which allows assessing functionally homogeneous machines of various types, models and modifications [1]. For this purpose, be the technical level we mean the truck quality at the stage of development of the design assignment.

According to the systematic approach, the functional criterion as the basis of this method appears to be the product of the truck performance Π_T by the specific energy of the mined bulk W transportation:

$$\lambda = W \cdot \Pi_T. \tag{1}$$

The specific energy of transportation can be calculated according to the correlation deduced by the authors (kJ/t·km):

$$W = \frac{g[f_c + i_{max}(2k_T + 1)]}{\sqrt{1 + i_{max}^2}}, \tag{2}$$

wheref_c – coefficient of rolling resistance for a truck on the road; i_{max}– possible slope for a truck to ride on; $k_T = m_a/m_{gr}$ – payload ratio (m_a, m_{gr} – truck payload and empty weight, t); g–gravity acceleration, m/s^2.

When assessing the technical level of the existing trucks and designing new ones in order to determine the functional criterion one should use the theoretically calculated performance values. Besides, for comparable results the influence of the payload upon the functional criterion is to be excluded, that is why it is reasonable to use 1 auto-ton performance (or truck specific performance) (t·km/ t·h):

$$\Pi_{T_y} = 0.56 N_{eng}^{ud}(1 + k_T)\frac{\sqrt{1 + i_{max}^2}}{g[i_{max} + f_c(2k_T + 1)]}, \tag{3}$$

where $N_{eng}^{ud} = N_{eng}/(m_a + m_{gr})$ – power-to-weight ratio of the engine (power-to-weight ratio of the truck), kW/t.

The final dependence for the functional criterion calculation is as follows (kW/t):

$$\lambda = 0.56 N_{eng}^{ud}(1 + k_T)\frac{f_c + i_{max}(2k_T + 1)}{i_{max} + f_c(2k_T + 1)}. \tag{4}$$

Another important issue of the assessment is the grounded selection of the parameters determining the technical level of haul trucks to have the essential and sufficient nomenclature made up. The parameters that determine the technical level to the most are selected from the initial nomenclature after a thorough analysis of the operational processes followed by a further verification of their matching (direct correlation) and representativity (the degree of influence upon the technical level depending, in its turn, on the purpose of its assessment). Eventually the following system of the parameters was obtained (Fig. 1).

The given parameters can be determined at the earliest stages of truck design. In compliance with the qualimetry principles extra parameters can be introduced depending on the specific conditions and demands. The integrated assessment of the technical level is carried out by means of unique (q_{ij}) and generalized (k_i) parameters according to the formulae of the method [1].

The developed method allows not only assessing the truck technical levels but also choosing the most promising models with its help and providing quantitative evaluations of the decisions to be madeat the design stage.

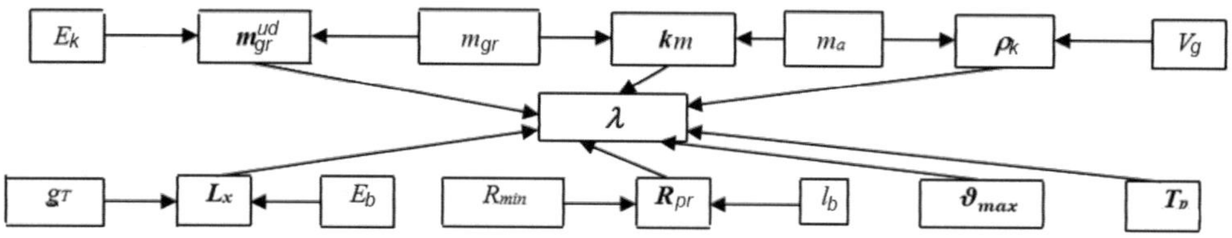

Fig. 1. Complex of haul trucks technical level parameters λ– functional criterion, kW/t; $m_{gr}^{ud} = m_{gr}/E_k$ – truck specific payload, t/m^3; $k_m = M/m_b$ – truck weight coefficient (comparative materials consumption); $\rho_k = m_a/V_g$ – component density, t/m^3; $L_x = 100E_b/g_T$ – distance to empty, km; $R_{pr} = R_{min}/l_b$ – specific turn radius; ϑ_{max}– maximum speed of the truck, km/h; T_p – predetermined service life (truck mileage), km; m_{gr} – truck payload capacity, t; E_k – body capacity, m^3; m_a – truck curb weight, t; V_g – total truck volume (product of length by width and height), m^3; E_b – fuel tanks capacity, l; g_T – linear fuel consumption, l/100 km; R_{min} – minimum turn radius, m; l_b – truck wheel base, m; $M = m_a + m_{gr}$ – gross weight, t; $m_b = (0.212N_{eng} - 8.4)$ – truck weighted average mass (corresponding to the installed engine capacity N_{eng})

In the paper according to the selected parameters, a comparative assessment of 22 trucks was made, 14 of which are BelAZ products, 4 – of Caterpillar, 1 of Komatsu and 1of Unit Rig. This group comprises the whole range of truck payloadcapacities(30 to 320 t). The trucks are compared in similar conditions.

After the generalized parameter of the truck technical level was calculated and analyzed it can be concluded that for all the 22 trucks this parameter exceeds 0.5 (Fig. 2), therefore, this result proves that the technical level of all the trucks is quite high.

The average technical level for all the machines in question is 0.695. Also, it should be noted that the level of BelAZ trucks and that of the trucks produced by foreign companies are virtually the same. The average value of the technical level generalized parameter for the trucks considered is 0.698 and that is only 0.7% higher than for BelAZ trucks, which means that as far as the vehicle structure perfection and new design ideas and solutions are concerned Belarusian Plant products show very little difference.

The results of the unique parameters calculations allow stating that BelAZ trucks surpass their counterparts in 4 parameters out of seven, namely in truck specific payload m_{gr}^{ud} (6.4% higher), component density ρ_k (11.2%), truck weight coefficient (comparative materials consumption) k_m (9.8%) and specific turn radius R_{pr} (8.7%); also BelAZ trucks are behind in distance to empty L_x (6.7% lower), maximum speed of the truckϑ_{max} (8.7%) and especially in predetermined service life (truck mileage) T_p (27.5%).

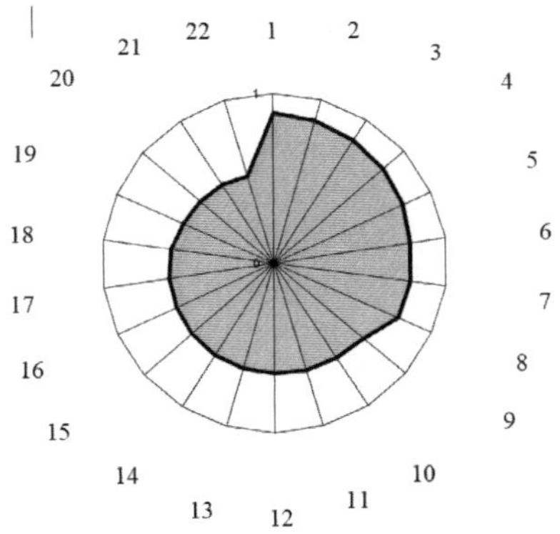

Fig. 2. The generalized parameter of the trucks technical level (in the brackets - the parameter value k)

On the basis of the analysis it can be concluded that power-to-weight ratio of new vehicles is to be increased, empty weight is to be lowered while reliability and service life are to be raised to a higher level; all truck parameters are to be balanced and reviewed as interrelated and interdependent. All this shows that the systematic approach to new wheeled open-pit mine machinery design is essential.

At the present moment, there is a distinctive deficiency of guidance materials and norms and specifications that could standardize products optimal design as an important stage of quality improvement. Systematic approach is to be regarded fundamental for optimal design as it allows perceiving the vehicle as a whole while its individual elements are being designed. The systematic approach to a vehicle parameters determining consists of the following: firstly, every parameter must correlate optimally with the others, and secondly, output operational performance established on the basis ofthese parameters is to comply with the specified values.

It is suggested that the mathematical model of truck parameters optimization should be presented as a set of regression analysis equations directly connecting output operational performance and the parameters which are to be optimized.

Haul truck as a system is characterized by its own structure and parameters. The truck structure predetermines its elements and their interrelationships; it is to ensure the truck optimal functioning and for this purpose the parameters characterizing the truck are to be optimized. It is reasonable to use generalized evaluations of the technical level as optimization criteria and its unique indicators as optimized parameters. Such approach enables to reduce the problem dimensions, to simplify its solution and also to ensure objective assessments of the suggested solutions.

After carrying out the analysis of the output operational performance (such as payload ratio k_T and 90% predetermined service life (truck mileage) T_p) specified in the regulations (including interstate GOST 30537-97) and specific truck performance Π_{T_y} as the main technical and operational parameter for mining haul trucks and taking into account their interrelationships(dependence (3) the essential and sufficient nomenclature of the optimized parameters was compiled. It is stated in Fig. 3.

The problem of finding out how much each of these parameters influences the technical level of the generalized parameter chosen as the optimization criterion and subsequent generating an objective function model on this basis is a problem of approximation type and can be solved with the help of the theory of experiment design [6, 7]. Having been carried out, the correlation analysis of the interrelationships between the parameters that are to be optimized determined the degree of their independence (Table 1) and allowed diminishing their number from 8 (Fig.3) to 3 (N^{ud}_{eng},k_T and T_p).

IOP Conf. Series: Materials Science and Engineering **253** (2017) 012006 doi:10.1088/1757-899X/253/1/012006

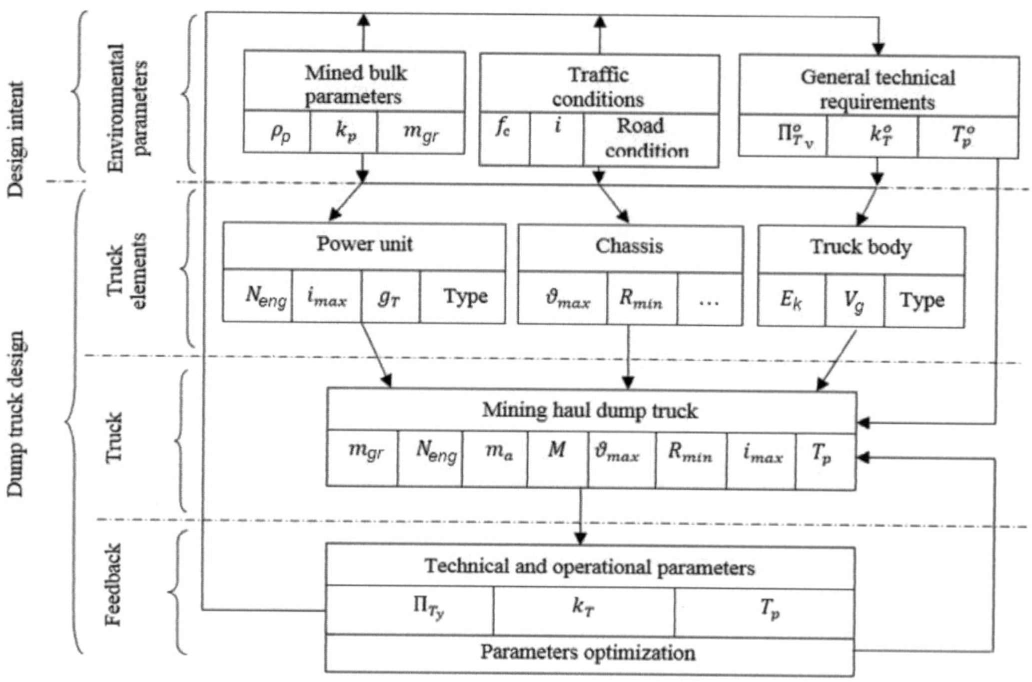

Fig. 3. Flow chart for haul truck design and truck parameters optimization ($\boldsymbol{\rho}_{п}$ – minedbulkdensity; $\boldsymbol{k_p}$ – mined bulk loosening factor)

Table 1.Results of the correlation analysis of the truck parameter inter relationships

Model	**Regression parameters**		**Regression parameter value, *F***	**Fisher criterion critical value, $F_{кр}$**	**Correlation index, *r***	**Correlation index reliability, *t***	**Student's t-test critical value, $t_{кр}$**
	a	***b***					
$N_{eng} = a + bm_{gr}$	70.9	7.950	1097.1	3.49	0.984	24.7	2.086
$M = a + bN_{eng}$	–8.40	0.212	1659.0	3.49	0.990	31.4	2.086
$\vartheta_{max} = a + bN_{eng}$	61.6	$-8.7 \cdot 10^{-3}$	784.4	3.49	0.695	4.32	2.086
$R_{min} = a + bM$	9.0	0.017	1588.3	3.49	0.919	10.4	2.086
$g_T = a + bN_{eng}$	–9.10	0.434	1384.8	3.49	0.988	28.6	2.086
$i_{max} = a + bN_{eng}$	0.294	$-6.3 \cdot 10^{-3}$	181.8	3.49	0.604	3.39	2.086
$i_{max} = a + bk_T$	0.870	–0.848	121.4	3.55	0.555	2.98	2.101
$k_T = a + bm_{gr}$	0.657	$1.04 \cdot 10^{-3}$	1031.5	9.55	0.890	3.38	3.182
$T_p = a + bm_{gr}$	0.378	$0.82 \cdot 10^{-3}$	660.6	19.0	0.959	4.79	4.303

When solving the problem of truck parameters optimization, the following constraints can be found.

1. Constraints on compliance of the output operational performance parameters Π_{T_y}, k_T, T_p with $\Pi^o_{T_y}$, k^o_T, T^o_p regulated by GOST 30537-97 (Fig. 3). The models of these constraints as well as the objective function model are formulated afterthe mathematics experiment by which we mean that the technical level generalized parameter k and the output operational performance parameters are calculated upon the condition of specific combinations of variables according to the experiment design matrix.

2. Constraints on the ratio of the k_T, T_p parameters to the truck payload capacity m_{gr}. The given interrelationships are shown in Table 1.

3. The boundary conditions describing the range of variations for the optimized parameters. The analysis shows that the only boundary condition that can be established for all these parameters is non-negativity ($N_{eng}^{ud} \geq 0$; $k_T \geq 0$; $T_p \geq 0$).

Eventuallya mathematical model for the optimization (MMO) of technical level parameters and truck parameters was derived which is a linear programming problem.

$$\begin{cases} k = a_0 + a_1 N_{eng}^{ud} + a_2 k_T + a_3 T_p \to max; \\ a_{11} N_{eng}^{ud} + a_{12} k_T + a_{13} T_p \geq \Pi_{T_y}^{o} - S_0^{\Pi} - a_{10}; \\ a_{11} N_{eng}^{ud} + a_{12} k_T + a_{13} T_p \geq \Pi_{T_y}^{o} + S_0^{\Pi} - a_{10}; \\ a_{21} N_{\text{дв}}^{\text{уд}} + a_{22} k_T + a_{23} T_p \geq T_p^0 - S_0^T - a_{20}; \\ a_{21} N_{eng}^{ud} + a_{22} k_T + a_{23} T_p \geq T_p^0 + S_0^T - a_{20}; \\ a_{31} N_{eng}^{ud} + a_{32} k_T + a_{33} T_p \geq k_T^0 - S_0^k - a_{30}; \\ a_{31} N_{eng}^{ud} + a_{32} k_T + a_{33} T_p \geq k_T^0 + S_0^k - a_{30}; \\ N_{eng}^{ud} \geq 0; k_T \geq 0; T_p \geq 0, \end{cases} \quad (5)$$

The evolved method is based upon the technical requirements regulated by GOST 30537-97 but it can be used for the purpose of optimization of haul truck parameters and other external conditions established by the customers. The method suggests a complete rethink of the establishment of critical parameters for truck specifications.

For the purpose of establishing parameters of the objective function model and constraints a mathematics experiment was carried out according to the Full Factorial Experiment 2^3 plan. Asaresult, a MMO was developed as follows:

$$\begin{cases} k = -0.422 + 0.172 N_{eng}^{ud} - 0.043 k_T + 0.391 T_p \to max; \\ 0.715 N_{eng}^{ud} + 9.077 k_T \geq 9.818; \\ 0.467 N_{eng}^{ud} + 9.077 k_T \leq 9.900; \\ 0.065 N_{eng}^{ud} - 1.250 k_T \geq -0.605; \\ 0.065 N_{eng}^{ud} - 1.250 k_T \leq -0.289; \\ 0.051 N_{eng}^{ud} - 0.195 k_T - T_p \geq -0.338; \\ 0.051 N_{eng}^{ud} - 0.195 k_T - T_p \leq -0.090; \\ N_{eng}^{ud} \geq 0; k_T \geq 0; T_p \geq 0. \end{cases} \quad (6)$$

The experiment confirmed linearity of the objective function models and all the constraints. The linearity is conditioned by the fact that all interrelations between the generalized and unique parameters of the technical level and between the truck parameters can be viewed as linear with a reasonable degree of accuracy as suggested by regression parameters high values and correlation indices reliability (Table 1).

At the same time, the analysis of the objective function model allows drawing the conclusion that the higher are the values of power-to engine ratio N_{eng}^{ud} and predetermined service life (truck mileage) T_p and the lower is payload ratio k_T, the higher the technical level generalized parameter (optimization criterion) k.

At that the optimal solution of the system will be the following combination of the optimized parameters and optimization criterion value:

$$N_{eng}^{ud*} = 8.308 \text{ kWt/t}; \; k_T^* = 0.663 \text{t/t}; \; T_p^* = 0.633 \text{ mln. km}; \; k^* = 1.227, \quad (7)$$

thus, for the technical level generalized parameter to reach its maximum, power-to-weight ratio is to be increase and make up no less than 8.3 kWt/t; payload ratio is not to exceed 0.660; 90% truck mileage is to be at least 630 thous. km. All this allows increasing the technical level by 38.5% in comparison with the best dump truck from the group under consideration and by 76.5% when compared to the group on average.

After the model (6) analysis the optimal values for power-to-engine ratio N_{eng}^{ud*}, payload ratio k_T^*, predetermined life service (truck mileage) T_p^* and the technical level generalized parameter k^*(7) were obtained.These composite indices can be synthesized into particular optimal parameters which are included into a truck specification. Optimal parameters synthesis is based on ensuring optimal functioning of a truck as a whole with due consideration of the existing interrelationships between these parameters.

Conclusion

The analysis of the conditions and practices in mining machinery operation enabled us to ascertain that the composition of BelAZ truck fleet is of relatively low quality. At present, the number of guidance papers and materials for mine trucks adequate assessment and recommendations aimed at their technical level increase is not sufficient.

Integrated comparative assessment of the technical level of mining machinery produced by both domestic and foreign manufacturers highlighted the prospect of the technology advances in mining machinery and their implementation as early as the design stage. Increase in the technical level of newly developed mining automotive machinery can be achieved through the usage of theoretically grounded methods of optimal design which allows balancing all parameters.

Optimal design is based upon the optimization mathematical modeling of machinery operations and parameters. The objective function and constraints of the optimization mathematical model are viewed as a set of regression analysis equations which connect corresponding figures with the optimized parameters directly. It is suggested to use the technical level generalized parameters as an optimization criteria and its unique indicators as optimized parameters.

Implementation of scientific methodological and technical recommendations will allow bringing newly developed mine machinery up to the level that matches world standards.

References

[1] S. Alarie, M. Gamache, Int. J. of Surface Mining, Reclamation and Environment, **16**, 1 (2002)
[2] C.H. Burt, L. Caccetta, Int. J. of Surface Mining, Reclamation and Environment, **21**, 4 (2007)
[3] C.H. Burt, L. Caccetta, Int. J. of Surface Mining, Reclamation and Environment, **22**, 1 (2008)
[4] R.A. Carter, Eng. and Mining Journal, (2012)
[5] R.A. Carter, Eng. and Mining Journal, **221**, 1 (2010)
[6] V.A. Temeng, F.O. Otunoye, J.O. Frendewey, Int. J. of Surface Mining, Reclamation and Environment, **11**, 4 (1997)
[7] V.A. Temeng, F.O. Otunoye, J.O. Frendewey, Mineral Resources Eng, **7**, 2 (1998)
[8] G. Lumley, *GBI Mining Intelligence white paper*, (2012)
[9 C.H. Ta, J.V. Kresta, J.F. Forbes, H.J. Marquez, Int. J. of Surface Mining, Reclamation and Environment, **19**, 3 (2005)
[10] A. Krause, C. Musingwini, The Journal of The Southern African Institute of Mining and Metallurgy, **107** (2007)
[11] S.V. Mkhatshwa, The Journal of The Southern African Institute of Mining and Metallurgy, **109** (2009)
[12] S. Tan,*Ph.D. Thesis*(Pennsylvania State University, USA, 1992)
[13] B. Kolonja, N. Vasiljevic,*Mine Planning and Equipment Selection, Rotterdam, Balkema*, (2000)
[14] J.W. White, J.P. Olson, S.I. Vohnout, CIM Bulletin, **86 (973)**, (1993)

[15] S.G. Ercelebi, A. Bascetin,The Journal of The Southern African Institute of Mining and Metallurgy, **109** (2009)
[16] C. Brown, Eng. and Mining Journal, (2012)
[17] Y. Kuo, *Ph.D. Thesis*(University of Texas, USA, 2004)

Manufacturing of New Welding Fluxes Using Silicomanganese Slag

Roman E Kryukov [1], Nikolay A Kozyrev [1], Olga A Kozyreva [2], Aleksander A Usoltsev [1]

[1] Siberian IndustrialStateUniversity
654007, Russia, Kemerovo region, Novokuznetsk, Kirova street, 42
[2] KemerovoStateUniversity, 6, Krasnay street, Kemerovo, 650000

E-mail: [a] Kozyrev_na@mtsp.sibsiu.ru

Abstract. There were developed the composition and manufacturing technology of new welding flux using industrial products such as slag of silicomanganese production. The effect of fractional composition on welding and fabrication characteristics was studied. It was found that using of small-sized fracture of welding flux at a rate of 30-40% decreases the oxide non-metallic impurity rating of the weld and herewith doesn't affect its constituents.
To increase technical-and-economic indexes it was suggested to mix small-sized fracture and water glass. The use of ceramic flux made by mixing the dust fraction of silicomanganese slug with fraction size less than 0.45 mm and water glass provides the decrease of non-metallic impurity rating of the weld. Herewith the increase of its content from 15 to 40% doesn't have a significant influence on the non-metallic impurity rating of the weld and its constituents.

Introduction

Great attention in the world is paid to the issues of production, research and development of new welding fluxes [1 – 18]. It was proposed to use the slag of silicomanganese production for welding flux manufacturing [19, 20], the technology is protected by patents [21, 22]. The possibility of effective use of slag of silicomanganese production for welding flux manufacturingis considered in this paper.

Methods and materials

For the flux manufacturing was used the slag of silicomanganese production which chemical composition is presented in table 1. Alsointhefirstseriesofexperimentsthe possibility of use of differentslag fractions ratio was examined (table 2). Double-sided flux plain butt welding was carried out on the samples of 500×75 mm size and 16 mm thick made of sheet and plate steel type 09G2S. The welding process was performed with wire Sv-08GA with the use of welding machineASAW-1250 in modes: I_{weld} = 700 A; U_{arc} = 30 V; V_{weld} = 35 m/h.

Table 1 – Chemical composition of the slag of silicomanganese production

Content, %										
Al_2O_3	CaO	SiO_2	FeO	MgO	MnO	F	Na_2O	K_2O	S	P
6.91-9.62	22.85-31.70	46.46-48.16	0.27-0.81	6.48-7.92	8.01-8.43	0.28-0.76	0.26-0.36	0.62	0.15-0.17	0.01

Samples were cut of the welded plates and the X-rays pectrometry of weld chemistry and metallographical tests of weld metal were conducted. Chemical composition of welding fluxes is presented in table 3. Chemical composition of slag crust is presented in table 4, chemical composition of weld metal is presented in table 5.

Table 2 –Fraction and component composition of fluxesin question

Sample	Ratio, %, of fractions, mm
1	100 % of fraction 0.45 – 2.5
2	95 % of fraction 0.45 – 2.5 + 5 % of fraction<0.45
3	90 % of fraction 0.45 – 2.5+ 10 % of fraction< 0.45
4	85 % of fraction 0.45 – 2.5 + 15 % of fraction< 0.45
5	80 % of fraction 0.45 – 2.5+ 20 % of fraction< 0.45
6	70 % of fraction 0.45 – 2.5+ 30 % of fraction< 0.45
7	60 % of fraction 0.45 – 2.5 + 40 % of fraction< 0.45
8	60 % silicomanganese slag + 40 % water glass
9	70 % silicomanganese slag + 30 % water glass
10	80 % silicomanganese slag + 20 % water glass
11	85 % silicomanganese slag + 15 % water glass

Table 3 – Chemical composition of welding fluxes

Sample	Content, %									
	Al_2O_3	CaO	SiO_2	FeO	MgO	MnO	F	Na_2O	S	P
8	5.29	25.84	51.75	0.55	5.02	7.39	0.36	4.66	0.12	0.01
9	5.48	26.68	51.73	0.57	5.16	7.59	0.39	4.19	0.13	0.01
10	5.88	25.53	52.53	0.56	5.07	7.75	0.31	4.07	0.13	0.01
11	6.55	26.81	51.14	0.56	5.78	8.10	0.35	2.62	0.14	0.01

Table 4 – Chemical composition of slug crusts

Sample	Content, %										
	MnO	SiO_2	CaO	MgO	Al_2O_3	FeO	Na_2O	K_2O	F	S	P
1	7.90	46.04	23.38	6.77	10.08	2.07	0.37	0.65	0.73	0.13	0.01
2	7.87	45.58	31.82	6.62	6.77	1.35	0.26	absent	0.32	0.11	0.01
3	7.83	44.54	23.84	6.43	9.64	3.59	0.37	0.65	0.69	0.12	0.008
4	8.09	45.91	31.15	6.60	6.79	1.39	0.27	absent	0.29	0.11	0.01
5	7.93	45.67	23.84	6.54	9.87	2.86	0.37	0.65	0.72	0.12	0.008
6	8.16	45.74	29.39	6.22	6.93	1.99	0.26	absent	0.36	0.12	0.01
7	8.23	45.52	29.12	6.29	6.65	1.88	0.28	absent	0.26	0.12	0.01
8	8.19	48.79	24.42	4.82	5.14	2.45	3.64	absent	0.35	0.09	0.01
9	8.29	49.92	26.12	5.37	5.60	2.64	3.25	absent	0.37	0.10	0.01
10	8.16	48.25	26.32	5.22	6.02	2.17	2.12	absent	0.33	0.12	0.01
11	8.18	48.09	27.24	5.67	6.36	1.97	1.64	absent	0.34	0.12	0.01

Table 5 – Chemical composition of weld metal

Sample	Content, %										
	C	Si	Mn	Cr	Ni	Cu	V	Nb	Al	S	P
1	0.09	0.71	0.51	0.03	0.10	0.11	0.001	0.014	0.023	0.018	0.012
2	0.08	0.54	1.33	0.04	0.05	0.08	0.003	0.014	0.015	0.008	0.008
3	0.09	0.61	1.49	0.04	0.11	0.11	0.01	0.013	0.018	0.016	0.010
4	0.07	0.45	1.24	0.02	0.05	0.07	0.002	0.014	0.014	0.006	0.007
5	0.08	0.66	1.42	0.03	0.10	0.11	0.002	0.015	0.023	0.018	0.012
6	0.08	0.61	1.42	0.02	0.06	0.08	0.003	0.014	0.029	0.010	0.011
7	0.08	0.59	1.39	0.02	0.02	0.05	0.004	0.018	0.091	0.014	0.009
8	0.05	0.52	1.25	0.02	0.04	0.05	0.003	0.017	0.020	0.005	0.007
9	0.03	0.51	1.23	0.02	0.04	0.06	0.002	0.017	0.017	0.007	0.008
10	0.06	0.53	1.31	0.02	0.04	0.06	0.004	0.016	0.018	0.012	0.009
11	0.09	0.52	1.31	0.02	0.04	0.06	0.003	0.015	0.013	0.010	0.008

Metallographical tests were conducted on micro sections without etching by means of optical microscope OLYMPUSGX-51 with optical magnification 100. Metallographical tests of metal structure in weld zone were conducted by means of optical microscopeOLYMPUSGX-51 in a bright-field in the magnification range of 500 after the etching of the samples surface in 4% nitricacid solution. Grain size was determined in accordance with GOST 5639 – 82.

Results

Theresultsofanalysisforpresenceofnon-metallicinclusions in the weld zonecarried out in accordance with GOST 1778-70, are presented in the figure 1 and in the table 6.

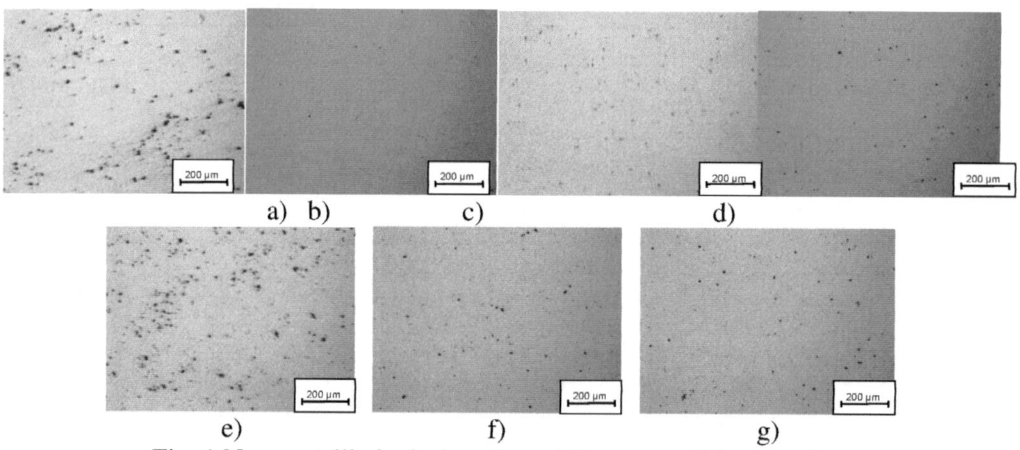

Fig. 1.Non-metallic inclusions in welding zone of the samples:
a – 1; b – 2; c – 3; d – 4; e – 5; f – 6; g – 7

Microstructures of weld metal are presented in the figure 2.

IOP Conf. Series: Materials Science and Engineering **253** (2017) 012007 doi:10.1088/1757-899X/253/1/012007

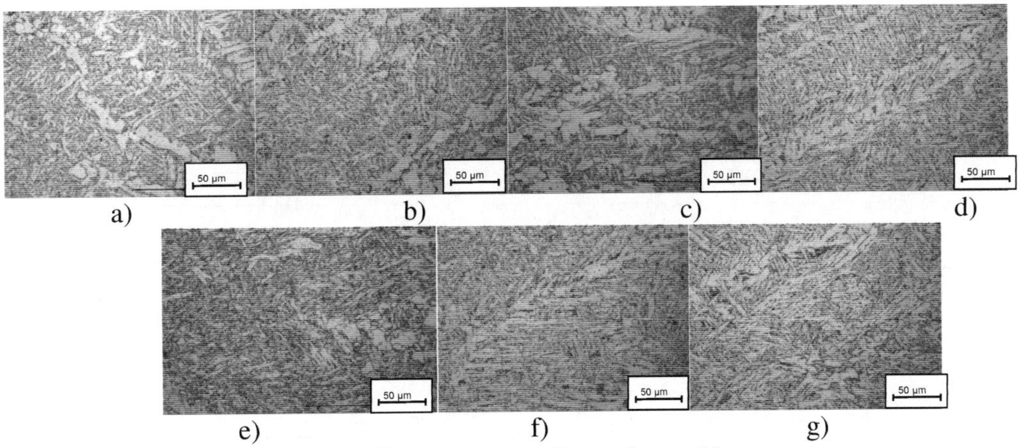

Fig. 2. Microstructures of samples welds :
a – sample 1; b –sample 2; c – sample 3; d – 4; e – 5; f – 6; g – 7

Samples were cut of the welded plates and their mechanical properties were determined. The results obtained by investigation of mechanical properties show the increase of impact strength (Figure 3).

The analysis of samples' mechanical properties shows that the optimal content of dust fraction with size less than 0.45 mm is 20-30%. This content of dust fraction with size less than 0.45 mm delivers the profitable set of mechanical properties of the samples cut of the welded plates.

In the weld metal structure of all samples' welds there is a ferrite presented in the form of unequiaxed grains elongated along the line of heat removal. There can be seen the transition from ferrite-pearlite uniform structure to the Widmanstatten ferrite-pearlite structure. However, relevant variation of grain sizes on a grain-size scale was not observed (table 6, 7).

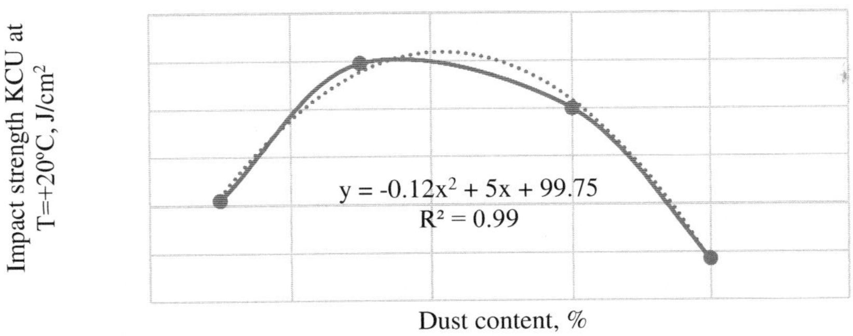

Fig. 3.The effect of dust fraction content (fraction size less than 0.45 mm) in the flux on the impact strength

In these cond series of experiments the possibility of use of ceramic flux made of water-glass-bonded silicomanganese slug's dust with fraction sizeless than 0.45 mm was studied. The production technique involved mixing silicomanganese slug and water glass in different proportions (table 2), drying, crush in gand size grading to get fractions 0.45-2.5 mm.

IOP Conf. Series: Materials Science and Engineering **253** (2017) 012007 doi:10.1088/1757-899X/253/1/012007

Table 6 – Non-metallicinclusions in weld zone

Sample	Non-metallic inclusions, scale number		
	Non deflecting silicates	brittle silicates	spot oxides
1	4b; 3b; 4a	3b	1a
2	2b; 1b; 3a; 4a	absent	1a; 2a
3	4b; 2b	absent	1a; 2a
4	2b; 4b	absent	1a; 2a
5	4b; 5b; 3b	absent	1a; 2a
6	2b; 1b; 2a; 2,5a	absent	1a; 2a
7	2b; 2a; 2,5a	absent	1a; 2a
8	2b; 1b; 2a; 2,5a	absent	1a
9	2b; 1b; 2a; 2,5a	absent	1a
10	2b;1b; 2a; 2,5a	absent	1a; 2a
11	2b; 2,5a	absent	1a, 2a

Table 7 – Welds' grain size according toGOST 5639-82

Sample	Grain size on grain size scale
1	№4, №5
2	№5, №4
3	№4, №5, №6
4	№4
5	№5, №4
6	№4
7	№4
8	№5, №4
9	№4, №5
10	№4
11	№4, №5

The analys is of the mechanical properties of the samples cut of welded plates allowed to determine the optimal content of water glass in the flux (up to 20-30%) to achieve the profitable set of mechanical properties of the samples cut of the welded plates (figures 4, 5).

However, the examined fluxes are oxidizing fluxes and are created on the principles of silicon-manganese-oxidation-reduction reactions, therefore resultants of such reactions are oxidic compound of silicon and manganese. Consequently the non-metallic impurity rating of weld increases and hence the physical and mechanical properties decreases, especially at low temperatures. To decrease the impurity rating of weld and to increase mechanical properties we investigated the possibility of introduction of previously developed carbon and fluorine containing admixture FD–UFS in the new flux.

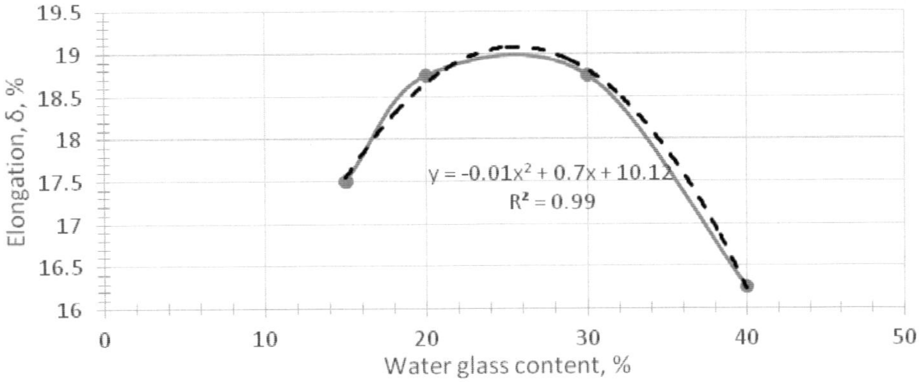

Fig. 4. The effect of water glass content in the flux on percentage elongation

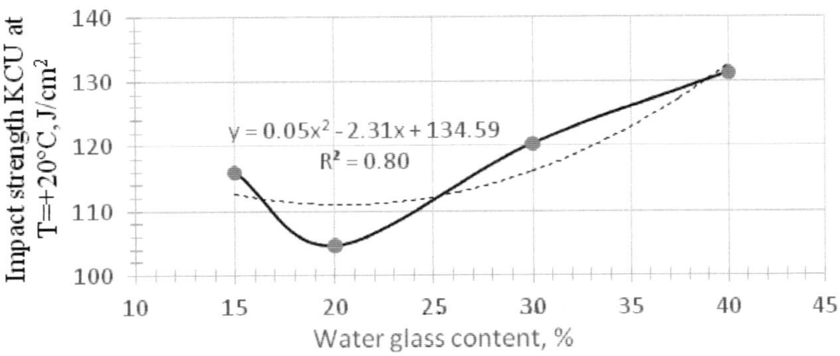

Fig. 5. The effect of water glass content in the flux on the impact strength

In the experiments flux-admixture was added at the ratio of 2, 4, 6, 8 % accordingly.

Chemicalcomposition of examined mixtures is presented in table 8, composition of slug crusts is presented in table 9, chemical composition of welds' metal is presented in table 10.

Table 8 – Chemical composition of examined flux mixtures, %

FD–UFS content in the flux, %	FeO	MnO	Ca	SiO_2	Al_2O_3	MgO	Na_2O	K_2O	S	P	ZnO	F
2	0.40	8.01	15.80	50.08	11.55	7.39	0.77	0.63	0.22	0.008	0.002	1.30
4	0.91	7.90	17.72	46.63	10.32	6.63	1.10	0.68	0.24	0.01	absent	1.95
6	0.81	7.68	16.79	43.64	11.27	5.71	2.25	0.65	0.34	0.01	0.003	4.04
8	0.46	7.46	16.00	43.64	11.86	5.56	2.30	0.60	0.33	0.01	0.002	3.96

Table 9 – Chemical composition of slug crusts, %

FD–UFS content in the flux, %	FeO	MnO	Ca	SiO_2	Al_2O_3	MgO	Na_2O	K_2O	S	P	ZnO	F
2	2.21	7.25	15.55	38.09	9.39	8.63	0.49	0.57	0.12	0.006	0.002	0.94
4	2.28	7.39	16.90	42.00	9.76	5.77	0.76	0.62	0.15	0.008	0.002	1.12
6	2.24	7.20	16.06	39.94	11.15	7.14	1.09	0.60	0.17	0.008	0.002	1.53
8	2.36	7.14	14.70	42.87	12.40	5.57	1.34	0.57	0.20	0.008	0.002	1.88

Therefore, the metal of the weld which has been made with flux without admixtures has the highest non-metallic impurity rating. The introduction of admixture FD–UFS decreases the non-metallic impurity rating and also decreases the impurities' size and number. Speaking of the examined ratio, the highest effect on the non-metallic impurity rating has the 8% amount of admixture.

Table 10 – Chemical composition of welds' metal

FD–UFS content in the flux, %	Mass fraction of the element %									
	C	Si	Mn	Cr	Ni	Cu	Nb	Al	S	P
2	0.09	0.62	1.40	0.02	0.06	0.09	0.014	0.023	0.020	0.008
4	0.10	0.60	1.34	0.02	0.07	0.08	0.010	0.013	0.023	0.009
6	0.12	0.66	1.43	0.02	0.06	0.10	0.011	0.012	0.027	0.008
8	0.13	0.65	1.36	0.03	0.06	0.09	0.013	0.013	0.024	0.008

The results of analys is for presence of non-metallic inclusions in the weld zone carried out in accordance withGOST 1778 – 70 are presented on figure 6 and in the table 11.

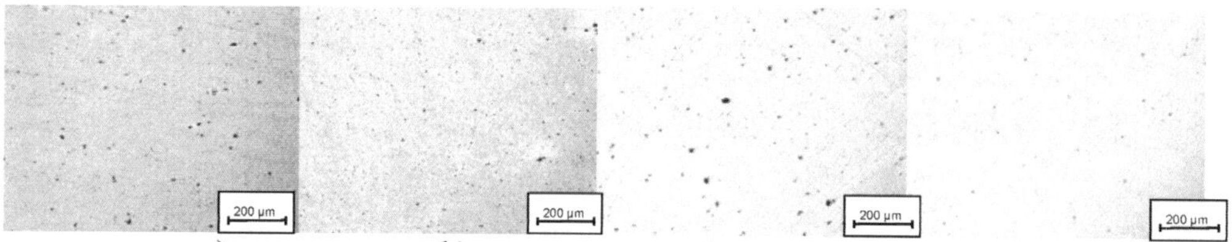

a) b) c) d)

Fig. 6.Non-metallicinclusions in the weld zone of the samples with admixture ratio, %:
a) 2; b) 4; c) 6; d) 8

Table 11 – Non-metallicinclusionsinweldzone

FD–UFS content in the flux, %	Non-metallic inclusions, scale number		
	Non deflecting silicates	brittle silicates	spot oxides
2	2b, 4b, 5a	absent	1a, 2a
4	2b, 4b	absent	1a, 2a
6	2b, 4b, 1b	absent	1a, 2a
8	2b	absent	1a, 2a

Microstructure of samples' welds is presented on figure 7. It was found out that introduction of admixture at a rate no more than 8% has no effect on the size and morphology of constituents.

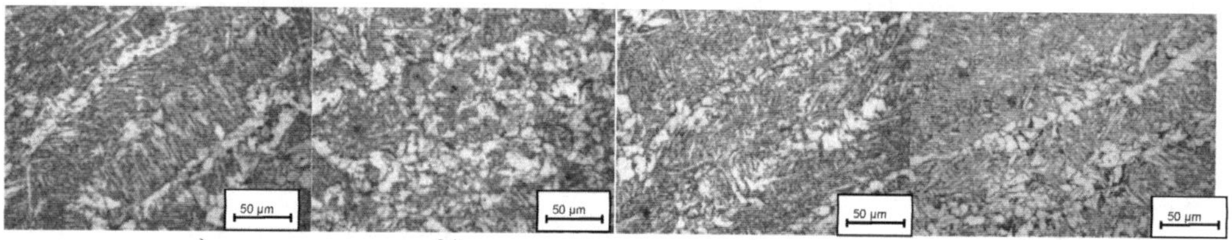

a) b) c) d)

Fig. 7.Microstructureofwelds of the samples with admixture ratio, %: a) 2; b) 4; c) 6; d) 8

The examination of mechanical properties has shown that as the amount of admixture increases the level of properties also increases (figure 8). The conducted researches formed the basis for RF patents [21, 22].

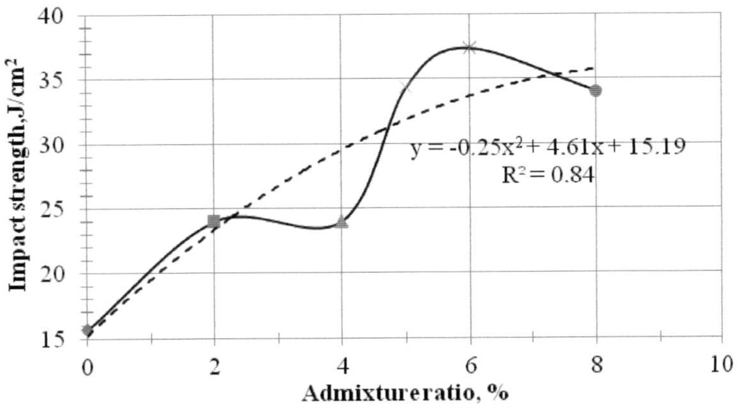

Fig. 8.Effect of admixture FD–UFS ratio in the flux on the impact strength (KCV at T=-20 °C)

Conclusions

1. The possibility in principle of use of slag of silicomanganese production for welding flux manufacturing was shown.
2. It is possible to use in the fluxes up to 30% of small-sized fracture(less than 0.45 mm). This content of dust fracture in the flux is optimal to achieve the profitable set of mechanical properties of the samples cut of the welded plates.
3. The optimal content of water glass in the flux allows to achieve the profitable set of mechanical properties is 20-30%.
4. To decrease the non-metallic impurity rating of weld and to increase mechanical properties of the weld it was suggested to introduce in the fluxes carbon-fluorine containing admixture FD–UFS at the ratio 2-8%. The introduction of admixture decreases the non-metallic impurity rating and also decreases the impurities' size.

References

[1] R. Q. Puchol, J. R. Blanco, L. P. Gonzalez, G. C. Hernández, C. R Gómez Pérez. The influence of the air occluded in the deposition layer of flux during automatic welding: a technological aspect to consider in the quality of the bead // Welding International. 2009. Vol. 23. № 2. P. 132 – 140.

[2] A.C.Crespo, R.Q.Puchol, L.P.Goncalez, L.G.Sanchez, C.R.Gomez Perez, E.D.Cedre, T.O.Mendez andJ.A. Pozol Obtaining a submerged arc welding flux of the MnO – SiO2 – CaO – Al2O3 – CaF2 system by fusion // Welding International. 2007. Vol. 21. № 7. P. 502 – 511.

[3] A.C.Crespo, R.Q.Puchol, L.P.Goncalez, L.G.Sanchez, C.R.Gomez Perez, G.Castellanos, E.D.Cedreand T.Ortíz. Study of the relationship between the composition of a fused flux and its structure and properties // Welding International. 2009. Vol. 23. № 2. P. 120 – 131.

[4] Golovko V.V., Potapov N.N. Special features of agglomerated (ceramic) fluxes in welding // Welding International. 2011. Vol. 25. № 11. P. 889 – 893.

[5] VolobuevYu.S., Volobuev O.S., Parkhomenko A.G., Dobrozhela E.I., Klimenchuk O.S. Using a new general-purpose ceramic flux SFM-101 in welding of beams // Welding International. 2012. Vol. 26. № 8. P. 649 – 653.

[6] VolobuevYu.S.,Surkov A.V.,Volobuev O.S., Kipiani P.N.,Shestov D.V., Pavlov N.V.,Savchenko A.I. The development and properties of a new ceramic flux used for reconditioning rolling stock components // Welding International. 2010. Vol. 24. № 4. P. 298 – 300.

[7] Potapov N.N.,Kurlanov S.A. A criterion for evaluating the activity of fused welding fluxes // Welding International. 1987. Vol. 1. № 10. P. 951 – 954.

[8] Babushkin P.L.,PersitsV.Yu. Determination of hydrogen in the form of moisture in basic electrode coatings and fluxing materials in metallurgical production // Welding International. 1991. Vol. 5. № 9. P. 741 – 742.

[9] Pavlov I.V.,Oleinichenko K.A. Regulating generation of CO by varying the composition of ceramic fluxes // Welding International. 1995. Vol. 9. № 4. P. 329 – 332.

[10] Chigarev V.V.,Kosenko A.A. Regulating the silicon-reduction process in welding under ceramic fluxes with an active deoxidising agent // Welding International. 1994. Vol. 8. № 10. P. 808 – 809.

[11] Kurlanov S.A., Potapov N.N,.Natapov O.B. Relationship of physical and welding-technological properties of fluxes for welding low-alloy steels // Welding International. 1993. Vol. 7. № 1. P. 65 – 68.

[12] Bublik O.V.,Chamov S.V. Advantages and shortcomings of ceramic (agglomerated) fluxes in comparison with fused fluxes used for the same applications // Welding International. 2010. Vol. 24. № 9. P. 730 – 733.

[13] Gur'ev S.V.,PletnevYu.M.,Murav'ev I.I. Investigation of the properties of welded joints produced by welding in a gas mixture and under a flux // Welding International. 2012. Vol. 26. № 8. P. 646 – 648.

[14] Parshin S.G. Using ultrafine particles of activating fluxes for increasing the productivity of MIG/MAG welding of steels // Welding International. 2012. Vol. 26. № 10. P. 800 – 804.

[15] A.C. Crespo, R.Q. Puchol, L.P. Goncalez, C. R Gómez Pérez, L.G. Sanchez, G.E. Vielsa, A.C. Sánchez Carbothermic reduction of pirolusite to obtain carbon-bearing ferromanganese and slag suited to the development of welding materials // Welding International. 2005. Vol. 19. № 7. P. 544 – 551.

[16] Barmin L.N. Effect of the composition of flux and welding wire on the properties of deposited metal of 05N4MYu type // Welding International. 1989. Vol. 3. № 2. P. 109 – 111.

[17] KazakovYu.V.,Koryagin K.B.,Potekhin V.P. Effect of activating fluxes on penetration in welding steels thicker than 8 mm // Welding International. 1991. Vol. 5. № 3. P. 202 – 205.

[18] Potapov N.N.,Feklistov S.I.,VolobuevYu.S.,PotekhinV.P. A method of selecting fused fluxes in welding pearlitic–ferritic steels // Welding International. 2009. Vol. 23. № 10. P. 800 – 803.

[19] Kozyrev N.A., Kryukov R.E., Kozyreva O.E., Lipatova U.I., Filonov A.V. Production of Welding Fluxes Using Waste Slag Formed in Silicomanganese Smelting // IOP Conference Series: Materials Science and Engineering. 2016. Vol. 125. P. 1 – 6: All-Russia Scientific and Practical Conference on Materials Treatment: Current Problems and Solutions 26 – 28 November 2015, Yurga, Russia.

[20] Kozyrev N.A., Kryukov R.E., Lipatova U.I., Kozyreva O.E. On the use of slag from silicomanganese production for welding flux manufacturing // IOP Conf. Series: Materials Science and Engineering. 2016. Vol. 150. P. 1 – 9.

[21] Kryukov N.E., Kryukov E.N., Kozyrev N.A., Kryukov R.E., Kozyreva O.A.; JSC «Novokuznetsk plant of reservoir metalware named after N.E. Kryukov»: «Welding flux», RF Patent 2576717, 05 June 2014.

[22] Kryukov N.E., Kryukov E.N., Kozyrev N.A., Kryukov R.E., Kozyreva O.A.; JSC «Novokuznetsk plant of reservoir metalware named after N.E. Kryukov»: «Welding flux»: RF Patent 2579412, 05 June 2014.

Quality of Metal Deposited Flux Cored Wire With the System Fe-C-Si-Mn-Cr-Mo-Ni-V-Co

Aleksander I Gusev [1], **Nikolay A Kozyrev** [1, a], **Ivan V Osetkovskiy** [1], **Roman E Kryukov** [1], **Olga A Kozyreva** [2]

[1] Siberian IndustrialStateUniversity
654007, Russia, Kemerovo region, Novokuznetsk, Kirova street, 42
[2] KemerovoStateUniversity, 6, Krasnay street, Kemerovo, 650000

E-mail: [a] Kozyrev_na@mtsp.sibsiu.ru

Abstract: Studied the effect of the introduction of vanadium and cobalt into the charge powder fused wire system Fe-C-Si-Mn-Cr-Ni-Mo-V, used in cladding assemblies and equipment parts and mechanisms operating under abrasive and abrasive shock loads. the cored wires samples were manufactured in the laboratory conditions and using appropriate powder materials and as a carbonfluoride contained material were used the dust from gas purification of aluminum production, with the following components composition, %: Al_2O_3 = 21-46.23; F = 18-27; Na_2O = 8-15; K_2O = 0.4-6; CaO = 0.7-2.3; Si_2O = 0.5-2.48; Fe_2O_3 = 2.1-3.27; C = 12.5-30.2; MnO = 0.07-0.9; MgO = 0.06-0.9; S = 0.09-0.19; P = 0.1-0.18. Surfacing was produced on the St3 metal plates in 6 layers under the AN-26C flux by welding truck ASAW-1250. Cutting and preparation of samples for research had been implemented. The chemical composition and the hydrogen content of the weld metal were determined by modern methods. The hardness and abrasion rate of weld metal had been measured. Conducted metallographic studies of weld metal: estimated microstructure, grain size, contamination of oxide non-metallic inclusions.

Metallographic studies showed that the microstructure of the surfaced layer by cored wire system Fe-C-Si-Mn-Cr-Mo-Ni-V-Co is uniform, thin dendrite branches are observed. The microstructure consists of martensite, which is formed inside the borders of the former austenite grain retained austenite present in small amounts in the form of separate islands, and thin layers of δ-ferrite, which is located on the borders of the former austenite grains.

Carried out an assessment the effect of the chemical composition of the deposited metal on the hardness and wear and hydrogen content. In consequence of multivariate correlation analysis, it was determined dependence to the hardness of the deposited layer and the wear resistance of the mass fraction of the elements included in the flux-cored wires of the system Fe-C-Si-Mn-Cr-Mo-Ni-V-Co. The calculated value of the average approximation error suggests that the dependence is adequate and can be used to determine the resulting indicators. These dependencies can be used to predict the hardness of the deposited layer and its wear resistance while changing the chemical composition of the weld metal.

Introduction

Mining equipment machines and mechanisms which work in abrasive and abrasive-shock conditions, breaks down early. Wear of working surface cause necessity for its recovering. That is why developing materials which could increase wear resistance of such details and technologies of their recovering is important purpose. Most advanced is the use of cored wires for depositing on the wearing out surface of details. For this reason, in our and foreign countries is conducted the development and manufacture of special surfacing cored wires [1-14]. Due to optimal selection of doping method, deposited coatings exhibit high level of hardness, abrasive and abrasive-shock wear resistance. Widespread flux cored wires of system Fe-C-Si-Mn-Cr-Ni-Mo classification type A and B by International Institute of Welding are for depositing coatings on wearing out items surface [15]. Currently flux cored wires of this system manufactured by the firm DRATEC (Germany) brand DT-SG 600 F and other cored wires firm ESAB brands OK Tubrodur 15.52, OK Tubrodur 58 O/G M are widely used in our country.

Materials and methods of research

This work continues the research started to develop new formulations of flux-cored wires for overlaying products working in conditions of abrasive wear in the mining industry [16-18], particularly researching the influence of vanadium and cobalt on the level of wear and hardness of surfaced coating once manufacturing test wire samples of system Fe-C-Si-Mn-Cr-Mo-Ni.

Manufacturing of powder wire was conducted on laboratory machine. Diameter of powder wire is 5mm, wire shell is made from metal ribbon of St3 grade. As a filler were used respective powder materials: iron powder brand GJW1 by GOST 9849-86, ferrosilicon powder brand ФС 75 by GOST 1415-93, powder of high carbon ferrochrome brand FH900A by GOST 4757-91, powder of carbon ferromanganese FMn 78(A) by GOST 4755-91, nickel powder PNK-1L5 by GOST 9722-97, ferromolybdenum powder brand FMo60 by GOST 4759-91, ferrovanadium powder brand FW50U 0,6 by GOST 27130-94, cobalt powder brand PK-1U by GOST 9721-79, plus as carbon-contained component dust was used from gas scrubbing of aluminum production, with the following component composition, wt.%: Al_2O_3 = 21-46.23; F = 18-27; Na_2O = 8-15; K_2O = 0.4-6; CaO = 0.7-2.3; Si_2O = 0.5-2.48; Fe_2O_3 = 2.1-3.27; C = 12.5-30.2; MnO = 0.07-0.9; MgO = 0.06-0.9; S = 0.09-0.19; P = 0.1-0.18.

Surfacing was conducted by manufactured wire under the AN-26S flux, on the metal plates of the brand St3 in 6 layers (to avoid mixing the weld metal with substrate) with the use of welding tractor ASAW-1250, with the following surfacing conditions: I=450 A, U=30 B, V=10 cm/min. Further plates were cut into appropriate test specimens. Chemical composition of the test specimens were defined by GOST 10-543-98 X-ray fluorescence method on spectrometer XRF-1800 and atomic emission method on spectrometer DFS-71. Chemical composition of the surfaced metal is shown in table 1. Hardness of the researched samples was measured with the use of hardness meter MET-DU. Wear resistance tests of samples was conducted on machine SMT-1 with the following conditions: capacity 30 mA, rate 20 rmp. Metallographic analysis was conducted on optical microscope OLYMPUS GX-51 in the light field in the limits of range ×100-1000 magnification after etching samples by 4% nital to determine the degree of influence changes in chemical composition on parameters of the microstructure. Martensite needles size was determined by GOST 8233-56 with × 1000 magnification. Studying of longitudinal deposited metal samples on nonmetallic inclusions existence was conducted in according with GOST 1778-70 with ×100 magnification.

Metallographic researches shown, that microstructure of surfaced layer by flux coder wire system Fe-C-Si-Mn-Cr-Mo-Ni-V-Co is uniform, there are thin dendritic branches. The microstructure consists of martensite, which is formed inside the borders of the former austenite grain, retained austenite, which is presented in small amounts in the form of separate islands, and thin layers of δ-ferrite, which is located on the borders of the former grain austenite.

It is found that the microstructure of the samples with a carbon content of 0.22 - 0.55% (samples № 1-6) observed fine-needled martensite with the needles size up to 6 microns (rating №4). The size of the former austenite grain conforms to scales ranges №6 - 7 (figure 1a, b, table 2).

By reducing the carbon content in the deposited layer to 0.19-0.2% while varying chromium, nickel, molybdenum and other present components in its composition, there is coarsening of the martensite needles and increasing the size of the former austenite grains (figure 1b, table 2).

IOP Conf. Series: Materials Science and Engineering **253** (2017) 012008 doi:10.1088/1757-899X/253/1/012008

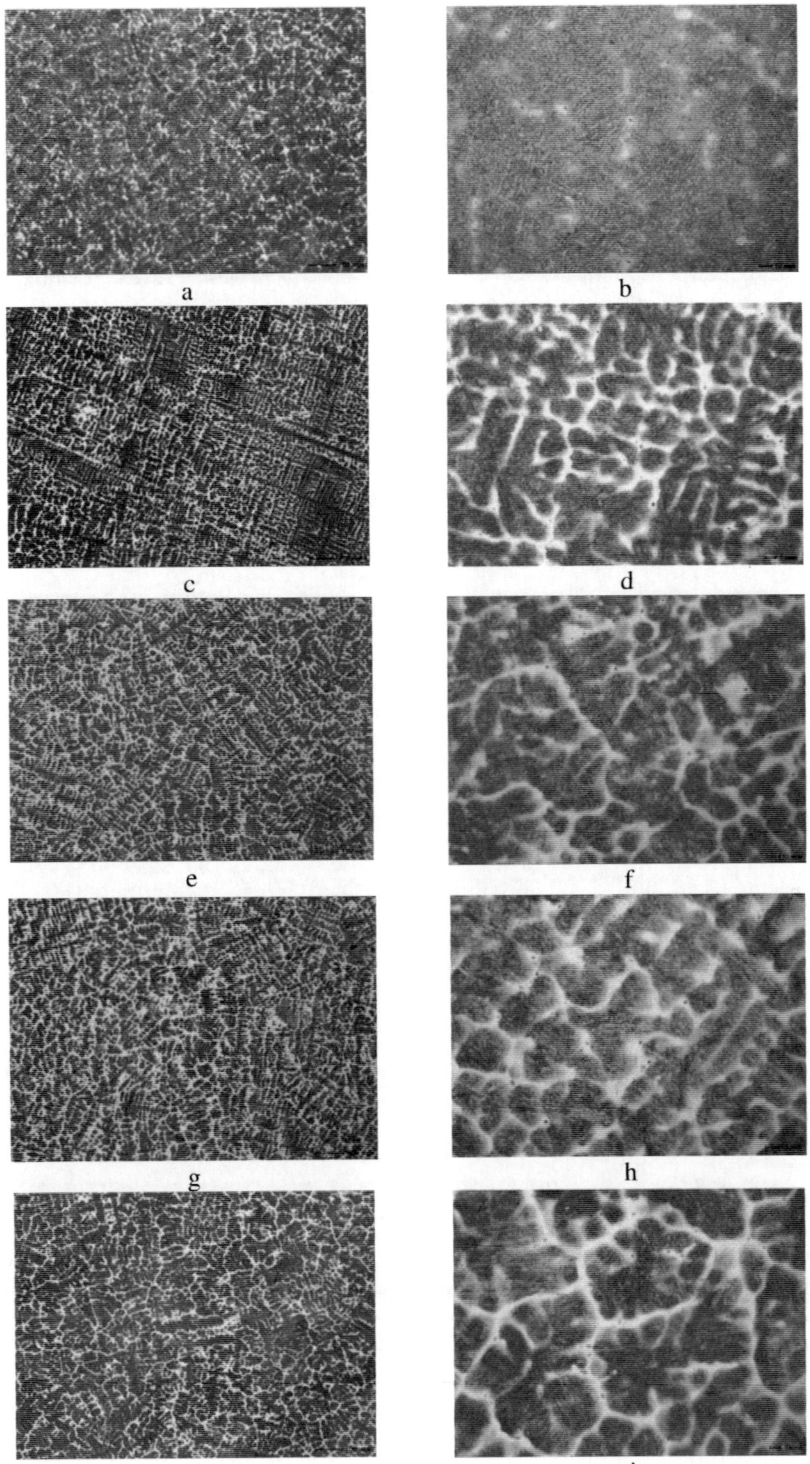

Fig. 1a. Microstructure of the researched samples, (a, c, e, g, i × 100), (b, d, f, h, j × 500):
a, b - №1; c, d-№2; e, f -№3; g, h-№4; i, j- №5

IOP Conf. Series: Materials Science and Engineering **253** (2017) 012008 doi:10.1088/1757-899X/253/1/012008

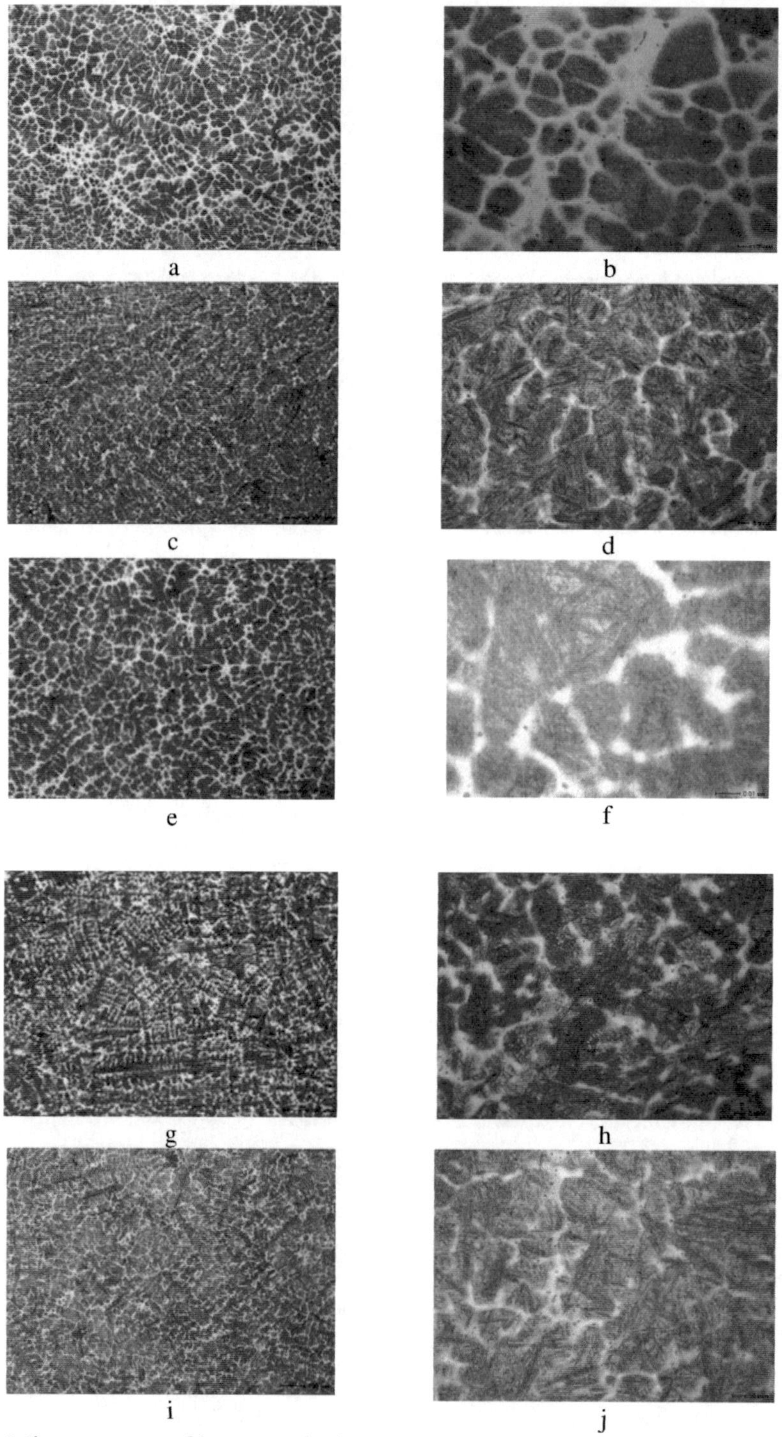

Fig. 2b. Microstructure ofthe researched samples, (a, c, e, g, i × 100), (b, d, f, h, j × 500): a, b-№6; c, d-№7; e, f -№8; g, h -№9; i, j-№10

Table 1 – Surfaced layers' chemical composition, hydrogen contain, wear and hardness

Sample number	Components weight content %														[H], cm^3/ 100 g	Samples hardness, HRC	Samples wear, grams per revolution.
	C	Si	Mn	Cr	Ni	Mo	B	V	Co	Al	Cu	Ti	S	P			
1	0.22	0.35	0.65	2.78	0.09	0.25	0	0.02	0.04	0.01	0.06	0.01	0.036	0.020	2.0	36	0.000040
2	0.43	0.37	0.84	7.04	0.42	0.49	0.01	0.03	0.06	0.02	0.08	0.01	0.038	0.020	2.0	56	0.000020
3	0.5	0.68	0.75	5.57	0.44	0.55	0.01	0.04	0.1	0.03	0.07	0.003	0.037	0.025	2.0	50	0.000015
4	0.55	0.81	0.7	5.59	0.6	0.58	0.01	0.03	0.11	0.07	0.07	0.005	0.044	0.023	2.4	53	0.000005
5	0.46	0.68	0.75	5.04	0.72	0.5	0.01	0.03	0.08	0.05	0.1	0.001	0.042	0.019	2.1	51	0.000071
6	0.44	0.74	0.73	5.59	0.86	0.53	0.01	0.03	0.09	0.06	0.07	0.02	0.038	0.020	2.3	52	0.000017
7	0.19	0.77	0.61	4.17	0.34	0.38	0.01	0.02	0.05	0.01	0.07	0.02	0.054	0.024	2.4	44.5	0.000071
8	0.19	0.63	0.65	4.06	0.3	0.38	0.01	0.03	0.06	0.01	0.08	0.03	0.056	0.019	1.7	43	0.000039
9	0.2	0.59	0.61	4.12	0.3	0.38	0.01	0.02	0.12	0.02	0.06	0.04	0.049	0.019	1.9	46	0.000044
10	0.2	0.64	0.6	4.03	0.3	0.39	0.01	0.03	0.2	0.01	0.08	0.03	0.058	0.021	2.0	30	0.000073

Table 2 – Characteristics of non-metallic inclusions and the structure of researched samples

Samplenumber	The contamination by nonmetallic inclusions, raiting		Size of austenite gaine, raiting	Size of martensite needles, mcm
	non-deformable silicates (fragile)	Oxides spot		
1	2b, 1b, 2a	1a	6, 7	3-6
2	2b, 1b, 3a	1a	7	2-6
3	1b, 2b, 3a	1a	7	2-6
4	2b, 1b, 2a, 3a	1a	6, 7	4-6
5	1b, 2b, 3a	1a	6, 7	3-6
6	1b, 2b, 3a	1a	7, 6	2-6
7	1b, 2a	1a	6	8-13
8	1b, 2b, 2a, 3a	1a, 2a	6	8-10
9	1b, 2b, 2a, 3a	1a, 2a	6	8-12
10	1b, 2b, 3a	1a, 2a	6	8-11

Assessing the influence of the flux-cored wires with chemical composition system Fe-C-Si-Mn-Cr-Mo-Ni-V-Co on wear and hardness of the deposited layer was carried out by means of multivariate correlation analysis, which allows to study the patterns of change in the resulting index, which depends on the behavior of various factors, the procedures set forth in the works [19, 20, 21].

To conduct the analysis were determined factors, that influence on resulting index, and then were selected most essential of them (table 1). After this initial check was made on the accuracy of the information, it's uniformity and conformity to the law of normal distribution. Further the model of factorial system was built. Deterministic factor analysis is used, because of presence of independent factorial signatures in the reduced system.

The connection type was determined using paired and partial correlation coefficients.It was found that the relationship has a rectilinear character, as evidenced by the same factor determinism, which is for the model of reduced factorial systems equal 1.

Calculation the key link indicators of correlation analysis was carried out in stages. Firstly, the first factor was considered, which influence is the most significant, then the second, third etc. At each stage the constraint equation and the parameters were calculated, by means of which was estimated its reliability.

Per the results of calculations obtained dependencies, the adequacy of which was checked by actual values in index of the average approximation error:

$$\bar{\varepsilon} = \frac{1}{m}\sum_{i=1}^{m}\left|\frac{Y_i - \tilde{Y}_i}{Y_i}\right| \cdot 100 , \quad (1)$$

m - the number of observations; $\tilde{Y}_i$ – calculatedvalue of resulted index; Y_i – real value of resulted index.

Hardness and wear resistance dependencies of deposited metal from its mass fraction of the elements included in the cored wire system Fe-C-Si-Mn-Cr-Mo-Ni-V-Co obtained in consequence of the analysis:

- Hardness of the deposited metal (with no regard for contained hydrogen): y=535.343 + 168.120·C - 276.437·Si - 890.442·Mn + 6.037·Cr + 108.957·Ni + 445.851·Mo - 433.688·Co - 1014.594·Al + 477.567·Cu (approximation error 0.01%);
- Hardness of the deposited metal (with the regard for contained hydrogen): y=214.819 - 163.253·C - 307.499·Si - 353.293·Mn - 21.239·Cr + 33.871·Ni + 858.517·Mo - 280.917·Co + 321.784·Cu + 13.830·H (approximation error 0.01%);
- Wear resistance of the samples (with no regard for contained hydrogen): y=0.002005 + 0.001110·C - 0.000804·Si - 0.003522·Mn + 0.000040·Cr + 0.000500·Ni + 0.000460·Mo - 0.001061·Co - 0.005525·Al + 0.003132·Cu (approximation error 0.32%);
- Wear resistance of the samples (with the regard for contained hydrogen): y=0.000260 - 0.000695·C - 0.000973·Si - 0.000597·Mn - 0.000108·Cr + 0.000091·Ni + 0.002707·Mo - 0.000229·Co + 0.002283·Cu +0.000075·H (approximation error 10.74%).

Calculated values of average approximation errorargue that obtained dependencies are adequate and they can be used to define the resulted indexes.

In consequence of multivariate correlation analysis, it was determined dependence to the hardness of the deposited layer and the wear resistance of the mass fraction of the elements included in the flux-cored wires of the system Fe-C-Si-Mn-Cr-Mo-Ni-V-Co.

Conclusions.

1. It is found that, by reducing carbon contained in the surfaced metal down to 0.19-0.2% while changing contain of chromium, nickel, molybdenum and other present components in its composition, promotes increasing in the size of martensite needles and former grain austenite.
2. In consequence of multivariate correlation analysis, it was determined dependence to the hardness of the deposited layer and the wear resistance of the mass fraction of the elements included in the flux-cored wires of the system Fe-C-Si-Mn-Cr-Mo-Ni-V-Co. These dependencies can be used to predict the hardness of the deposited layer and its wear resistance while changing the chemical composition of the weld metal.

References

[1] Kirchgaßner M, Badisch E, Franek F. Behaviour of iron-based hardfacing alloys under abrasion and impact. / Wear Journal. 2008. Vol.265. P. 772–779.
[2] Azzoni M. Directions and developments in the types of hard phases to be applied in abrase deposits against abrasion. \ Weld International. 2009. Vol.23. P. 706–716.
[3] Klimpel A. Abrasion resistance of GMA metal cored wires surfaced deposits. / Klimpel A, Dobrzanski LA, Janicki D, Lisiecki A. // Materials Processing Technology. 2005. Vol.164 165. P. 1056–1061.
[4] Wang Q, Li X. Effects of Nb, V, and W on microstructure and abrasion resistance of Fe–Cr–C hardfacingalloys. / Welding. 2010. Vol 89. P. 133–139.
[5] Metlitskii V.A. Flux-cored wires for arc welding and surfacing of cast iron / Welding International. 2008. Vol 22. P. 796-800.
[6] R. Kejžar & J. Grum.Hardfacing of Wear-Resistant Deposits by MAG Welding with a Flux-Cored Wire Having Graphite in Its Filling \ Welding International. 2005. Vol 20. P. 961-976.
[7] R. Li. Wear and high temperature oxidation behavior of wire arc sprayed iron based coatings / R. Li, D. Y. He, Z. Zhou, Z. J. Wang & X. Y. Song. // Surface Engineering. 2014. Vol 30. P. 784-790.
[8] H. R. Ma. Fe-based amorphous coating with high corrosion and wear resistance / H. R. Ma, X. Y. Chen, J. W. Li, C. T. Chang, G. Wang, H. Li, X. M. Wang & R. W. Li // Surface Engineering. 2016. Vol 46. P. 1-7.
[9] M.A. Filippov. Structure and wear resistance of deposited alloys based on metastable chromium–carbon austenite / M.A. Filippov, V.I. Shumyakov, S.A. Balin, A.S. Zhilin, V.V. Lehchilo&G.A. Rimer. // Welding International. 2015. Vol 29. P. 819-822.
[10] D.S. Liu, R.P. Liu&Y.H. Wei.Influence of tungsten on microstructure and wear resistance of iron base hardfacing alloy / Materials Science and Technology. 2013. Vol 30. P. 316-322.
[11] S. C. Lim. WearresistantWC–Co composite hard coatings // S.C. Lim, M. Gupta, Y.S. Goh&K.C. Seow. // Surface Engineering. 1997. Vol 13. P. 247-250.
[12] Zhuk Yu. Super-Hard Wear-Resistant Coating Systems / Materials Technology. 1999. Vol 14. P. 126-129.

IOP Conf. Series: Materials Science and Engineering 253 (2017) 012008 doi:10.1088/1757-899X/253/1/012008

[13] J. Hardell. Abrasive wear behaviour of hardened high strength boron steel / J. Hardell, A. Yousfi, M. Lund, L. Pelcastre&B. Prakash // Tribology - Materials, Surfaces & Interfaces. 2014. Vol 8. P. 90-97.
[14] X.T. Deng. Epsilon carbide precipitation and wear behaviour of low alloy wear resistant steels / X.T. Deng, T.L. Fu, Z.D. Wang, R.D.K. Misra&G. D. Wang // Materials Science and Technology. 2016. Vol 32. P. 320-327.
[15] Technology of electro-fusion welding steels and alloys / Publishing by B.E.Paton. M.: Metallurgy, 1974. 768p.
[16] GusevA.I. Studyingpropertiesofmetaldepositedbypowderwireofsystem C-Mn-Si-Cr-V-Mo-Co / GusevA.I., OsetkovskiyI.V.// Progressivetechnologiesandeconomicin mechanical engineering: WorkscollectionofVIINational Scientific-Practical conference for students and young students in 2volumes. Vol. 1.UrgaTechnologicalInstitute. –Tomsk: Publishing of TomskPolytechnicUniversity, 2016. – P. 91-94.
[17] A study on the properties of the deposited metal by flux cored wires 40GMFR and 40H3G2MF/ A I Gusev, N V Kibko, N A Kozyrev, M V Popova, I V Osetkovsky// IOP Conf. Series: Materials Science and Engineering 150 (2016) 012033- pages 1-9| doi:10.1088/1757-899X/150/1/012033
[18] Structure and properties of deposited layers, surfaced by using powder wires 40GMFR and 40H3G2MF / Gusev A.I., Kibko N.V., Popova M.V., Kozyrev N.A., Osetkovskiy I.V. // Bulletin of mining-metallurgy section in Russian Academy of Natural Science. Metallurgy department: Collection of Scientific works - Vol. 36.- Novokuznetsk: Publishing. SibISU, 2016 P. 174-181.
[19] WentzelE.S. Thetheoryofprobabilities / E.S. Wentzel. M.: High School, 1999. 576 p.
[20] WentzelE.S. Thetheoryofprobabilities anditsengineering applications / E.S. Wentzel, L.A. Ovcharov. – M.: Publishing center «Academy», 2003. 464 p.
[21] Gmurman V.E. Thetheoryofprobabilityandmathematical statistics / V.E. Gmurman. – M.: High School, 2003. 479 p.

The Surface Layer Mechanical Condition and Residual Stress Forming Model in Surface Plastic Deformation Process with the Hardened Body Effect Consideration

M S Mahalov [a], V Yu Blumenstein [b]

T.F. Gorbachev Kuzbass State Technical University, 28 street Vesennyaya, Kemerovo, 650000, Russian Federation

E-mail: [a] maxim_ste@mail.ru, [b] blumenstein.vu@gmail.com

Abstract. The mechanical condition and residual stresses (RS) research and computational algorithms creation in complex types of loading on the product lifecycle stages relevance is shown. The mechanical state and RS forming finite element model at surface plastic deformation strengthening machining, including technological inheritance effect, is presented. A model feature is the production previous stages obtained transformation properties consideration, as well as these properties evolution during metal particles displacement through the deformation space in the present loading step.

Introduction

It is known that the mechanical condition, such as deformation degree and first kind residual stresses (RS) - one of important metal condition parameters, which largely determines product operational life, especially in different types of alternating loads [1].

However, accurate stress-strain calculation, acting in a particular part of the workpiece, it is not always possible, most of the computational methods and algorithms, as a rule, cover only simple (one-, two-axle) loading types and don't allow to perform calculations in the complex power and thermal loading types, taking place in manufacturing and operation stages processes [2].

Along with it, the technological inheritance (TI) phenomenon significantly affects to the mechanical condition formation and transformation, because the workpiece stress-strain condition (SSC), obtained at the previous loading stages, influence to the flow of the processes at each subsequent stage.

Theory

Among the machining methods high potential for operational life increase by generating in the surface layer (SL) hardening effect and favorable compressive RS are similar to the metal yield stress values have surface plastic deformation (SPD) finishing-hardening treatment methods (Fig. 1) [3].

During SPD treatment in the deforming tool and workpiece contact area occurs asymmetrical deformation zone (DZ) ABCDEFG, limited contour lines front non-contact ABC (zone 1), front contact CD (zone 2), back contact DE (zone 3) and a rear non-contact EF (zone 4) surfaces and also FGA curve, describing the boundary of metal plastic flow zone. Due to the stress and strain fields impact, metal particles are displaced in the area of wave along the some metal plastic flow lines, forming a workpiece SL (fig. 1). The initial condition parameters, which the metal particles have before enter the deformation zone (line AG), are transformed into the accumulated on DZ exit (line GF).

Solution of the deformation and RS analytical determination after complex loading types, including SPD treatment, taking into account TI effects is possible through the technological inheritance mechanics instruments, in which [3-5]:

IOP Conf. Series: Materials Science and Engineering **253** (2017) 012009 doi:10.1088/1757-899X/253/1/012009

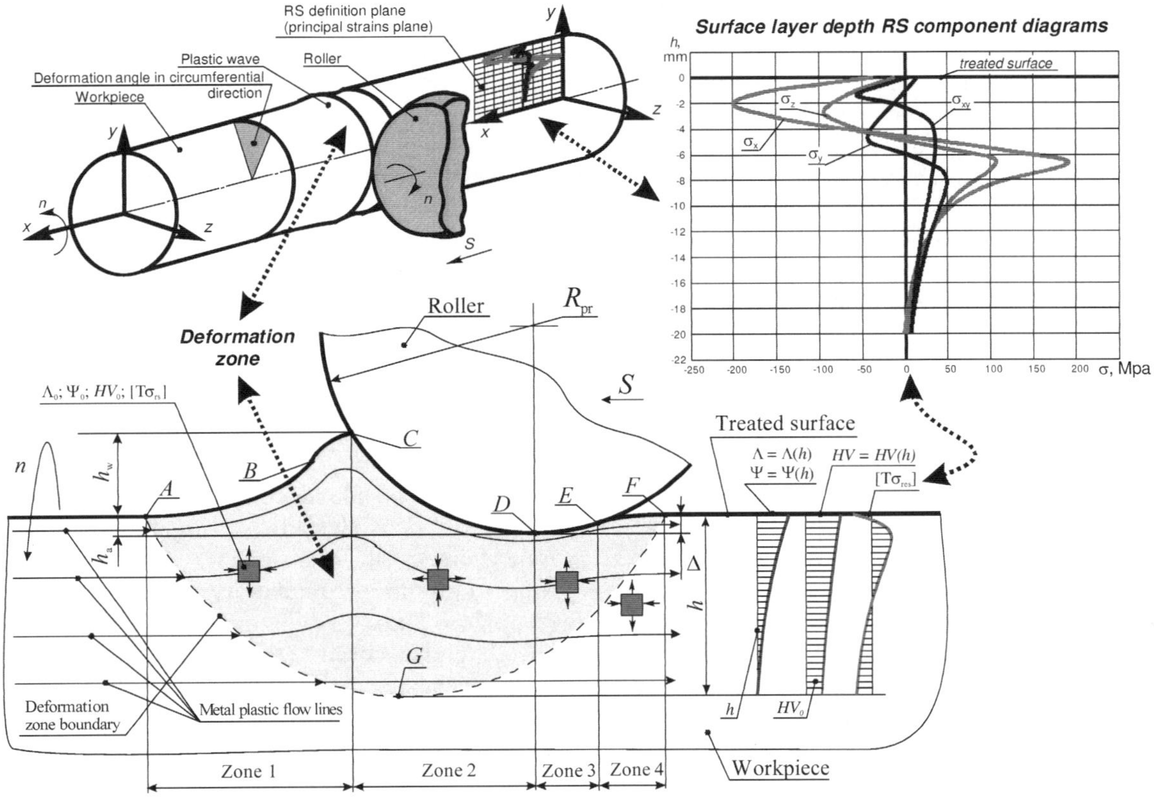

Fig. 1. Deformation zone scheme in the feed plane during processing SPD

- the SL mechanical condition formation and transformation in the machining stages and subsequent operational loading are considered as a uniform continuous process of strain accumulation, plasticity reserve exhaustion and RS transformation in surface layer metal, resulting complex non-monotonic metal loading in the DZ. Thus, along with traditional quality parameters, for describing SL properties respectively using: the cumulative shear strain level (CSSL) Λ, the plasticity reserve exhaustion level (PREL) Ψ and residual stresses components tensor $[T\sigma_{rs}]$ in coordinate system, associated with the workpiece or construction element form;
- the mechanical condition and RS formation occurs in the metal mechanical properties conditions changing due to plastic deformation: loading history affects to the processes flow at every trajectory point inside DZ;
- final mechanical condition and RS manifest themselves during the operation, transforming in each operating loading cycle; durability defines entire load prehistory.
- in the operational fatigue loading, the strain accumulation and plasticity reserve exhaustion processes continue, flowing under continuous residual stresses relaxation. With the strain limit accumulation ($\Lambda = \Lambda_p$), there is a complete plasticity reserve exhaustion occurs (Ψ = 1). This condition corresponds to the first kind residual stress tensor relaxation to negligible values ($[T\sigma_{ocm}] \approx 0$) and the fatigue crack appearance. A strain limit accumulation, a complete plasticity reserve exhaustion and fatigue crack nucleation occurs at a probable failure point, which may be located both on the surface and at some distance from it. Under cyclic loading compressive residual stress increases and tensile - reduce fatigue life.

Thus, for a complete TI phenomenon account and its influence on the SL metal final properties formation necessary to create a model representations about the metal stress-strain condition is not only taking into account the properties transformation, obtained in previous production and operation stages, but also taking into account changes in these properties during the metal particles move through DZ space at the current load stage.

To solve this problem the RS formation at roller tool SPD hardening SL treatment finite element model was created, taking into account inside DZ metal properties transformation.

During the modeling the following initial conditions and assumptions was taken:

1. Simulated isotropic material with the mechanical condition parameters corresponding to parameters of steel 45 (GOST 1050-88, hardness 160-180 HV) in the delivery state.
2. The occuring strain had only mechanical nature due to the small metal heating in the loading process.
3. Flow curve linear approximation, taking into account the metal hardening, is used.
4. In view of the relatively low error value, to simplify the calculations, the Baushinger effect were not taken into account.
5. The workpiece releases unload elastic stresses tensors was adopted zero because a set of conditions, specified in [6] for SPD treatment, was performed, and also because workpiece model was large enough compared to the appearing deformation zone geometrical dimensions.
6. The plane-strain formulation was used, during which the strain accumulation, plasticity reserve exhaustion and RS formation processes was modeled in the feed plane, passing through the workpiece rotation axis. According to [7], since the plastic wave height and, accordingly, the strain values in the longitudinal direction much larger than in the cross, from a mechanical point of view, feed plane better suits the principal strains plane concept.

As the treated surface cylindrical workpiece fragment of length L = 50 mm and height h = 20 mm was simulated (fig. 2). The fragment lower and side boundaries rigidly fixed on both axes. Stress-strain state zone size, which occurs in the processing, is much less enough than the simulated fragment, therefore, the edge effects, Appearing in this formulation, make the simulation results very small error.

Deforming indenter simulated as an absolutely rigid body and was represented by a circle with a determined profile radius arc. This assumption is valid, because numerous researches have shown, that during the SPD treatment, deforming tool has very small abrasion, thus any significant tool deformation is not fixed.

At the same time, if we consider the principal strains plane (fig. 1-2), the DZ displacement in tool feed direction has a discrete character: next DZ appears with the feed value offset from the previous, which upon SPD treatment usually varies between 0.05-0.3 mm/rotation.

This circumstance accounting is a model novelty. So in the early finite element SPD cutting and dimensional joint running-in process models, proposed in [4-5, 8], said discrete not modeled, that on the one hand simplifies the model and subsequent calculations, on the other hand - it is a disadvantage.

In the initial position the tool has some clearance from the surface (fig. 3).

On the first and any subsequent odd simulation step was performed loading - the surface direction tool displacement with the some value, presupposing indention into the surface with the specified tightness value.

On the second and any subsequent even simulation step was performed unloading - the surface direction tool retraction to the initial distance, with its simultaneous displacement along the surface with the feed value.

A total 300 steps was simulated - 150 loading steps and 150 unloading steps. The treated surface length in the presented model was l = S*n/2 = 15 mm.

The basic modeling idea is that, the certain modeled fragment section AB is positioned so, that at the first load step is not yet in the DZ space. At the same time at the penultimate load step this section has already left DZ space. In the feed direction distance l1 from the tool starting position to the section AB was 5 mm, distance l2 from section AB until the end tool position - 10 mm (fig. 2-3).

IOP Conf. Series: Materials Science and Engineering **253** (2017) 012009 doi:10.1088/1757-899X/253/1/012009

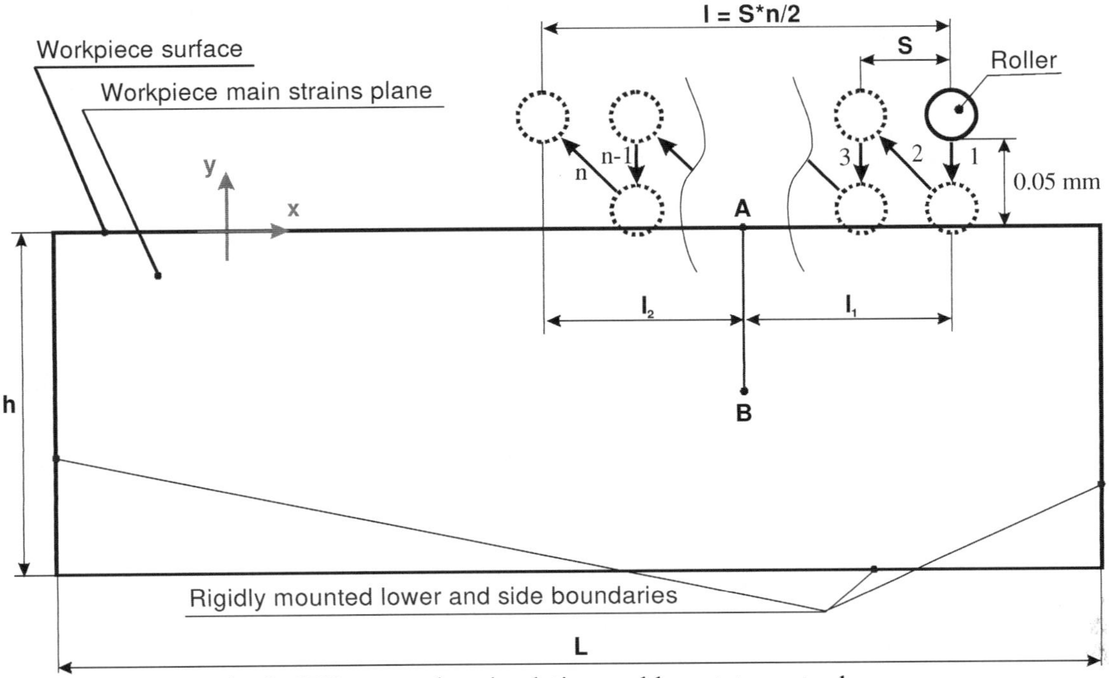

Fig. 2. SPD processing simulation problem statement scheme

Thus, in the all modeling steps implementation said section passes through a DZ space, this section is the treated SL depth trajectory, in which strain accumulation, partial plasticity reserve exhaustion and RS tensor forming based on changing the SL properties was happened. Fig. 3 is an example of the axial stresses distribution after the 1-st, 2-nd, 299-th and 300-th load steps. Marked section in which SSC parameters at each simulation step are fixed.

After solving the model in the selected section AB for each simulation step the following values were fixed: node coordinates, nodes vector displacement components, stress tensor components, elastic, plastic and total elastic-plastic strain tensor components.

For further mechanical condition parameters calculations according to [3, 7] the circumferential direction deformation angle was taken to be 100. Workpiece rotation frequency is taken equal to 300 rpm.

Thus, one workpiece rotation time is 0.2 s. Because taken deformation (DZ space) in circumferential direction is 100, cycle time, in which SL loading and unloading occurs, is 0.0054 s. It was taken, that half of this time (0.0027 s) loading occurs, and as many - SL unloading.

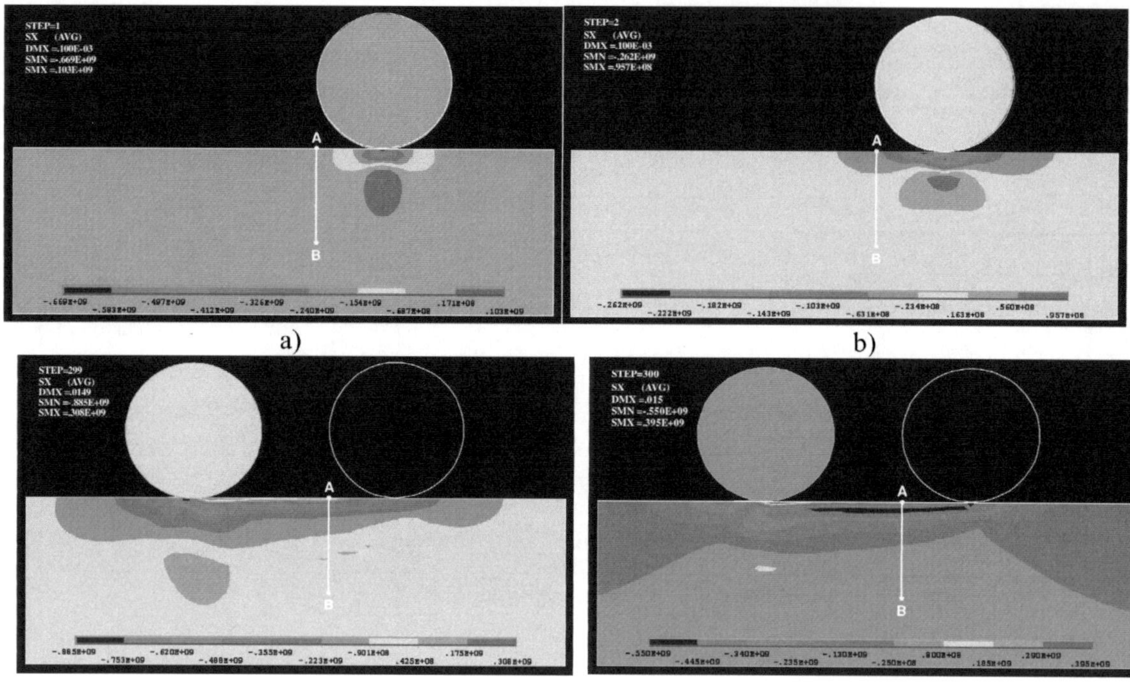

a) b)

c) d)

Fig. 3. The axial stresses distribution simulation after: a) 1-st step (loading); b) 2-nd step (unloading); c) 299-th step (loading); d) 300-th step (unloading)

According to taken assumptions, as the workpiece SL loading and unloading, accompanied by strain accumulation and plasticity reserve exhaustion, which occur in, stress condition scheme index changing conditions П [9-10]:

$$\Pi = \frac{\sigma}{T}, \tag{1}$$

where σ - is the hydrostatic pressure (defined through the stress tensor first invariant)

$$\sigma = \frac{I_1(T_\sigma)}{3} = \frac{(\sigma_{xx} + \sigma_{yy} + \sigma_{zz})}{3}; \tag{2}$$

T - the shear stresses intensity (defined through the stress deviator second invariant)

$$T = \sqrt{|I_2(D_\sigma)|} =$$

$$= \frac{1}{\sqrt{6}}\sqrt{(\sigma_{xx} - \sigma_{yy})^2 + (\sigma_{zz} - \sigma_{xx})^2 + (\sigma_{yy} - \sigma_{zz})^2 + 6(\sigma_{xy}^2 + \sigma_{yz}^2 + \sigma_{zx}^2)}. \tag{3}$$

The cumulative shear strain level Λ characterizes the strain condition and is defined as:

$$\Lambda = \int_0^t H(\tau) d\tau, \tag{4}$$

where H - is the shear strains velocity intensity (defined through the strain velocity deviator second invariant)

$$H = 2\sqrt{\left|I_2(D_\xi)\right|} =$$

$$= \sqrt{\left((\xi_{xx} - \xi_{yy})^2 + (\xi_{yy} - \xi_{zz})^2 + (\xi_{zz} - \xi_{xx})^2\right)\frac{2}{3} + 4(\xi_{xy}^2 + \xi_{yz}^2 + \xi_{zx}^2)} \quad (5)$$

Stress condition scheme index values defines each loading or unloading step maximum cumulative shear strain level Λp at which the metal fatigue crack nucleation will occur:

$$\Lambda_p = f(\Pi) \quad (6)$$

The plasticity reserve exhaustion level (PREL) Ψ is a complex parameter, that describes as stress and strain material condition. According to the Kolmogorov linear model the plasticity reserve exhaustion level is defined as [9]:

$$\Psi = \int_0^t \frac{\Lambda}{\Lambda_p} d\tau \quad (7)$$

Results and discussion

As an example of obtained results the distributions of RS tensor components, cumulative shear strain level Λ and plasticity reserve exhaustion level Ψ at the SPD treated surface layer depth are shown (after 300 simulation step) (fig. 4).

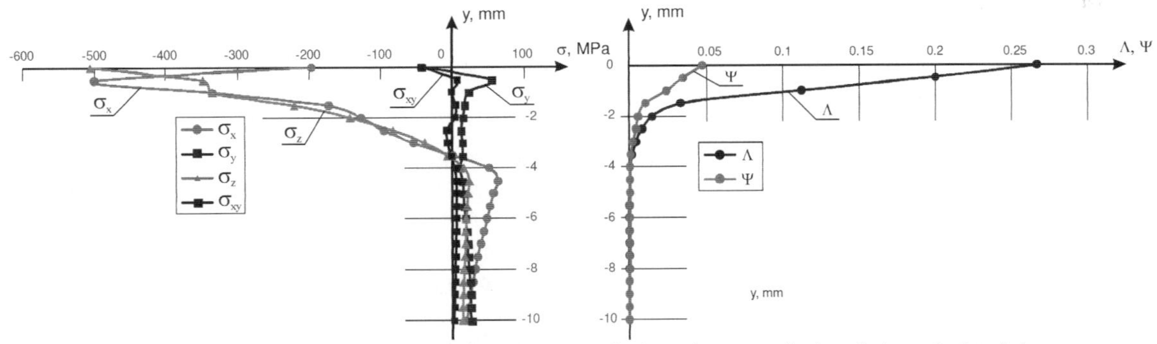

Fig. 4. The residual stress tensor components, cumulative shear strain level Λ and plasticity reserve exhaustion level Ψ distribution in the surface layer depth after the 300-th simulation step (roller profile radius Rpr = 10 mm; preset roller tightness ha = 0.05 mm; feed S = 0.1 mm/rot; workpiece rotation frequency n = 300 rpm)

The largest compressive stress values (up to -500 MPa) has an axial component σx. These fact is favorable because of a significant influence of this component to the workpiece (machine part) cyclic durability during it operational (fatigue) loading at circuit bending with rotation. Also the axial component has the compressive stress extremum located at a depth of about 0.5 mm under the workpiece surface, the compressive stress distribution depth reaches 4 mm.

These values correspond to the experimental data, obtained in [7], where the axial RS value at SPD reaches -800 MPa, and the circumferential RS - up to -400 MPa.

RS formation picture is also correspond to the results obtained in [11] for the ball SPD treatment. The author marks the possibility of compressive RS extremum formation on the surface and at some depth.

The surface layer depth cumulative shear strain level (CSSL) distribution has a maximum value Λ = 0.267 at the surface and continuously decreases as the distance from the surface. The plasticity reserve exhaustion level has a similar distribution character, the maximum value for this treatment mode is Ψ = 0.047. The Λ and Ψ significant values depth distribution is 2-3 mm.

IOP Conf. Series: Materials Science and Engineering **253** (2017) 012009 doi:10.1088/1757-899X/253/1/012009

Conclusions

Overall, the picture is in agreement with the analytical and experimental results obtained by the earlier works authors.

As a modeling result the stress-strain distribution parameters and calculated surface layer mechanical condition and residual stresses, formed during SPD hardening treatment, parameters were obtained.

A proposed model feature is the TI phenomenon consideration not only at the the previous loading (treatment or operation) stages level, but at the current loading stage level, during metal particles displacement through the deformation zone space (the effect of the hardened body effect).

Accounting this, for hardened body effect percentage estimate (TI in current loading stage), using the proposed approach further the SPD treatment simulation without hardened body phenomenon consideration requires.

References

[1] S.I. Ivanov, V.F. Pavlov, Influence of residual stresses and hardening on fatigue strength, Strength Problems. 5 (1976) 25-27.

[2] M.S. Mahalov, V.Y. Blumenstein, Residual stresses simulation on the products life cycle stages, Mechanical Engineering Journal. 12 (2014) 21-25.

[3] V.Y. Blumenstein, V.M. Smelyanskiy, Technological Inheritance Mechanics in the Machine Parts Treatment and Operation Stages, Mechanical engineering-1, Moscow, 2007.

[4] V.Y. Blumenstein, M.S. Mahalov, The surface layer Mechanics in dimensional joint running-in treatment, Hardening technologies and coatings. 2 (2006) 18-26.

[5] V.Y. Blumenstein, M.S. Mahalov, The surface layer mechanical condition analytical model of the hardened parts under cyclic loading after dimensional joint running-in treatment, Hardening technologies and coatings. 2 (2006) 18-26.

[6] A.N. Ovseenko M.M. Gajek, V.I. Serebryakov, Formation of the Machine Parts Surface Layer Condition by Technological Methods, Politechnika Opolska, Warsaw, 2001.

[7] V.M. Smelyanskiy, Machine Parts Surface Plastic Deformation Hardening Mechanics, Mechanical engineering, Moscow, 2002.

[8] A.A. Krechetov, The machine parts surface layer mechanical condition parameters calculation method, The T.F. Gorbachev Kuzbass State Technical University bulletin. 5 (2001) 27-31.

[9] V.L. Kolmogorov, B. A. Migachev, V. G. Burdukovsky, The Phenomenological model of damage accumulation and fracture under different loading conditions, Ural branch of RAS, Ekaterinburg, 1994.

[10] G.A Smirnov-Alyaev, Mechanical fundamentals of metals plastic treatment, Mechanical engineering, Moscow, 1968.

[11] D.D. Papshev Machine Parts Running-in Balls-Hardening, Mechanical engineering, Moscow, 1968.

IOP Conf. Series: Materials Science and Engineering **253** (2017) 012010 doi:10.1088/1757-899X/253/1/012010

Definition of the Mathematical Model Coefficients on the Weld Size of Butt Joint Without Edge Preparation

Vladimir P Sidorov, Anna V Melzitdinova

Togliatti State University
445020, Volga Federal District, Samara region, Togliatti, Belorusskaya str., 14

E-mail: vladimir.sidorov.2012@list.ru

Abstract. This paper represents the definition methods for thermal constants according to the data of the weld width under the normal-circular heat source. The method is based on isoline contouring of "effective power – temperature conductivity coefficient". The definition of coefficients provides setting requirements to the precision of welding parameters support with the enough accuracy for an engineering practice.

Introduction

Automatic welding of thicknesses without edge preparation takes an important place in total welding work content. The issues of welding conditions optimization and improvement of quality for these joints are in priority. The most prospective way is to solve the arising problems with a help of mathematical methods, available in engineering practice.

Research Methods

This paper studies the prospects of use the analytic model of the normal-circular heat source on the surface of flat layer [1, 2] for improving the technology of argon arc welding. There were studied the experimental data for weld width in a work [3], in which it is described the welding for butt joints made of steel 12X18H9T of 1.5; 2; 3 mm thick on a cooper backing with a slot. The facing weld width $E1$ and the turning roll width $E2$ were measured. Experiments in the work [3] were conducted under the changing of welding speed V_w for several arc current I.

As a basis for analysis we study the results for the thickness of δ = 3mm with the arc length of L_a = 3 mm. Despite of intensive experiment performance, there was a significant unit variation of results, which rises up to 10% from the average values. This instability can be described by the influence of surface phenomena on the metal spreading of welding pool on the weld border.

It was found that the dependencies of weld width from weld speed are hyperboles. According to this fact, the data of [3] were processed by known software [4], and coefficients of hyperbolic dependencies were defined. Each dependency is for one current value and arc length value. It turned out that in absolute, the average relative deviations of the approximated data do not exceed 1% for facing weld width $E1$. For turning roll width $E2$ the deviations are slightly higher, but they do not exceed 4%. Figure 1 shows the example of approximating dependencies. These curves allow using not only experimental values, but any intermediate values for the analysis.

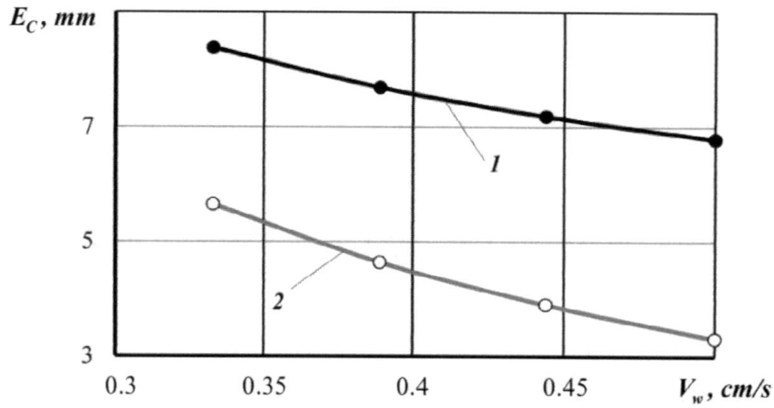

Fig.1. Calculated data of weld width based on the experimental points
1. For $E1$; 2. For $E2$. I_a= 190 A.l_a= 2 mm

It is natural to assume that approximating data, averaging the whole sets of experiments, in general better describe the real dependencies than the experimental points. Therefore, approximating data were further used for the analysis.

In [3] the weld parameter was calculated by numerical method considering the change of thermal constants on the temperature and a separation of heat source into two sources – the normal-circular heat source and the linear heat source with correspondence of their power of 0.8/0.3-0.8/0.2. According to the data in [3], the change of temperature conductivity coefficient ***a*** from the temperature was calculated. As a result, in temperature interval from 100ºC to 900ºC the coefficient ***a*** was rose from 0.04 cm²/s to 0.07 cm²/s. Figure 2 shows the calculation results.

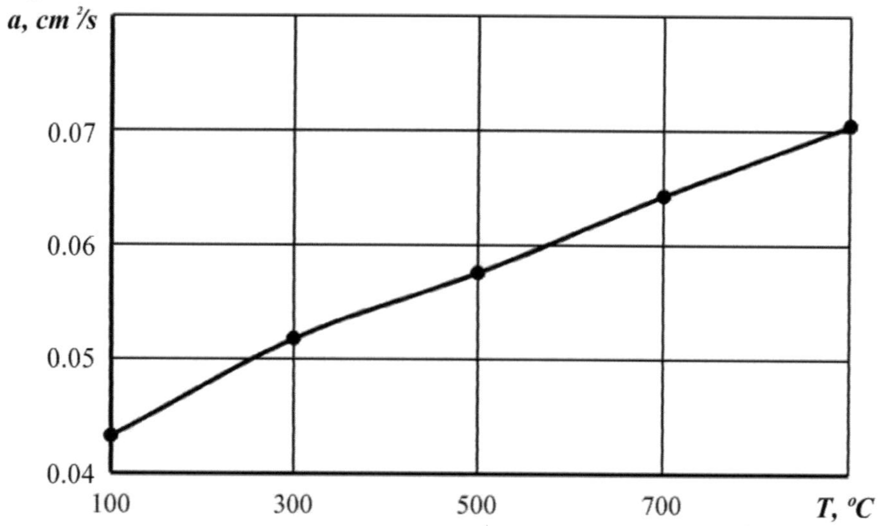

Fig.2. Dependency of ***a*** from the temperature for steel 12X18H9T

The temperature expression from the normal-circular heat source shows that the correspondence of effective power q_e and volumetric heat capacity $c\rho$ can be as one of the parameters of $Q = q_e / c\rho$. Then the influence of q_e and the temperature conductivity coefficient ***a*** on the weld size were calculated with $c\rho$ =4.5 J/(cm³ºC).

To calculate the weld width, we developed a BASIC software program. A settling time for welding was 20 seconds, each of which was divided into 50 parts. A dichotomy method was used as a numerical one [4,5]. The weld width along the X axis was defined with a space of 1 mm, and the calculation precision was 0.1 mm.

Results and Discussion

There were built dependencies of two experimental values for weld size *E*1 and *E*2 from q_e for several coefficients ***a***, which varied within the limits of dependency that is shown in Fig. 2. The values of *E*1 and *E*2 were chosen for average value of variation interval V_w. The example of such dependencies for *E*1 is shown in Figure 3.

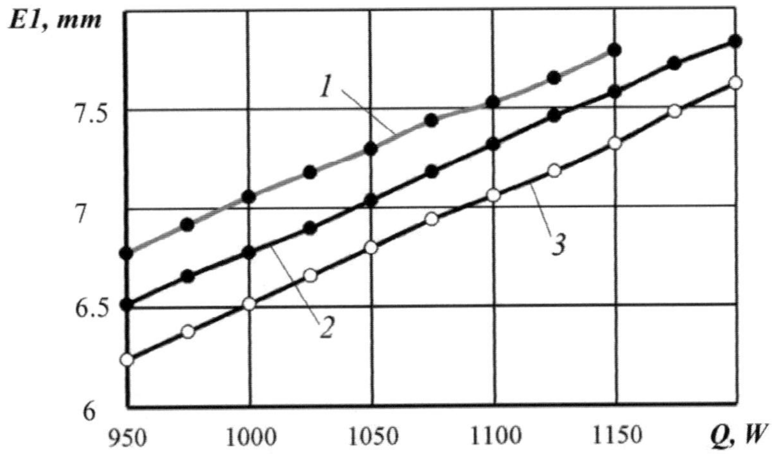

Fig.3. Series of width dependencies *E*1 from effective power
1. $a = 0.04$ cm²/s; 2. $a = 0.05$ cm²/s; 3. $a = 0.06$ cm²/s.

Drawing a line, parallel to the *Q* axis, we obtain points to draw the isoline *Q*-***a***.

The dependence of the calculated weld width *E*1 and *E*2 from q_e was plotted for three ***a*** values consequently for three values of axial heat flow q_O. Welding speed remained constant $V_w = 0.444$ cm/s. Fig. 4 shows the example of dependencies for q_O =3400 W/cm².

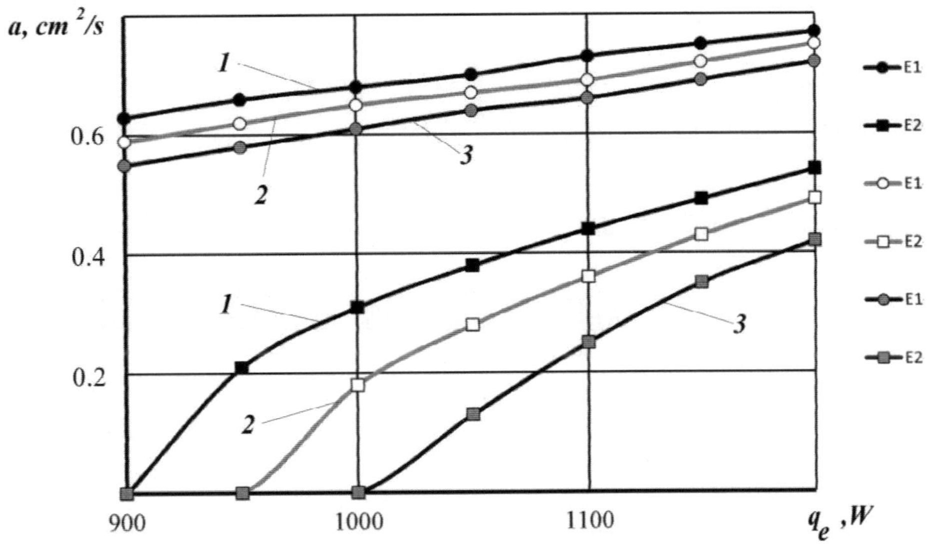

Fig. 4. Series of width dependencies *E*1 and *E*2 from effective power
1. $a = 0.04$ cm²/s; 2. $a = 0.05$ cm²/s; 3. $a = 0.06$ cm²/s.

Curve for facing weld width and for turning roll width are completely different. On the studied conditions, the derivative dE/dq_e for the turning roll is much bigger, than for facing weld width. The feature is more pronounced in low power. In high power, the gradients become similar.

Experimentally taking the weld width values of $E1$=7.1 mm, $E2$=3.8 mm, we have isolines "q_e – temperature conductivity ***a***" for two weld sizes (Fig.5, 6).

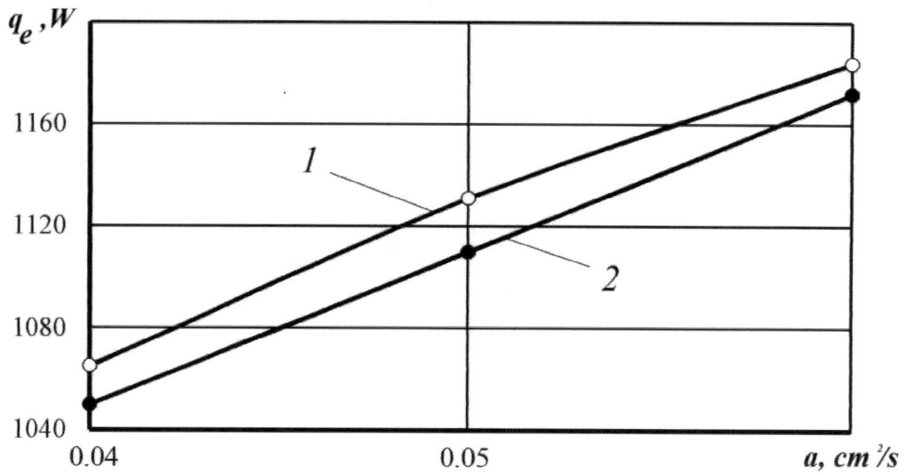

Fig. 5. Isolines for weld sizes. 1 - $E1$; 2 - $E2$. q_O =3400 W/cm^2.

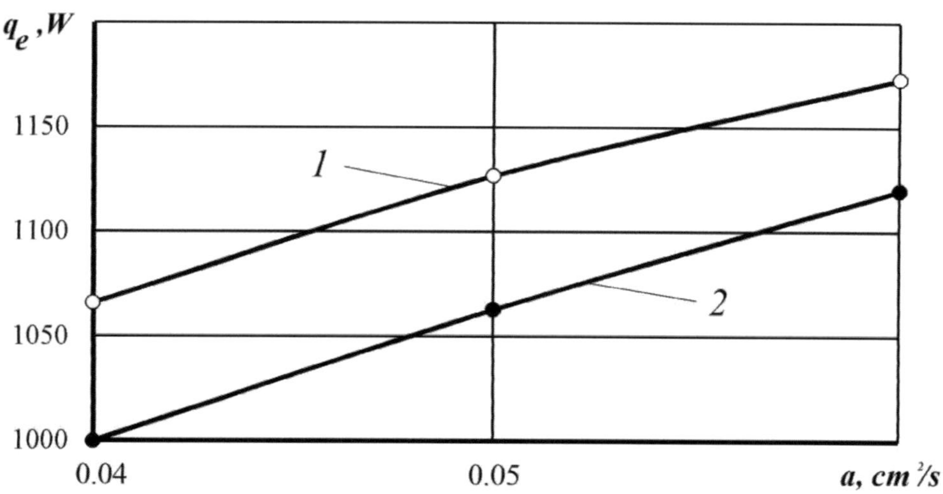

Fig. 6. Isolines for q_O =4200 W/cm^2. 1 - $E1$; 2 - $E2$.

Isolines on Fig. 3 for q_O =3400 W/cm^2 are closer to each other than isolines for q_O =4200 W/cm^2. Isolines for q_O =5000 W/cm^2 are much farer from each other. To define the model parameters, the isolines on Fig.5 were chosen. As there was not their cross point, the optimal parameter values were the average coordinate values of the segment of the minimal distant between the isolines. For Fig. 5 the value q_e = 1170 W, ***a*** = 0.059 cm^2/s. The Volt equivalent of effective power is 1170 W/190 A = 6.2 W/A. Effective efficiency $\eta_e \approx 0.62$. Diameter of heating spot on the studied condition was D_H = 1.15 cm, concentration coefficient of heat flow k=9.12 cm^{-2}.

According to these values, the weld sizes on nominal condition I=190 A, V_w =0.444 cm/s was calculated. The values of $E1$= 7.0 mm, $E2$= 3.9 mm were taken. The difference with experimental values is 0.1 mm that concurs with the calculation precision of the weld width. It may be concluded that the model parameters were right.

After this, the resulting values q_O; $\boldsymbol{a}$ were used to calculate the weld width on the others arc current. The effective power was changed in proportion with a help of the Volt equivalent of effective power, which practically does not depend on the arc current. [5, 6, 7] Herewith, there were changes of the heating spot diameter D_H and the concentration coefficient of heat flow *k*. [8] The comparison of experimental and calculated data are shown in the Table.

Absolute and relative deviations for current values are shown in the Table.

Table

Weld width , deviation	*I*=170 [A]		*I*=210 [A]		*I*=225 [A]	
	Experiment	Result	Experiment	Result	Experiment	Result
*E*1, [mm]	6.5	6.2	8.2	7.7	8.2	7.6
Δ*E*1/*E*1, [%]	-4.6		-6.1		-7.3	
*E*2, [mm]	2.5	1.6	4.4	5.4	4.7	4.8
Δ*E*2/*E*2, [%]	-36.0		+22.7		+2.1	

Note: for current I = 225 A the welding speed V_w = 0.5 cm/s.

Data from the Table show that resulted values q_e and $\boldsymbol{a}$ are well description of the facing roll width in a range of ±15%, that differ from the turning roll width. Obviously, it is connected to the fact that the welding was on a cooper backing. Also an intensive heat removal from the back surface influences on the result.

Conclusion

1. The methods of definition for three mathematical model coefficients of normal-circuit heat source on two weld width values were developed. These coefficients are the correspondence of effective power to volumetric heat capacity of parts, temperature conductivity coefficient and axial heat flow.
2. Calculated data for turning roll width are two times inaccurate then the difference of experimental data. Herewith, to define the relative deviations of welding conditions for turning roll, this method should be used for experimental data on the turning roll width.

References

[1] Denisov P.V., Mirlin G.A. The calculation of heating temperature of sheet metal by normal distributed source for spot arc pulse welding // *Svarochnoe proizvodstvo* [Welding Engineering]. 1974, No 1, PP. 3-6. (in Russian)
[2] Sidorov V.P., Khurin S.A. Penetration modeling for arc welding of butt joint without edge preparation // *Svarka i Diagnostika* [Welding and Diagnostic]. 2011, No 6, PP. 36-40. (in Russian)
[3] Nikiforov R.V. *Sovershenstvovanie tekhnologii avtomaticheskoiy argonodugovoiy svarki neplavyaschimsya electrodom stykovykh soedineniy iz korrozionno-stoyikikh staleiy s uchetom termodeformatsionnykh protsessov v izdelii.* Kand. Tech., Diss. [Technology improvement of automatic argon-arc welding by nonconsumable electrode for butt joints made of corrosion-resistant steel considering the processes of thermal deformation in the unit. Tech. Sc. Cand., Diss.]. Ufa, 2014, 181 p.
[4] Dyakonov V.P. *Spravochnik po algoritmam i programmam na yazyke Beisik dlya personalnykh EVM* [Hand book for algorithms and programs on BASIC for personal computers]. Moscow, Nauka, 1987, 240 p.

IOP Conf. Series: Materials Science and Engineering **253** (2017) 012010 doi:10.1088/1757-899X/253/1/012010

[5] Gladkov E.A. *Avtomatizatsiya svarochnykh protsessov* [Automation of welding processes] / E.A. Gladkov, V.N. Brodyagin, R.A. Perkovskiy. – Moscow, Izdatelstvo MGTU im. N.E. Baumana, 2014, 421 p.
[6] Sidorov V.P. The research of allowed deviation of parameters for arc double-side welding / V.P. Sidorov, A.V. Melzitdinova // *Svarochnoe proizvodstvo* [Welding Engineering]. 2016, No 3, PP. 11-15. (in Russian)
[7] *Svarka v mashinostroenii: spravochnik.* [Welding in machine building engineering: a hand book]. In 4 volumes. Vol.4 / under the editorship of Yu.N. Zorin. Moscow, Mashinostroenie, 1979, 512 p.
[8] Lenivkin V.A. *Avtomatizatsiya svarochnykh protsessov* [Automation of welding processes] / V.A. Lenivkin. Rostov-on-Don, Izdatelskiy tsentr DGTU, 2008, 205 p.

The Analysis of Welding Conditions Under the Flux of Double-Sides Joints Without Edge Preparation

Vladimir P Sidorov

Togliatti State University
445020, Volga Federal District, Samara region, Togliatti, Belorusskaya str., 14

E-mail: vladimir.sidorov.2012@list.ru

Abstract: This paper represents the results of analysis for automatic welding conditions under the flux of double-sides butt joint without edge preparation. As the process characteristics, a specific energy of welding, joint formation rate, average weld width, fusion rate of base metal and other parameters were used. It is determined an optimal joint rate of about 1 cm^2/s, that can be used to calculate welding conditions. The paper founds the use of linear dependence between specific energy of welding and cross-section area of base metal's fusion.

Introduction

The analysis of automatic welding conditions under the flux of the butt joints is of interest to create adequate methods for calculating of optimal welding conditions. As it is noted in [1], a learned method, which is represented in [2], is not meet modern requirements, because it is based on the calculation of the heat transmission from a point heat source, which is active on the surface of semi-infinite body. The conditions of double-sides welding do not fully correspond to this scheme, because they do not consider the influence of the unit thickness and the distribution of a heat source to the fusion of the unit. In our opinion, the calculating method of welding conditions should consider patterns to energetic characteristics of the process. From these positions, this paper researches the welding conditions, which are typical for double-sides joint described in [1, 3].

Research Methods

We study energetic characteristics and parameters of fusion efficiency of base metal described in [1, 3]. Source data for welding conditions are shown in the Table 1. Parameter marks of common use for welding engineering are used in this table.

Table 1

Reference	δ, [mm]	d_e, [mm]	U_a, [V]	I_a, [A]	V_w, [cm/s]	V_e, [cm/s]
[1]	20	3.0	39.5	680	1.1	5.25
[1]	20	3.0	44.0	850	0.69	7.83
[3]	16	4.0	32.0	712	0.67	2.80

Experimental geometrical characteristics of welds, defined by the photos of macro sections with a help of today's software, are shown in the Table 2.

IOP Conf. Series: Materials Science and Engineering **253** (2017) 012011 doi:10.1088/1757-899X/253/1/012011

Table 2

I_a, [A]	Weld width B, [mm]	Penetration H, [mm]	F_O, [mm^2]	F_f, [mm^2]
712	20.0	9.5	115	52
680	17.0	8.5	58	29
850	22.0	13.9	142	78

Herewith, found cross section areas of base metal F_O and filled metal F_f are completely different from data in [1]. The area of filled metal on arc current I_a= 680 A was concurred both for the calculation on the speed of electrode fusion, and on the dimension of areas on micro-sections. The following data were calculated by experimental values:

1. Average width of fusion for base metal

$$B_a = \frac{F_O}{H}, \quad (1)$$

where H – penetration, mm.

2. Joint formation rate J

$$J = V_W \cdot H. \quad (2)$$

3. Specific energy of welding E [4].
4. Welding frequency [5]

$$f = \frac{F_O}{J}. \quad (3)$$

5. Specific consumption of fused base metal for joint formation

$$g_O = B_a \cdot \rho, \quad (4)$$

where ρ – density of base metal, for steel ρ=7.8 g/cm^3.

Besides, we calculate coefficients for fusion of base metal α_O, and electrode metal α_P, and thermic efficiency of the process η_t. For η_t a heat content of welding pool was taken as 7800 J/cm^3.

Results and Discussion

The calculation results are shown in the Tables 3 and 4.

Table 3

I_a, [A]	B_a, [cm]	J,[cm^2/s]	E, [kJ/cm^2]	f, [Hz]
712	1.21	0.64	2.95	1.8
680	0.68	0.94	2.37	0.62
850	1.02	0.96	3.23	1.48

Table 4

I_a, [A]	g_O, [g/cm^2]	α_O, [g/AHr]	α_P, [g/AHr]	η_t
712	9.44	30.9	14.5	0.32
680	5.30	26.3	13.9	0.23
850	7.96	32.4	18.7	0.25

In the Tables 3 and 4, data show that the most optimal condition is the second one for current I_a=680 A. Regarding to the first condition, the second one provides 1.5 times more joint rate J under the higher energy cost reduction E. It is proved by the minimal welding frequency f, which characterizes the efficiency of use of fused base metal on joint formation, as g_O does. The fusion coefficient of base metal α_O is approximately the same for all conditions. It is in average two times more then α_P, that is caused by the similar heat content reduction of welding pool's metal in comparison with the heat content of electrode metal. Herewith, it should be considered, that the second condition is not fully provide requirements for double-sides welding – the penetration of base metal should be at least 50% of thick. The first and the third conditions provide such requirement. For the first condition, the correlation of penetration to the thickness is $H/\delta = 0.59$, and for the third condition it is $H/\delta = 0.65$. The joint rate $J \approx 1.0$ cm²/s for the second and the third conditions can be a reference point for optimal welding conditions of different thickness.

The important index of the weld is a fusion area of base metal, which is difficult to be defined. Together with the cross-section area of filled metal, F_O influences on a part of base metal in a weld ψ_O. For the data shown in the Table 1, a linear dependence of fused area of base metal was selected

$$F_O = K \frac{q_e}{V_w \delta}, \tag{5}$$

where q_e – effective arc power, kW, K – coefficient.

A quantity q_e/δ was called a specific effective power, i.e. effective power, which suits on the welding thickness unit. Its calculation can be useful for welding of double-sides welds. For example, with the help of specific effective power a fused area of base metal can be estimated with a help of the calculated scheme of linear heat source. So, $q_e/V_w\delta$ can be called as specific running energy of welding.

An effective power of arc was calculated by the values of effective efficiency $\eta_e=0.829$, described in [6].

The values of K, comparison of calculated F_{OC} and experimental F_O cross-section areas of base metal, which were found by the formula (1), are shown in the Table 5.

Table 5

q_e, [kW]	K, [cm⁴/kJ]	F_O, [cm²]	F_{OC}, [cm²]	Δ,[%]
15.11	$8.3\cdot10^{-2}$	1.17	1.06	-9.4
18.66	$6.8\cdot10^{-2}$	0.58	0.64	+10.3
26.50	$7.4\cdot10^{-2}$	1.42	1.40	-1.4

The average value of K_a=7.5 cm⁴/kJ. The average on absolute value and relative deviation of calculated data from the experimental ones is 7% that is fully acceptable for engineering calculations. To use the formula (5) for calculating of welding conditions, it is necessary to study the dependence of specific welding energy E of one weld from the welded thickness.

To switch to the particular welding conditions, it is necessary to use the formula of arc voltage in [2]

$$U_a = 20 + 0.05 \frac{I_a}{(d_e)^{0.5}} \pm 1. \tag{6}$$

For the studied welding conditions, the calculation of arc voltage on formula (6) shows good coincidence of these conditions according to [2] and deviation of + 18% according to [3].

To determine the penetration area of the base metal, the most prospective is the use of the formulas of heat transfer in welding [7]. For arc welding by a consumable electrode, the penetration of the base metal depends on the polarity of the arc. At the same time, the effective arc power does not depend on the polarity. In [8] this is explained by the influence of the liquid interlayer under the arc. Our hypothesis is that the power, transmitted by a consumable electrode to the product, affects the penetration differently than the arc power [9].

Based on this, a technique for calculating the penetration area of the base metal is proposed. This technique is the following. From the total effective arc power, it is necessary to take the power, transferred to the welding pool by the electrode metal. For testing were used the experimental data on the determination of the penetration while welding under the flux, given in [3]. To calculate thermal cycles using the equivalent heat source method, the volume heat capacity $c\rho$=4.5J/(cm³°C), the thermal conductivity λ=0.32 W/(cm °C) was assumed in [3]. It gives the temperature conductivity coefficient a=0.071cm²/s. In this mode, it has been obtained a cross-section of the penetration on the macro section with a weld width of B=20 mm and a penetration of h=9.5 mm. With the help of modern software, the areas of penetration of the base metal F_O=1.17 cm^2 and the filled metal F_f= 0.52 cm^2 were measured.

The effective power was determined by adopting the effective arc efficiency under the flux with the same mode, η_e=0.829, according to the data of [6]

$$q_e = 0.829 \cdot 22784 = 18888 \text{ W}.$$

Using the formula for a normal-circular heat source [7], the penetration area of the base metal was first determined by a known technique, using the total effective arc power. To do this, the maximum width of the welding pool in sections along the thickness of the plate with a step of 1 mm was calculated with a help of special software in BASIC. Based on the obtained dimensions, a transverse profile of penetration of the base metal was built.

By the value of the effective power q_e and the experimental value of the penetration, the diameter of the heating spot of a normal-circular heat source D_H=1.34cm was determined. Therewith, the axial heat flow was q_M=40000 W/cm^2, and the heat flow concentration factor was k=6.63 cm^{-2}. The maximum calculating penetration at y=0 was H_M=9.43 mm, which differs from the experimental value by only 0.07 mm. The large value of q_M corresponds to the data obtained in [10], and it is due to the immersion of the arc in the welding pool under the arc pressure.

Fig. 1 (curve 1) shows the total calculated profile of penetration of the base metal. According to it, the average penetration was H_a=0.66 cm, and the calculated penetration area of the base metal is F_O=146 mm^2. The deviation of the calculated value of the area from the experimental value is 146-117 = 29 mm^2, which is 25%.

As before, by the proposed technique for reversing polarity of the arc, the penetration area of the base metal was determined. The power q_{MA}, transferred to the weld by the electrode metal anode, was determined too.

According to the [9], the voltage equivalent of the electrode power of the anode with the heating of the stick-out, is U_e= 5.3 W/A. So,

$$q_{MA} = 5.3 \cdot 712 = 3774 \text{ W}.$$

The power for usage in calculating of the penetration of base metal is

$$q_O = q_e - q_{MA} = 18888 - 3774 = 15114 \text{ W}.$$

When calculating the penetration of the base metal, the diameter of the heating spot D_H=1.34 cm was kept. For this effective power, an axial heat flow q_M =32000 W/cm^2, and a concentration coefficient k=6.63 cm^{-2} was obtained. The maximum calculated penetration at y=0 was H_M=8.18 mm, the average penetration was H_a = 0.57 cm, the cross-section area of the base metal was F_O=113 mm^2 (curve 2 in Fig. 1). The deviation from the experimental value of the area is 113-117= 4 mm^2, which is -3.4%, that is approximately 7 times more accurate than the known method provides.

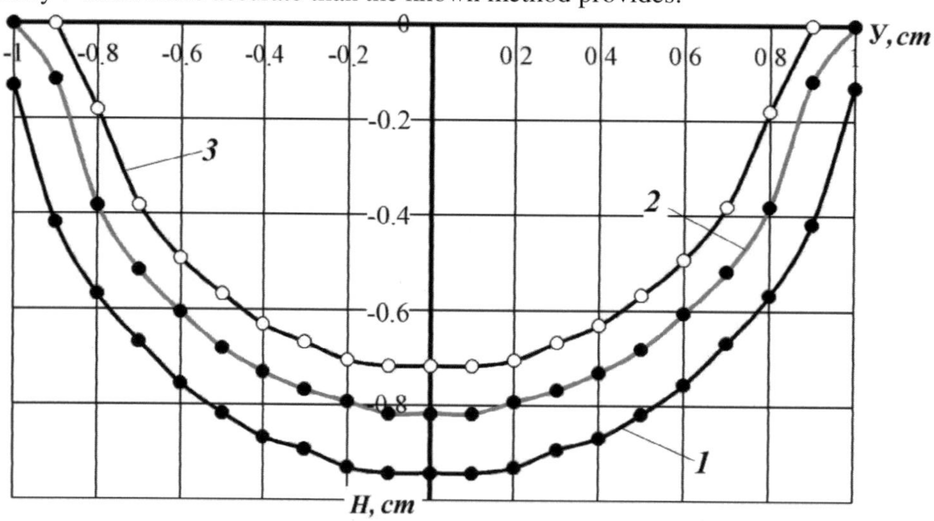

Fig.1. Calculated profiles of the penetration:

1 – without polarity; 2 – reversing polarity; 3 – normal polarity

As before, the penetration area of the base metal was determined by the proposed technique for normal polarity of the arc. The power q_{MA}, transferred to the weld by the electrode metal anode, was determined too.

According to the [6], the voltage equivalent of the electrode power of the cathode with the heating of the stick-out, is U_e= 9.3 W/A. So, the power, transferred by electrode metal to the weld is

$$q_{MA}= 9.3\cdot 712 = 6622 \text{ W}.$$

The power for usage in calculating of the penetration of base metal is

$$q_O= q_e- q_{MA} = 18888 - 6622 = 12266 \text{ W}.$$

When calculating the penetration of the base metal, the diameter of the heating spot D_H=1.34 cm was kept. For this effective power, an axial heat flow q_M=25000 W/cm^2 was obtained. The maximum calculated penetration at y=0 was H_M=7.18 mm, the average penetration was H_a = 0.50 cm, the calculated penetration area of the base metal was F_O= 91 mm^2 (curve 3 in Fig. 1). The deviation from the experimental value of the area, obtained on the reversing polarity is 91-117= -26 mm^2, which is -22%.

The ratio of the obtained penetration areas of the base metal corresponds to the relations of penetrations in [2], depending on the polarity of the arc.

Conclusion

1. Optimal welding condition for one run from double-sides weld should be chosen considering available data about optimal joint rate, which is about 1.0 cm^2/s, and considering characteristics of efficient use of fused base metal.
2. The idea of dependence of fusion area for base metal of double-sides weld's one run, which is proportional to specific running energy of welding, is prospective.
3. The known methods for calculating the penetration of the base metal on welding by a consumable electrode, do not take into account the effect features on the electrode power process. It is most clearly seen when considering arc welding by a consumable electrode at different polarities.
4. Calculation of the penetration area of the base metal on welding by a consumable electrode should be carried out by reducing the effective power to power transferred to the welding pool by the electrode metal.

References

[1] Zaiytsev N.L. Calculation and experimental estimation method of weld configuration / N.L. Zaiytsev, A.M. Osipov, D.V. Gladkovskiy, I.A. Ilin // *Sovremennye problemy svarochnogo proizvodstva* [Today's problems of welding engineering] : collection of studies / under the editorship of M.A. Ivanov, I.A. Ilin. – Chelyabinsk, Izdatelskiy tsentr YuUrGU, 2016, PP. 161- 169. (in Russian)

[2] Akulov A.I. *Tekhnologiya i oborudovanie svarki plavleniem* [Technology and equipment of fusion welding] / A.I. Akulov, G.A. Belchuk, V.P. Demyantsevich. Moscow, Mashinostroenie, 1977, 432 p.

[3] Tsvelev R.A. Error estimate of reproduction of temperature field under the welding on flux basing on equivalent heat source / R.A. Tsvelev // *Svarochnoe proizvodstvo* [Welding Engineering], 2013, No 6, PP.11-15. (in Russian)

[4] *Teoriya svarochnykh protsessov* [Theory of welding processes] / A.V. Konovalov, A.S. Kurkin, E.L. Makarov, V.M. Nerovnyiy, B.F. Yakushin; under the editorship of V.M. Nerovnyiy. Moscow, Izdatelstvo MGTU im. N.E. Baumana, 2007, 752 p.

[5] Sidorov V.P. The criteria of arc fusion on welding / V.P. Sidorov, N.A. Borisov // *Svarka i Diagnostika* [Welding and Diagnostic], 2013, No 6, PP.24-27.

[6] Tsvelev R.V. and al. Physical and mathematical modeling of process of arc welding under the flux / R.V. Tsvelev, D.V. Slezkin, V.A. Sudnik and al.// *Sbornik dokladov 6oiy mezhdunarodnoiy*

konferentsii Matematicheskoe modelirovanie i informatsionnye tekhnologii v svarke i rodstvennykh protsessakh [Proc. 6th International Conf. Mathematical modeling and Informational Technologies in Welding and Allied Processes] 29.05–1.06.2012. Kiev, 2012, PP. 156-163.
[7] Karkhin V.A. *Teplovye protsessy pri svarke* [Thermal processes on welding] / V.A. Karkhin. Saint-Petersburg, Izdatelstvo Politekhnicheskogo universiteta, 2015, 572 p.
[8] Kobernik N.V. Influence of the type and polarity of the current on the penetration of the electrode and base metal for arc welding under flux / N.V. Kobernik, G.G. Chernyshov, P.P. Gvozdev and al. // *Svarka i diagnostika* [Welding and Diagnostic], 2011, No 5, PP. 24-27. (in Russian)
[9] Sidorov V.P. Influence of the type and polarity of the current on the penetration of the electrode and base metal for arc welding under flux / V.P. Sidorov // *Svarka i diagnostika* [Welding and diagnostic]. 2013, No 3, PP. 20-23. (in Russian)
[10] Berezovskiy B.M. Calculation of the parameters of the heat flow distribution of the surface welding arc / B.M. Berezovskiy, V.A. Stikhin // *Svarochnoe proizvodstvo* [Welding Engineering], 1980, No 2, PP. 1-4. (in Russian)

IOP Conf. Series: Materials Science and Engineering **253** (2017) 012012 doi:10.1088/1757-899X/253/1/012012

Diagnostics of the Technical State of Bearings of Mining Machines Base Assemblies

Boris L Gerike [a], **Andrey A Mokrushev**

Federal State Budget Scientific Institution «The Federal Research Center of Coal and Coal Chemistry of Siberian Branch of the Russian Academy of Sciences», Institute of Coal., Leningradsky av., 10, Kemerovo, 650065, Russia

E-mail: [a] gbl_42@mail.ru

Abstract: The article reviews the methods of technical diagnostics of equipment used during maintenance of mining machines in accordance with their actual technical state, and considers the basics of vibration parameters measuring. The classification of existing methods for diagnosing the technical condition of rolling bearings is given. The advantages and disadvantages of these methods are considered. The main defects of rolling bearings arising during manufacturing, transportation, storage, and operation are considered.

1. INTRODUCTION

Vibration diagnostics – technical discipline that considers the theory and methods of recognition of technical state of machines and mechanisms at the source of the information contained in vibro-acoustic signal [1]. Is the primary method of functional diagnostics of mining machines and equipment. Methods of vibration diagnostics aimed at the detection and identification of malfunctions of units of mining machinery and equipment, affecting their vibration – defects of the rotors, components of the stator reference system, plain bearings and rolling bearings, toothed and belt transmissions experiencing or generating dynamic loads [1, 2].

The objectives of monitoring are:

- Warning the development of defects in the assembly and reducing the cost of its restoration
- Determining the optimal technology to restore the unit if the defect eliminates the possibility of normal operation

Carrier of information about the technical state of elements of working equipment in the vibration diagnostics is a vibro-acoustic signal – a collective term that includes information about oscillatory processes (vibration, hydro - gasdynamic, etc.) and acoustic noise mechanism [3, 4].

The peculiarity of the process of vibroacoustic signal almost instantly responds to changes in the condition of the equipment that determines the speed of diagnosis and decision making.

Vibration diagnostics solves the following practical tasks of maintenance of machinery and equipment:

- the division of the many possible technical status of the unit into two subsets: good versus bad;
- detection of a possible defect at an early stage and predict its development in time;
- the diagnosis consists in determining the nature and localization of one or group of defects corresponding to the vibrating condition of the unit;

- detection of hidden defects;
- absence, as a rule, installation and dismantling works of the equipment and reducing the time of diagnosis
- the reduction of the emergency situation when the equipment is operated;

However, vibration analysis is not without disadvantages, which include:

- special requirements to the method of mounting the vibration sensor;
- the dependence of vibration parameters from a large number of factors and the complexity of the allocation of the vibration waveform due to the presence of a fault that requires a deep application of methods of correlation and regression analysis;
- the accuracy of diagnosis in most cases depends on the number of smoothed parameters, for example, the number of estimated SPM [1].

From the point of view of creation of technical regulations vibro-acoustic diagnosis is necessary at the stage of creating the product, in the process of testing to solve a number of tasks aimed at the formation of technical features and values of the bands corresponding to the technical requirements or safe operation. For the formation of ranges of appropriateness in vibration diagnostics is one of the most important tasks is the definition of own and forced frequencies of the products [6-8].

2. FAULT DIAGNOSIS METHODS

Vibro-diagnostic approach is effective if it includes direct methods of spectral analysis, envelope and synchronous accumulation that allows you to obtain the necessary information about the state of gears, as well as to identify signs of defects and criteria of the maximum allowable technical condition [9].

One of the most common problems of vibration diagnostics is to determine the technical condition of rolling bearings installed on the shafts of the nodes of mining machinery. In many cases trouble-free operation of this equipment largely depends on the condition of rolling bearings and has an impact not only on the technological process, but also the safety of the unit, unit, area mining operations. Thus, the question of diagnosis of these nodes is a very important task [10, 11].

During the rolling of rolling elements, mechanical defects on the bearing races, as in the defects on the rolling elements in the bearing vibration processes occur. The cause of the vibration is a single pulse excitation. The duration of the excitation efforts is very small and is often of the shares or units of milliseconds. Each pulse force is from a defect occurring in the bearing elements and the elements of the mechanical design of the unit shall be considered as free fading fluctuations. Since the exciting force is of short duration, the ranges of frequencies of free oscillations occur is very wide and could occupy a frequency band from hundreds of Hertz to hundreds of kilohertz. This explains the successful application for diagnostics of rolling bearings of various diagnostic methods that analyze vibration in the audible frequencies and in the area of ultrasound and higher [12].

3. RESULTS

There are the following types of bearing damage – primary and secondary [13, 14]:

Primary injuries include:

- Wear occurs when bearing penetrate debris or insufficient lubrication. It can also be the result of vibrations of non-rotating bearing. Figure 1. Wear of a rolling bearing ;

IOP Conf. Series: Materials Science and Engineering **253** (2017) 012012 doi:10.1088/1757-899X/253/1/012012

Figure 1. Wear of a rolling bearing

- Dents on the raceways and rolling elements can occur in cases when the forces are transmitted to the mounting ring via the rolling elements. Equally dents occur when excessively large loads on bearings at a time when the bearings do not rotate. The cause of the dents can be penetration into the bearing of foreign particles. Figure 2. Dents on the raceways of a rolling bearing ;

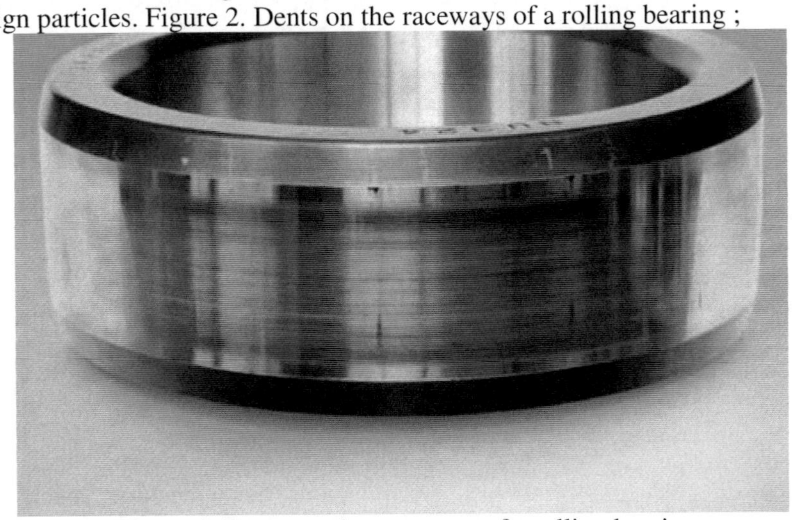
Figure 2. Dents on the raceways of a rolling bearing

- Burrs occur when there is insufficient lubrication of sliding surfaces under load, when there is a transfer of metal particles from one surface to another. The surface look rough. If you encounter scuffing of the bearing material is heated to a temperature at which there is vacation. Occurs local stress concentration that resulted in the formation of cracks and cavities. Also bullies can arise when the rolling elements enter the loaded zone with great acceleration. Figure 3. Burrs on the raceways of a rolling bearing;

IOP Conf. Series: Materials Science and Engineering **253** (2017) 012012 doi:10.1088/1757-899X/253/1/012012

Figure 3. Burrs on the raceways of a rolling bearing

- Surface destruction occur in case of too thin lubricating layer between the tracks and the rolling elements when the top of the roughness briefly touch each other. While on the surface minute cracks occur. In the process of surface destruction of the initially microscopically small, then increase rapidly and, in the end, prevent a smooth rotation of the bearing. Cracks of the described type can speed up the process of formation of fatigue cracks under the surface of the raceways and thereby reduce the durability of the bearing. With adequate lubrication there is no risk of destruction of the described type as long as the lubricating layer becomes too thin, either due to changes in oil viscosity due to temperature increase or excessive increase in workload. Figure 4. Surface destruction on the raceways of a rolling bearing;

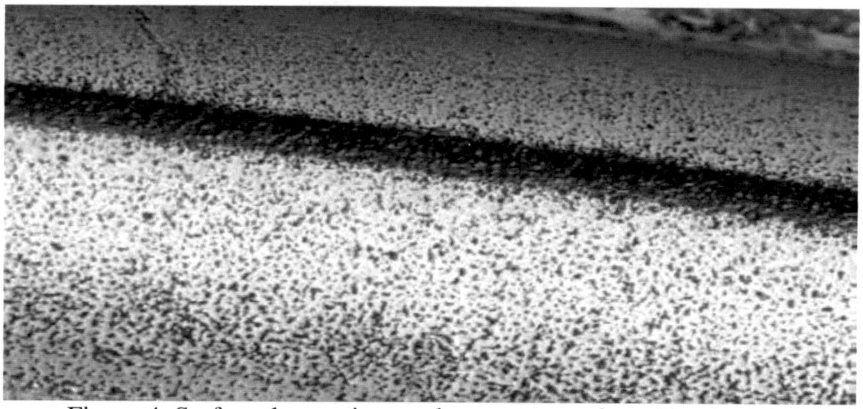

Figure 4. Surface destruction on the raceways of a rolling bearing

- Atmospheric corrosion occurs in the case of penetration into the bearing to atmospheric moisture or an aggressive environment in such volume, that the break (dilute) lubricating film in the contact of bodies and raceways. Figure 5. Atmospheric corrosion on the raceways of a rolling bearing ;

IOP Conf. Series: Materials Science and Engineering **253** (2017) 012012 doi:10.1088/1757-899X/253/1/012012

Figure 5. Atmospheric corrosion on the raceways of a rolling bearing

- The passage of electric current through the bearing from one ring through the rolling elements to another ring causing bearing damage. In places of transition process similar to electric arc welding. The bearing material may be heated to the melting temperature. In this form the coated areas of different size in which the material is annealed and hardened again, and sometimes even melted. In places where the metal has been melted, there may also be small moons. Figure 6. The passage of electric current through the bearing;

- Figure 6. The passage of electric current through the bearing

Secondary damage can be attributed:

- Fatigue sinks – which are the result of normal fatigue damage that occurs at the end of the normal long-term operation of the bearing. This, however, is not the most frequent cause of bearing failure. In most cases, can have causes different from the process of fatigue fracture of the material. If the shell detected at an early stage, when their sizes are not too large, it often becomes possible to identify the cause of the damage, allowing to develop appropriate cautionary measures, and thereby to prevent re-injury. If the shell has reached a certain stage of development, the presence of damage to detect noise and vibrations. Figure 7. Fatigue sinks on the raceways of a rolling bearing;

Figure 7. Fatigue sinks on the raceways of a rolling bearing

- Cracks that occur in the rings due to improper Assembly or disassembly of the bearing, too tight fit of the ring on the tapered shaft neck, can also occur due to burrs. Figure 8. Cracks on the raceways of a rolling bearing;

Figure 8. Cracks on the raceways of a rolling bearing

4. CONCLUSIONS

Thus, with the help of modern software tools it is possible to conduct a spectral analysis of the signal and determination of possible damage to the bearing assemblies, and in the case of condition monitoring key nodes mining machines – prediction of residual life of bearings [12, 15].

The recording of vibrations in a digital format allows you to more accurately and without the use of special instruments to assess the level of vibration and noise, to build a spectrogram and a qualitative assessment of the reliability of the mechanism by increasing the degree of vibration and noise from the original or regulated values

References

[1] Diagnostics of mining machines and equipment. Textbook. B.L. Gerike, P.B. Gerike, V.S. Kvaginidze. Moscow «Nikitskygate», 2012. 400 p.

[2] Kelly, S. Graham. Advanced vibration analysis. 2013. 637 p. (Dekker mechanical engineering).
[3] Kostyukov V.N, Naumenko A.P. Fundamentals of vibro-acoustic diagnostics and monitoring of machines. A tutorial. Omsk. 2011. 368 p.
[4] Salnikov A. F. Vibro-acoustic diagnostics of technical objects. A tutorial. Perm. 2011. 246 p.
[5] Kravchenko V. M., Sidorov V. A., Sedush V. J. Technical diagnostics of mechanical equipment. The textbook. Donetsk. 2009. 459 p.
[6] Kuzin E. G., Gericke, B. L. condition Monitoring of gearboxes frequency-regulated electric drive of the mine belt conveyors. // Bulletin of the research center for work safety in the coal industry, no 1, 2016. p. 82-88
[7] Delvecchio, S.; Delia, G.; Mucchi, E.; Dalpiaz, G. Advanced signal processing tools for the vibratory surveillance of assembly faults in diesel engine cold tests, Journal of Vibration and Acoustics, April 2010, Vol. 132, Issue2, 021008-10, ISSN: 1048-9002, doi:10.1115/1.4000807.
[8] Zakharov A. Yu., Shiryamov D. V. determination of the critical value of the resistance to the rotation of the conveyor rollers.// Mining equipment and electromechanics, 2016. No. 1. p. 3-8.
[9] State standard R ISO 13373-2-2009. Status monitoring and diagnostics of machines. Vibratory condition monitoring of machines. Part 2. Processing, analysis and presentation of the results of the vibration measurement. M.: STANDARTINFORM, 2009. p. 33.
[10] Nondestructive testing: Handbook: In 8 volumes / Under the General editorship of V. V. Klyuev. Vol. 7: In 2kN. KN. 1: V. I. Ivanov, I. E. Vlasov. The method of acoustic emission. KN. 2: F. Ya. Balitsky, A. V. Barkov, N.. Barkov etc. Vibration diagnostics. – 2nd ed. Rev. M.: Mashinostroenie, 2006. 829 p.
[11] One Estimation of the Technical Condition of the Selective Headers./ V. Kovalev, A. Khoreshok, B. Gerike, A. Meshkov. Regarding One Estimation of the Technical Condition of the Selective Headers. //The conferecewas funded by RFBR according to the research project # 16-05-20506. October, 10-12, 2016, Kemerovo, Russia. – Atlantis Press. Amsterdam, Paris, Beijing. 2016. P. 291-296.
[12] Rudloff L., Arghir M., Bonneau O., Guingo S., Chemla G., Renard E. Experimental Analysis of the Dynamic Characteristics of A Hybrid Aerostatic Bearing. Journal of Engineering for Gas Turbines and Power, vol. 134 (18). 2012.
[13] Troubleshooting bearings ./ NSK Motion & Control, 2009. 42 p.
[14] Damage of rolling bearings and their causes./ SKF AB, 2002. 46 p.
[15] B. L. Gerike, P. B. Gerike. one assessment of residual life of main fans. // Examination of industrial safety in the Siberian Federal district: the collection of proceedings/ under the editorship of Smirnova A. N. T. 2. M.: Mashinostroenie, 2015. P. 85-89.

Diagnostics of Technical Condition of Gear Units of Belt Conveyors for the Aggregate of Methods of Nondestructive Testing

Evgeny G Kuzin [1, a], Boris L Gerike [2], Yuriy V Drozdenko [2], Michael G Lupiy [1], Natalya V Grigoryeva [1]

[1] Regional Campus of Kuzbass State Technical University after T. F. Gorbachev, 653039, Russia, Prokopievsk, Nogradskaya str. 19a.
[2] Kuzbass State Technical University after T. F. Gorbachev, 650000, Russia, Kemerovo, Vesennyaya str. 28

E-mail: [a] kuzinevgen@gmail.com

Abstract: The article reviews the issues of complex use of methods of technical diagnostics of gearboxes for belt conveyors, with the aim of creating an effective system of maintenance. The article is showing the results of the evaluation of the technical condition of the drives of belt conveyors based on vibration monitoring and thermal parameters, and analysis of lubricating oil.

Introduction.

Transportation of coal from the coalface to the surface is an important process, the cost of which ranges from 70 to 80% of the total costs of production.

Thus, on one hand the spent time on non-productive pauses in the transportation chain should be kept to a minimum, and on the other hand it should be minimizing energy costs.

The quality of the work conveyor transportation to a large extent depends on the reliability and efficiency of drive units and control equipment.

Technical diagnostics allows to estimate and predict the state change of mining equipment, timely maintenance, and to more effective plan repairs. In addition an aggregate diagnosis of diagnostic characteristics allows to reveal defects at an early stage and to prolong the service life of equipment due to proactive (proactive) maintenance [1].

Statement of the problem and research methods choices.

Today vibration analysis is the strongest and most common method of detecting and determining fault in gears of the gearbox [2]. It should be noted that fluctuations related to the defect of the rolling bearing have a much lower amplitude than the variations associated with many other lesions, such as imbalance, misalignment or defective gear. A large variety of bearing designs and usage conditions, operation speeds and loads makes it very difficult to use the total allowable level of vibration, which would have worked satisfactory in most cases [3, 5, 6].

These observations determined the set of technical diagnostics methods for gearboxes of coal mine belt conveyors in the parameters of vibration, lubricants and thermal control by infrared thermography.

In order to diagnose drives mine belt conveyors the results of the monitoring of vibration parameters, lubricants and thermal control were used. The studies were conducted periodically starting with from the time of installation of conveyor lines. The drive station is equipped with gearboxes Moventas

Santasalo, Morley motors of 500 kW and a frequency Converter station, which provides smooth starting, torque control and speed of the motors.

The vibro-diagnostics specifics that vibration alerts almost instantly react to changes in the condition of the equipment. Analysis of vibration parameters allows to determine the imbalance of the ligament of the engine to the gearbox, the weakening of the landing bearings, defective gears, bearings and defects attaching to the foundation [4, 7, 8]. Thus, reducing the speed of the tape significantly reduces the level of vibration.

The analysis of the parameters of the lubricating oil in addition to accumulation speed wear and General degradation of the oil helps to assess which elements of the gear wear in the first place.

Thermal control of drive components by infrared thermography complements, enhances the diagnostic abilities and allows to reveal defects of cooling systems including mechanical and electrical parts of transmission stations.

Results.

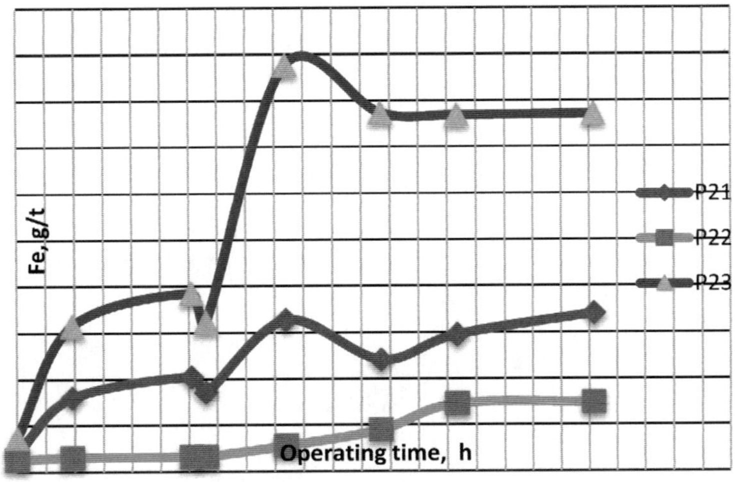

Fig. 1. The change of iron content in oil of gearbox from the time in operating

This paper describes the monitoring results of parameters of the lubricating oil, vibration, and thermal control gearboxes Moventas Santasalo mining belt conveyors LL1600. Fig. 1. shows the result of the iron accumulation, [g/t] in the oil for exactly the same gear from the moment of installation untill 21800 hours are reached. The results differ from the minimum accumulation of iron about 150 [g/t] in the drive P22 to a maximum of about 780 [g/t] in the drive P23. The maximum allowed iron content in the lubrication oil should be not exceed 200 [g/t].

IOP Conf. Series: Materials Science and Engineering **253** (2017) 012013 doi:10.1088/1757-899X/253/1/012013

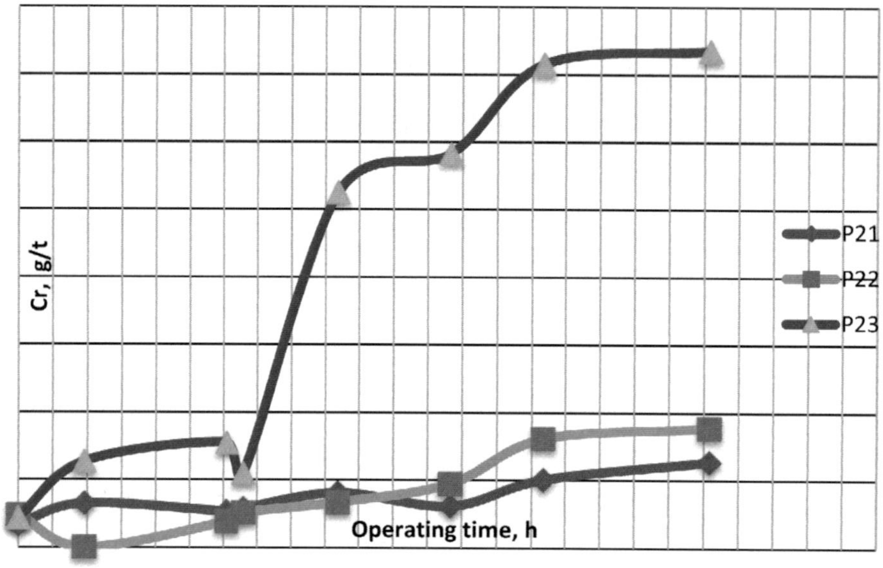

Fig. 2. The change in the content of chromium in oil of gearbox from the time in operating

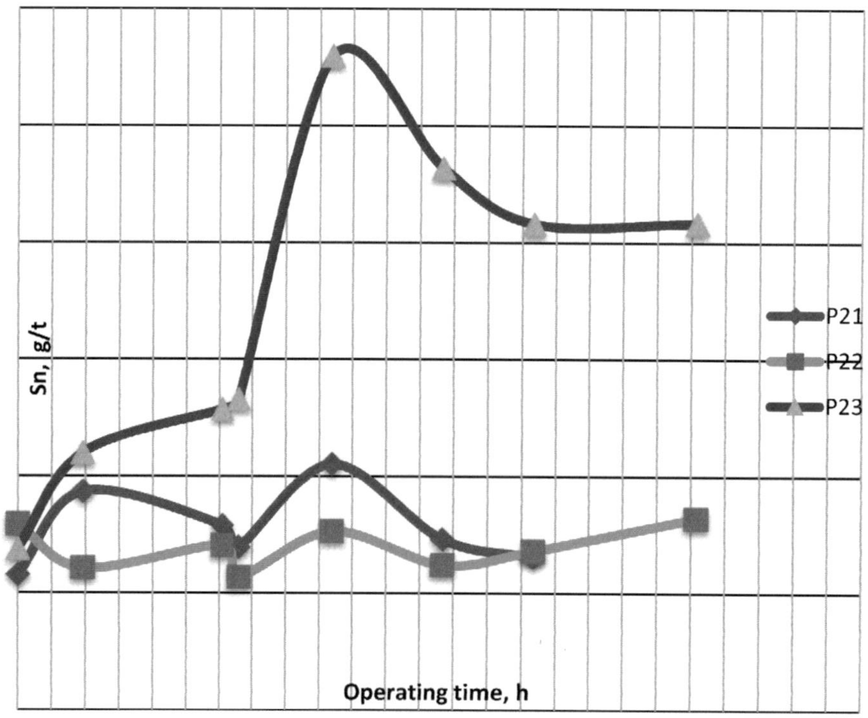

Fig. 3. The change in the content of tin in oil of gearbox from the time in operating

IOP Conf. Series: Materials Science and Engineering **253** (2017) 012013 doi:10.1088/1757-899X/253/1/012013

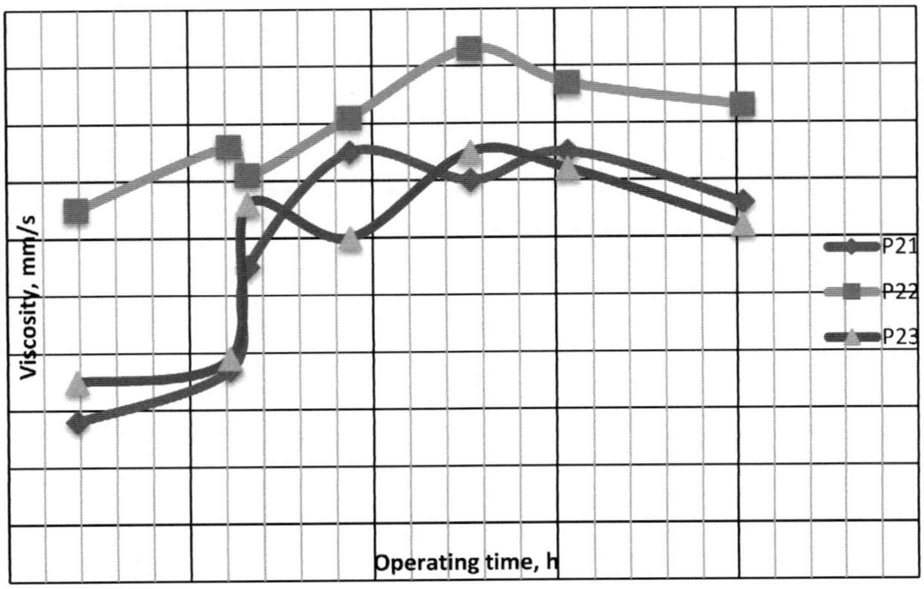

Fig. 4. The change in viscosity of the oil in a gearbox of a belt conveyor

After the analyzing the parameters of the oil, it can be concluded that more viscous oil (like in the gearbox P22) helps to reduce the accumulation of mechanical impurities. In the parameters of the lubricating oil reducer P22 that is in good condition, the gears P21 and P23 require oil changes.

Additionally monitoring of vibration parameters was performed. The effect of the overall RMS value of vibration velocity is shown in Fig. 5. The parameters of the vibration reducers P22 and P23 are in good condition, and satisfactory P21.

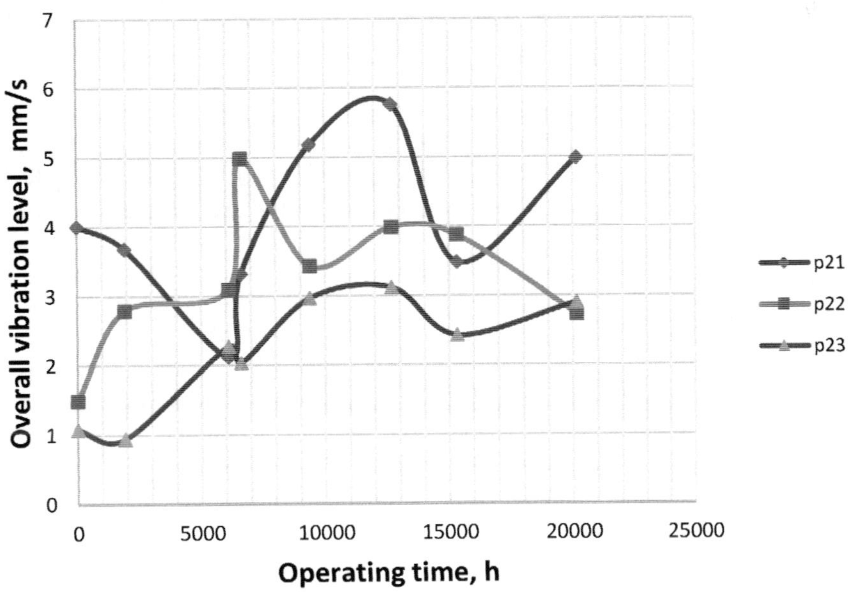

Fig. 5. The change of vibration parameters from the time in operating

IOP Conf. Series: Materials Science and Engineering **253** (2017) 012013 doi:10.1088/1757-899X/253/1/012013

Spectral analysis helps to assess the condition of gearbox components. Fig. 6 shows the spectrum of velocity vibration in one of the feature points P21 reducer which increases the level of vibration at odd sub-harmonics of the reverse frequency. This range characterizes a certain stiffness of the system.

The amount and nature of vibration in the weakening of the rigidity depends on the degree of defect, the force excitation magnitude of (as a result it depends on a load to the actuator) and the properties of a particular unit of anizotrop and nonlinearity of the reference frame. The vibration signal of gears mounted on a jet-powered (no capital base) usually has a complex nature. It usually presents oscillations in a wide frequency range: vibration with excitation frequency, its harmonics and subharmonics, and sometimes in the other frequency components.

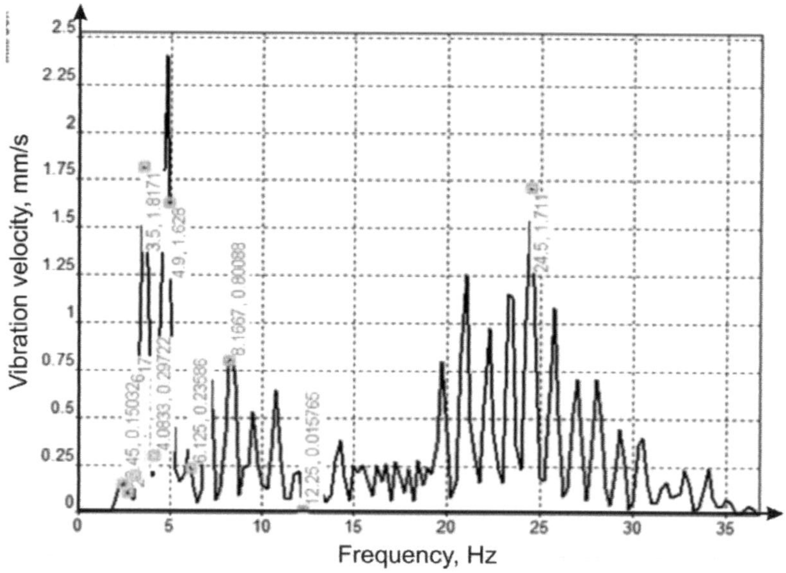

Fig. 6 Spectrum of the vibration velocity P21 point 3A appear odd sub-harmonics, which is the frequency of circulating

Thermal control of actuators via infrared thermography allows more accurate control of the faulty nodes, such as, to identify one local overheating of the bearing and failure of the cooling system. Fig. 7. presents the thermograms of gears having different temperature. This difference may be caused by the parameters of the oil and vibration and different load on the drive drums depending on their place of installation in the scheme of the pipeline.

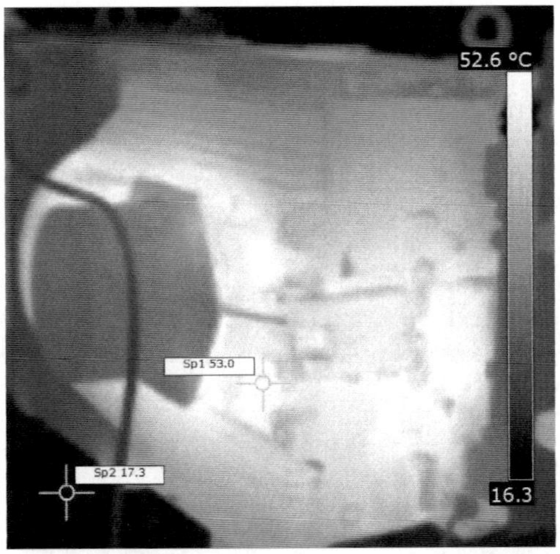

a) reducer P21 the highest temperature 53 °C

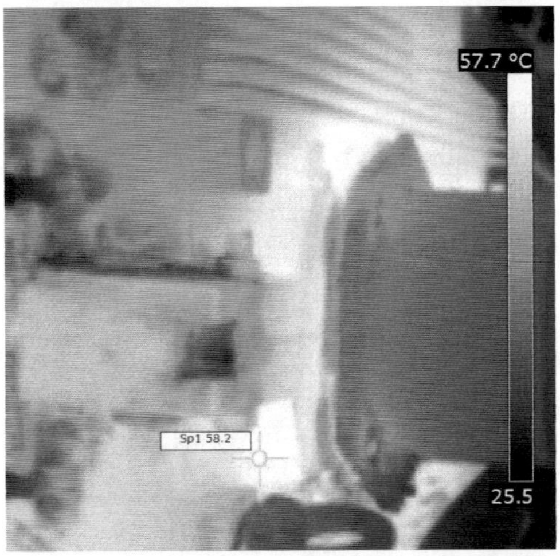

b) reducer P22 highest temperature 58.2 °C

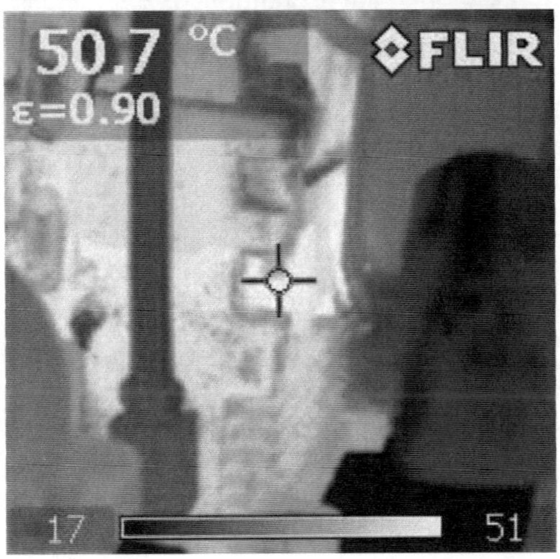

c) reducer P23 the highest temperature is 50.7 °C

Fig. 7. The thermograms of the respective gears at 15100 hours of operating

Conclusion

With decreasing oil viscosity the level of vibration increases and wear occurs more intensively. At higher oil temperature the evaporation of low boiling fractions and the viscosity slightly increases. At higher vibration it is recommended to increase the viscosity of the oil extending the life of the drives.

For reducer P21 it is recommended to inspect a mechanical fastener.

In reducer P23 it is necessary change the oil. In general the gearboxes are in a satisfactory technical condition and the projected life of these gearboxes is at least 20000 hours.

Systematic monitoring will allow to take measures in advance for repair preparation or replacement of defective gearbox, such as rewiring the equipment of the longwall next time. The results of the monitoring will help to create a database on the nature of the processes of wear and set proper evaluation criteria for the gears mentioned above.

References

[1] E. G. Kuzin, B. L. Gericke, the peculiarities of the vibration diagnostics of gearboxes of the mine belt conveyors. Proceedings of the V International scientific and practical conference prospects of innovative development of coal regions of Russia. Prokopyevsk, 2016, pp. 137 – 142.

[2] Ali Rostami, Rohallah Panahi Leavoli, Jalaleddin Ghezavati, Mohommad Homaei. Gearbox fault Detection using vibration signals based on the method of Hilbert. Indian Journal of Scientific Research 1(2): p. 675-689, 2014.

[3] Daniel Lynn. The detection of defects in rolling element bearings using vibration analysis/ TRANS. from English. I. R. Chanak, edited by V. A. Smirnov. - http://www.vibration.ru/v_defekt.shtml

[4] B. L. Gerike, P. B. Gerike Detection of operational defects of belt conveyors methods of vibration diagnostics. M. Mountain information-analytical Bulletin (scientific and technical journal) Issue 12/ volume 10/2009.

[5] Rudloff L., Arghir M., Bonneau O., Guingo S., Chemla G., Renard E. Experimental Analysis of the Dynamic Characteristics of A Hybrid Aerostatic Bearing. Journal of Engineering for Gas Turbines and Power, vol. 134 (18). 2012.

[6] Delvecchio, S.; Delia, G.; Mucchi, E.; Dalpiaz, G. Advanced signal processing tools for the vibratory surveillance of assembly faults in diesel engine cold tests, Journal of Vibration and Acoustics, April 2010, Vol. 132, Issue2.

[7] V. Kovalev, A. Khoreshok, B. Gerike, A. Meshkov. One Estimation of the Technical Condition of the Selective Headers./ Regarding One Estimation of the Technical Condition of the Selective Headers. Atlantis Press. Amsterdam, Paris, Beijing. 2016. pp. 291-296.

[8] Siju K. Abraham, V. Sugumaran, M. Amarnath. Acoustic Signal Based Condition Monitoring of Gearbox using Wavelets and Decision Tree Classifier. Indian Journal of Science and Technology, Vol 9 (33), September 2016.

Problems of Industrial Safety Examination in Modern Conditions and New Resource Evaluation Methodology

Alexander N Smirnov [1], Nikolay V Ababkov [1, a], Alexander I Petrachkov [2], Vladimir M Radchenko [3]

[1] T.F. Gorbachev Kuzbass State Technical University, 28 str. Vesennyaya, Kemerovo, 650000
[2] OOO PTP "Sibenergochermet", 13 str. Ordzhonikidze, Novokuznetsk, 654005
[3] Kemerovo State Agricultural Institute, 5 str. Markovtseva 5, Kemerovo, 650056

E-mail: [a] n.ababkov@rambler.ru

Abstract: A paper presents a fundamentally new approach to the evaluation of health and metal resource potentially dangerous equipment of thermal power stations, based on the identification of patterns of change in the structural-phase state, and the internal stress fields in the long-running metal and welded joints spectrally-acoustic method, taking into account the structural factor. Information about changing the structural state of the metal in the process of long-term operation of heat-resistant steels is given.

Introduction

The technical condition of industrial equipment at many enterprises in Russia does not meet the requirements of the current regulatory documentation. So, according to the latest data of Rustekhnadzor, about 70% of the technical devices of hazardous industrial facilities (TDHIF) have worked out the estimated period. And further operation of such equipment is unsafe. At the end of the last century, an industrial safety examination system (ISE) was created in Russia, using the regulatory framework of which the real state of the TDHIF was assessed and its performance was predicted. However, starting in 2014, a new ISE system was created, which rejected all the methodological developments created earlier (almost 20 years), recognizing them as illegitimate.

According to the requirements of the current regulatory documents of Rustekhnadzor [1, 2] for all technical devices, when performing an expert examination of industrial safety, it is necessary to calculate (estimate) the residual resource. However, nowhere is the methodological approach to this work written out. At present, Rustekhnadzor has not decided on the question of how to consider this resource (estimate). In addition, there is practically no basis for legitimate methodological documents for the technical diagnosis and examination of industrial safety of technical devices, buildings and structures. There are no standards for the dismantling of the base metal and welded joints for various equipment after a long service life. Methodical instructions (normative base) are necessary in order that after carrying out certain regulated diagnostic work, it would be possible to estimate the period of further operation of the equipment (according to the algorithm of work to be performed) and, if necessary, to check (by the customer or the controlling body) the scope of work performed by the expert, That is, at least in the first approximation, the quality of technical diagnosis. In addition, the state criminalized the

IOP Conf. Series: Materials Science and Engineering **253** (2017) 012014 doi:10.1088/1757-899X/253/1/012014

expert for the quality of the expertise, not providing it, to a sufficient extent, with legitimate regulatory documentation.

From the foregoing it is clear that the new ISE system is in a "clean" state and, in the absence of a regulatory framework, it is not possible to fulfill the RF laws and the requirements of Rustekhnadzor in the required volume. At present, there is no single terminology of definitions in Russia either, nor is there a single methodological approach to assessing the residual resource, while Rustekhnadzor documents require such work. There are also several definitions of the limiting state [3].

Theory

Let us draw your attention to the developments of the leading scientist in the field of strength, safety and the resource of Corresponding Member of the Russian Academy of Sciences prof., Doctor of Technical Sciences Makhutov N.A. and his school in the field of residual resource assessment [4]. Two methods for estimating the residual resource are proposed: - one by the criterion for the formation of cracks; - another according to the criteria of crack resistance.

Method for assessing the resource by the criterion for the formation of cracks. The determination of the residual resource is carried out using data on the technical state, obtained by experimental and calculated methods, for the following limit states:

- formation of cracks under cyclic loading;
- development of cracks under cyclic loading;
- the appearance of viscous or brittle fracture in the presence of initial technological and operational cracks.

When calculating the remaining resource, the cyclical, temporary, corrosive and other damages accumulated during the previous operation are taken into account here, as well as the main design, technological and operational factors that change the characteristics of the limiting states. The evaluation of the technical state for the subsequent determination of the residual life of the equipment elements is made by combined experimental and calculated (standard and non-standard) methods. When assessing the technical state by destructive and non-destructive methods, physicomechanical characteristics, the characteristics of the stress-strain state and the characteristics of the state of the crack defects (their depth, length, location and orientation) should be obtained.

Based on the analysis of all the above characteristics, the residual resource is evaluated for the resistance to cyclic destruction.

Another technique for assessing the technical condition and residual life by crack resistance criteria is used for critical equipment elements in which defects in the assessment of the technical condition are found to be unacceptable and allowed by the control standards (if any). With a single loading, the remaining life and survivability are determined:

- on critical temperatures of brittleness and minimum temperatures of metal of bearing elements during operation;
- by the critical stress intensity or strain rate coefficients by stress intensity or strain rates for the corresponding design defects and operating stresses.

When determining the residual life, the following criteria are taken into account for the crack resistance criteria: mechanical loads (internal and external pressure, own weight of the product, weight of other elements, reaction of supports, etc.); Temperature effects; Vibration; Seismic loads; Loads of emergency modes; Residual stresses; Nominal strain stresses from mechanical loads, etc.

To assess the residual life of the equipment in accordance with the above methods, it is necessary to have a powerful testing laboratory and a staff of highly qualified specialists. This work is highly specialized and has a high cost. Hence it follows that **it is not feasible to assess the residual resource by these methods for all technical devices**, both technically and economically. These techniques are acceptable for expensive equipment such as oil and gas complexes, where the damage from accidents is maximized.

For the equipment of 2-4 hazard class categories, a large number of resource estimation techniques developed by both industry institutes and individual expert organizations were proposed. Most of them are not approved by the bodies of Rustechnadzor.

One of the most important drawbacks of all these techniques is the lack of analysis of the structural-phase state of the metal being diagnosed. Nowhere, except energy, do not take into account the changes that occur in the structure of the metal in the process of long-term operation.

We propose a new methodology for estimating the resource, based on an integrated approach taking into account the structural factor.

Obviously, the total time of safe operation of any equipment consists of the time of operating the equipment up to the moment of examining $\tau_{c.s}$ (current state) of the current and the further operating time of the equipment until the limit state is reached τ_{res} (residual life).

$$\tau_{l.s} = \tau_{c.s} + \tau_{res}\,, \tag{1}$$

In the metal of equipment that has worked for a long time under difficult stress conditions, often in aggressive environments (before the time of inspection), complex physicochemical and structural changes occur. The structure is the most important index of metal, characterizing its performance. **In the memory of the metal**, at the time of the diagnosis, all types of thermal effects and loads were detected, to which the metal was exposed, beginning with its manufacture and ending with a certain period of operation (in our case – $\tau_{c.s}$). **The main thing is to correctly understand and evaluate these changes.** However, the structural-phase state of the metal after running-in ($\tau_{c.s}$) Is one side of the question. On the other hand, to assess the real resource, it is also necessary to take into account macro-defects (sometimes they are permissible, they are often not eliminated, but simply missed) that occurred during the manufacture, repair, operation of equipment during the flow $\tau_{c.s}$. These defects have a significant effect on the reliability of the resource estimate.

At the same time, the current state of the equipment at the time of diagnosis characterizes the coefficient of its current technical condition K_C, which can be represented as

$$K_C = f(K_{str.c}, K_{reg}, K_{def}) \rightarrow \tau_{c.s}, \tag{2}$$

where, $K_{str.c}$ – the structural coefficient of the metal in the current state, K_{reg} – the coefficient, which takes into account the operating modes (temperature, pressure (on-load), cyclicity, medium), K_{def} – factor, taking into account the presence of defects in manufacturing, installation and Repair in the main metal and welded joints. $K_{str.c}$ is determined by the nature of the change in the structural-phase state and internal stress fields, the redistribution of the sources of the stress fields and their density. $K_{str.c}$ can be expressed as a function of the structure parameters in the form

$$K_{str.c} = f(\sigma_{in}, \tau, \rho_{\pm}, \rho, \mu, d_{gr}, \gamma_c), \tag{3}$$

where σ_{in} is the amplitude of the internal stress fields, MPa; τ – tangential stresses; $\rho_{\pm}$ – the excess density of dislocations; ρ – the scalar density of dislocations; μ – the density of sources of internal stress fields; d_{gr} is the grain size, mkm; γ_c – a coefficient that takes into account the size and nature of the distribution of carbide phases in the metal. For each class of materials, engineering methods for calculating the structural coefficient are developed. For the steels most studied by us, a structural criterion is proposed that is worked out on long-term heat-resistant steels at all stages of the life cycle (from the initial state to reaching the limiting state and destruction), which has the following form

$$K_{str} = \left(\frac{\sigma_{in}}{\mu} + \frac{\tau}{\mu}\right) \times \frac{\rho}{\rho\pm} \gamma_c \cdot d_{gr}^{-1/2}, \tag{4}$$

The structural criteria must be determined on equipment that is operated under difficult stress conditions at high voltages and temperatures, under cyclic and shock loads, in corrosive environments during periods of routine maintenance (the date of the diagnostic is specified, and a certain period is worked out).

For the operating organizations, this will be an additional cost, "but the game is worth the candle," because according to this (studied) current state of the equipment, it is possible to estimate the resource and the deadline for reaching the limit state with sufficient reliability.

To assess the real state of equipment, in addition to the structural coefficient, it is necessary to take into account both operating modes, and mounting, repair and metallurgical defects, and accumulated operational defects, and on their basis to choose coefficients. Therefore, to determine K_C, it is necessary to determine the values of K_{def} and with K_{reg}.

From the operating conditions of the equipment, it can be represented as a coefficient depending on the number of work cycles (N), pressure, or load (P), temperature (T), aggressiveness of the medium (L).

Hence, $K_{reg} = K_N \cdot K_P \cdot K_T \cdot K_L$, the numerical values of the coefficients are developed for a specific type of equipment.

$K_{def} = K_{met} \cdot K_{rep} \cdot K_{ins}$, numerical values of coefficients are developed for the equipment being diagnosed.

As previously stated – K_C, the coefficient characterizing the technical state of the object of control at the time of control. Given the preservation of at least the relative constancy of K_{reg} and K_{def} throughout the life of the equipment, in the first approximation can be written

$$\frac{\tau_{res}}{\tau_{c.s}} = \frac{K_{l.s.} - K_C}{K_C}, \qquad (5)$$

Where $K_{l.s}$ – a coefficient characterizing the technical condition of the equipment at the moment of reaching the limit state.

Here the approach is necessary purely individual, but taking into account the level of exploitation and accumulation of defects in similar equipment with close terms of operation. The determination of this coefficient was carried out on the basis of an analysis of operating conditions (the coefficients K_{reg} and K_{def} were determined). The structural coefficient of the limiting state is estimated from the results of a study of the base metal and welded joints of equipment that has exhausted its life. From (5), after carrying out small transformations and substituting the results of calculations and studies into the formula, we obtain the desired result – the residual operation time of the equipment.

$$\tau_{res} = \frac{\tau_{c.s}(K_{l.s} - K_C)}{K_C}. \qquad (6)$$

At present, a database of equipment for the fuel and energy complex of Kuzbass is being created, and a large amount of research is needed to study the mining equipment. The proposed methodology has a significant drawback, consisting **in the difficulty of defining structural criteria**. It is necessary to perform non-expensive electron microscopic studies.

Results and its discussion

There are several ways to free yourself from this shortcoming. The most promising is the development of non-destructive physical methods for assessing structural changes. For a number of steels (such as Cr-Mo-Va), we have developed acoustic and magnetic structural criteria for the evaluation of damage. So, for example, for a base metal and welded joints of steam pipelines, a complex acoustic criterion of the limiting state has been developed, which allows to determine the resource of a technical device with respect to the delay time of surface acoustic waves (SAW). The criterion is based on the regularities of the change in acoustic characteristics as a function of the structural-phase state of the metal and the amplitudes of the internal stress fields. On the example of two destroyed bends (bent sections of a steam pipe) made of 12H1MF steel (Fig. 1), curves for the distribution of the SAW delay time and the amplitudes of the internal stress fields were experimentally studied. As a non-destructive testing equipment (NDT), a multifunctional highly sensitive information and computing complex ASTRON was used. The study of the nature of the microstructure, measurement of internal stress fields was carried out in the same areas where the SAW delay measurement was performed by the spectral-acoustic method.

In the bend "A" at a distance of 28–33 mm from the main crack from the outer surface revealed a secondary crack parallel to the main. The study was carried out in the zone of the end of the secondary crack. The average amplitude of curvature-torsion and the density of the extinction contours decrease with increasing distance from the shores of the main cracks, and in the "A" bug again increase at the secondary crack. Local internal stress fields (bending-torsion) are large, approaching the limit of strength and exceeding it; They can (with prolonged use) lead to the initiation of microcracks on the surface of the metal, their fusion, and the formation of main cracks. In the zone of the end of the secondary crack (bend "A"), an increased level of internal stresses and delay time of the surfactant (curve 1, 3, fig. 2) is recorded. In the "B" bend, these characteristics change monotonically from the rupture zone (curve 2, 4, fig. 2). It can be seen from the graphs (Fig. 2) that with increasing local internal stress fields (average curvature-torsion of the crystal lattice), the delay time of the surfactant increases, which is explained by the weakening of ultrasonic oscillations at the sources (concentrators) of internal stress fields [13, 14], Sources, a decrease in the curvature-torsion of the crystal lattice, and the magnitude of local internal stress fields, the degree of attenuation of ultrasonic oscillations decreases and the delay time of the SAW drops [5–7]. The uniformity of the curves for changing the delay time of surfactants and changes in the amplitudes of internal stresses indicates the existence of a connection between these characteristics and the high sensitivity of the method of acoustic structuroscopy.

a)

b)

Fig. 1. General type of destruction:
a – the character of crack opening – bend A (gap length – 2.1 m);
b – end of a crack in the section of the tube – bend B

It is established (Fig. 3) that, as the long-term strength of heat-resistant steels decreases, the delay time of the surfactant increases due to an increase in the density of sources of internal stress fields and their amplitudes. The maximum tensile strength of the surfactant corresponds to the minimum strength characteristics and the minimum long-term strength.

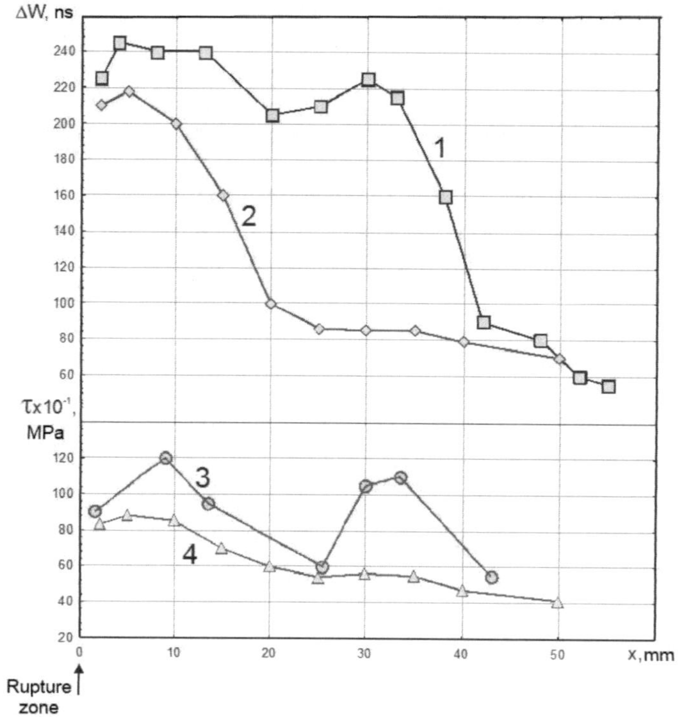

Fig. 2. Change of SAW delay time (ΔW – curve 1, 2) and bending-torsional stresses (τ – curve 3, 4), depending on the distance to the shores of main cracks in two destroyed bends:
Curve 1, 3 – bend "A"; Curve 2, 4 – bend "B"

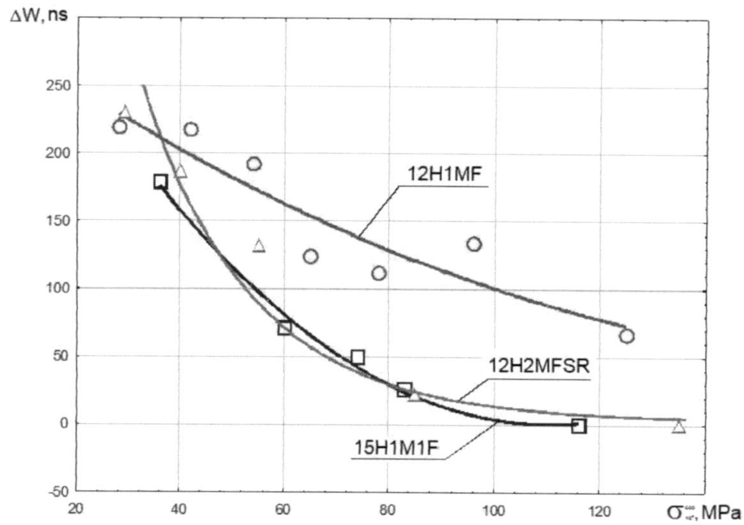

Fig. 3. Influence of long-term strength of heat-resistant steels
For the delay time of SAW

The results of the studies [8–12] confirm the high sensitivity of the spectral-acoustic method to the determination of changes in the parameters of the microstructure (dislocation density, mean curvature-torsion amplitude, and density of extinction loops).

Thus, application of the spectral-acoustic method allowed to develop complex criterion limiting condition long working metal.

$$K_f = \frac{W_\tau - W_0}{W_f - W_0} \cdot \frac{W_f}{W_\tau} \cdot \gamma, \qquad (7)$$

Which is determined by the delay time of the surfactant in the metal with the initial state of the structure (W_0), in the metal that has exhausted its service life (W_f) And in the controlled metal (W_τ), where γ – the coefficient that takes into account the material of the element being monitored [12]. The complex criterion of the limiting state (CCLS) has been tested on a number of destroyed elements of power equipment. It has been experimentally proved that at $K_f \geq 0.7$ the metal reaches the limiting state.

The received criterion was tested in the study of long-working 238 bends and straight sections of steam and steam pass pipes and four destroyed bends of power plants in Siberia. The data of the measurements of the acoustic characteristics give good agreement with the results of metallographic studies carried out on the cuts from the metal scraps. In addition, the criterion was tested in the diagnosis of technical devices and their elements made of steel 20, 15H1M1F, 12X2MFSR, 17GS, 09G2S, where results were also obtained, allowing to assess the structural state and life of the metal.

An acoustic criterion for evaluating the working capacity of high-pressure boiler drums has been developed. To assess the resource of the boilers of boilers made from boiler steels, a set of studies (steel 20M, 22K, 16GNM). The delay time for surfactants in the main metal of six drums of high-pressure boilers (66 or more bridges between the holes of the water pipes on each) and in the weld metal after repairs of the molybdenum-steel boiler boilers was measured by the acoustic-acoustic method.

Electron-microscopic studies showed that the most serious structural changes occurred in the HAZ and in the vicinity of the crack fringe, where microcracks were detected and the amplitude of local internal stress fields increased (Fig. 4). The maximum SAW delay time (Fig. 4) was observed at a distance of 1 mm from the fracture zone (crack), the maximum amplitude of local internal stress fields (1000 MPa) was recorded here. The increase in the volume fraction of the material structure dangerous to fracture (segments of the defective α phase Led to an increase in the average local internal stress fields on the one hand, and on the other hand, the growth of internal stresses (300–900 MPa in the main metal, 900–1100 MPa near the rupture zone) leads to an increase in the SAW delay time (30–45 ns in the main metal, 120–150 ns near the crack fringes.) Such regularities are explained by the weakening of ultrasonic oscillations at the sources of internal stress fields and the presence of inhomogeneities in the structure of the metal.

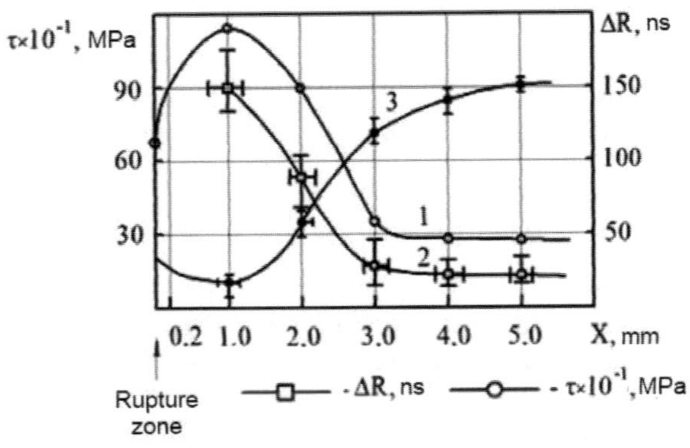

Fig. 4. Change in the mean value of local internal stresses (τ) (curve 1) and the delay time of the surfactant (ΔR) (curve 2) and the intensity of the MN (ΔIMN) (curve 3) when moving over the sample from the fracture zone to the base metal (sample long working metal)

The established dependencies show the principal possibility of identifying zones with high values of local internal stress fields in the long working and welded metal of welded drums of high-pressure boilers made from special molybdenum steels (type 20M) by the spectral-acoustic method.

The criteria has the following form

$$K_S = \frac{\Delta R_1}{\Delta R_2}, \tag{8}$$

where ΔR_1 and ΔR_2 – the anisotropy of the acoustic properties in the metal after repair and in the long-running metal before repair, respectively. These values are calculated as the difference in propagation times of Rayleigh waves in the metal perpendicular and parallel to the weld:

$$\Delta R_1 = \left|\Delta \bar{R}_1^{perp} - \Delta \bar{R}_1^{par}\right|, \quad \Delta R_2 = \left|\Delta \bar{R}_2^{perp} - \Delta \bar{R}_2^{par}\right| \tag{9}$$

The developed criterion is tested on the long-working and melted metal of a number of high-pressure boiler drums. Calculations of K_S showed that at $K_S > 0.22$, the metal of the drum of the boiler under study is in the stage of pre-destruction and repair and recovery measures are necessary [12] (Fig. 5) [12].

But in any case (to avoid large measurement errors), there is a need to perform reference electron microscope studies for each grade of steel.

And finally, for engineers-practitioners, atlas of histograms is developed, with the help of equipment (after carrying out the analysis of repair and operational documentation), which, with the knowledge of the structural criteria, the operating conditions of the specialist will find by the histogram the service life of the given technical device.

This methodology does not exclude the implementation of standard NDT and TD methods, they complement the results of our tests, at their expense refinement corrections to the corresponding coefficients are introduced.

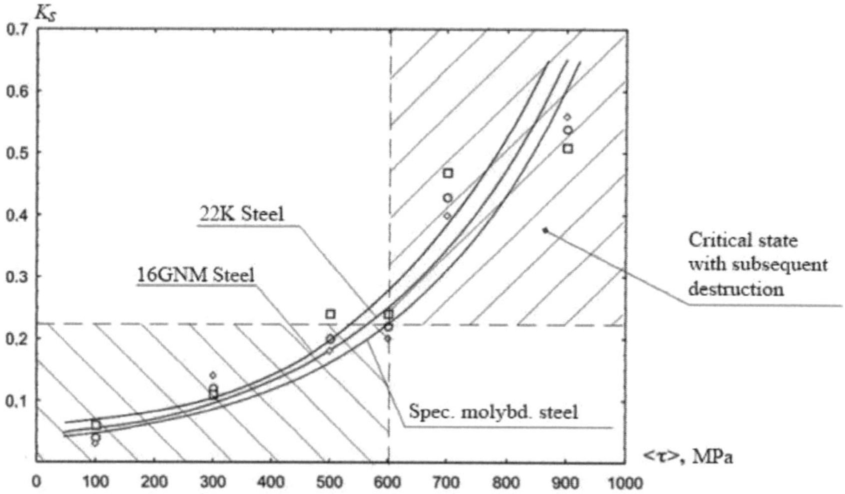

Fig. 5. Relationship between the criterion for estimating the resource of welded drums of K_S boilers with a change in the magnitude of local internal stress fields <τ>

Conclusions

1. At the present stage, the solution of the above-mentioned problems of industrial safety expertise (the lack of legitimate methods of technical diagnosis and resource evaluation, the absence of standards for the categorization of NDT defects, the development and introduction of new methods and equipment for technical diagnostics, etc.) is possible with the use of self-regulation mechanism. Only the community of specialists can solve the problems of ensuring the safety of potentially dangerous equipment, its diagnosis and evaluation of the resource. Neither the officials nor the bodies of Rustekhnadzor will solve these problems, either because of their workload, or the lack of relevant specialists.

2. It is shown that most of the existing various methods for estimating the resource do not take into account the nature of the change in the structural-phase state of the metal during long-term operation, which leads to a decrease in the reliability of the results obtained.

3. A new methodology for estimating the resource is proposed, based on the character of the structural parameters change during the long-term operation under difficult stress conditions and on the interrelation of these parameters with the acoustic characteristics. For various FEC equipment, work packages are carried out, including electron microscopic and non-destructive studies.

The work was supported by the RNF grant agreement 14-09-00724 and Russian President's grant for young candidates, MK-1341.2017.8.

References

[1] FNP v oblasti promyshlennoy bezopasnosti Pravila provedeniya promyshlennoy bezopasnosti p.28 Prikaz Rostekhnadzora ot 14.11.2013 №538 «Ob utverzhdenii federal'nykh norm i pravil v oblasti promyshlennoy bezopasnosti» (Zaregistrirovano v Minyuste Rossii 26.12.2013 №30855); In Russian.

[2] FNP v oblasti promyshlennoy bezopasnosti Pravila promyshlennoy bezopasnosti opasnykh proizvodstvennykh ob"yektov, na kotorykh ispol'zuyetsya oborudovaniye, rabotayushcheye pod izbytochnym davleniyem p. 415 pp. g. Seriya 20.Vypusk16.-M.: Zakrytoye aktsionernoye obshchestvo «Nauchno-tekhnicheskiy tsentr issledovaniy problem promyshlennoy bezopasnosti», 2014. 254 p. ISBN 978-5-9687-0609-6.; In Russian.

[3] Smirnov A.N., Petrachkov A.I. Problemy otsenki ostatochnogo resursa tekhnicheskikh ustroystv opasnykh proizvodstvennykh ob"yektov / A.N. Smirnov, A.I. Petrachkov //. Tekhnadzor. 2016. 1. p.25–28.; In Russian.

[4] Makhutov N.A., Gadenin M.M. Tekhnicheskaya diagnostika ostatochnogo resursa i bezopasnosti. M.: Spektr, 2011. 185 p.; In Russian.

[5] Koneva N.A., Kozlov E.V. Dislokatsionnaya struktura i fizicheskiye mekhanizmy uprochneniya metallicheskikh materialov / Pod red. D.L. Mersona // Perspektivnyye materialy (uchebnoye posobiye). Tula: Izd-vo TGU, MISiS, 2006. p. 267–320.; In Russian.

[6] Kozlov E.V., Lychagin D.V., Popova N.A. i dr. Dal'nodeystvuyushchiye polya napryazheniy i ikh rol' v deformatsii strukturno-neodnorodnykh materialov // Fizika prochnosti geterogennykh materialov. L.: FTI, 1988. p. 3–13.; In Russian.

[7] Aleshin, N. P. Fizicheskiye metody nerazrushayushchego kontrolya svarnykh soyedineniy: Uchebnoye posobiye. M.: Mashinostroyeniye, 2006. 368 p.; In Russian.

[8] Bugai, N. V. Working capacity and durability of metal power equipment [Text] / N. V. Bugai, T. G. Berezina, I. I. Trunin. M., Energoatomizdat, 1994. 272 p.; In Russian.

[9] Criteria for the evaluation of the technical state of the long-lived metal of HPP equipment based on acoustic structuroscopy / Smirnov A. N., Ababkov N. V., Muravev V. V., Folmer S. V. // Russian Journal of Nondestructive Testing. 2015. T. 51. 2. p. 94-100.

[10] Substructure, grain boundaries, and microcracks in long-operating metal / Smirnov A. N., Kozlov E. V., Koneva N. A., Popova N. A. // Metal Science and Heat Treatment. 2005. T. 47. 3-4. p. 155-161.

[11] N.A. Koneva, E.V. Kozlov, Regularities of Substructural Hardening Rus. Phys. J. 1991. V. 34. No. 3. pp. 224-236.

[12] Analysis of acoustic and physical-mechanical parameters Of Cr-Mo-Va steel in the vicinity of the crack / Smirnov A., Logov A., Ababkov N. // AIP Conference Proceedings 2. "Advanced Materials in Technology and Construction, AMTC 2015: Proceedings of the II All-Russian Scientific Conference of Young Scientists "Advanced Materials in Technology and Construction"" 2016. Ch. 020009.

[13] E.V. Kozlov, N.A. Popova, N.A. Koneva, Fragmented Substructure in BCC Steels at Deformations, Izv. RAN. Seriya fizicheskaya [in Russian]. 2004. V.68. No. 10. pp. 1419-1427.

[14] L.I. Tushinskii, A.A. Bataev, L.B. Tikhomirova, Steel Structure and Strength. Novosibirsk: Nauka, 1993. 280 p.

Influence of Microdamage Metal of Energy Long Working Objects on Acoustic and Magnetic Characteristics

Nikolay V Ababkov, Alexander N Smirnov

T.F. Gorbachev Kuzbass State Technical University, 28 str. Vesennyaya, Kemerovo, 650000

E-mail: n.ababkov@rambler.ru

Abstract: The samples of damaged metal steam pipes are studied by acoustic and magnetic methods in the paper. In particular, there are studied sample without damage, samples with creep cracks and samples of with microdamage. All samples were made of 12H1MF steel and have different operating time. A graph of acoustic and magnetic characteristics of the operating time and score of microdamage are constructed. A values of acoustic and magnetic characteristics, corresponding to the maximum, defective score of micro steam pipes of metal specimens made of steel 12H1MF are established.

Introduction

In the process of operating elements of technical devices of hazardous industrial facilities (TDOPO) of the energy industry, a number of problems arise that can lead to man-made disasters and human casualties. First and foremost, this is due to the fact that most of the equipment (85-90%) has developed its resource and needs a complete or partial replacement. However, the current state of the economy in the heat and power industry does not allow achieving the set goals. Therefore, the issue of developing new, effective technologies, methods and methods for assessing the operability and restoration of the resource of the elements of technical devices becomes very urgent. And to ensure the safe operation of critical structures, the concept based on "prediction and prevention" instead of the old concept of "detection and elimination" [1] is becoming increasingly widespread.

The object of the research of this article are direct sections with welded joints of the main steam pipelines, bends of the steam-through pipes from the output collectors of the convective superheater of the IV stage to the steam chamber of the boiler, made of 12Kh1MF steel and which were operated at different power stations and the Kemerovo region heat and power plant (Fig. 1).

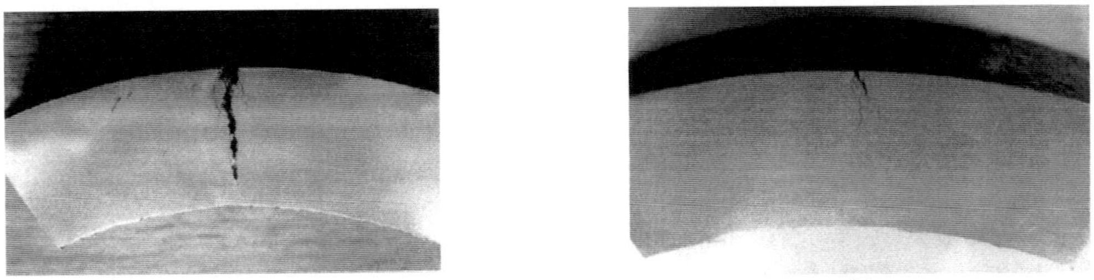

a) *b)*

Fig. 1. Type of test samples of bends of steam-through pipes from the output collectors of the convective superheater of the IV stage into the boiler's steam chamber; size Ø133.0 × 17.0:
a – after 230 000 hours of operation; *b* – after 182 621 h of operation

For the material under investigation, destruction by the creep mechanism is characteristic. According to this mechanism, the destruction of metal occurs due to the gradual accumulation of micropores, which in the subsequent develop into microcracks. The presence of micropores in the volume of the metal is called microdamage [1]. Microdamage is assessed on a point scale in accordance with Annexes G and A of the OST 34-70-960-96. An image of the structure of 12H1MF steel with various microdamage points is shown in Fig. 2, *a* and *b*.

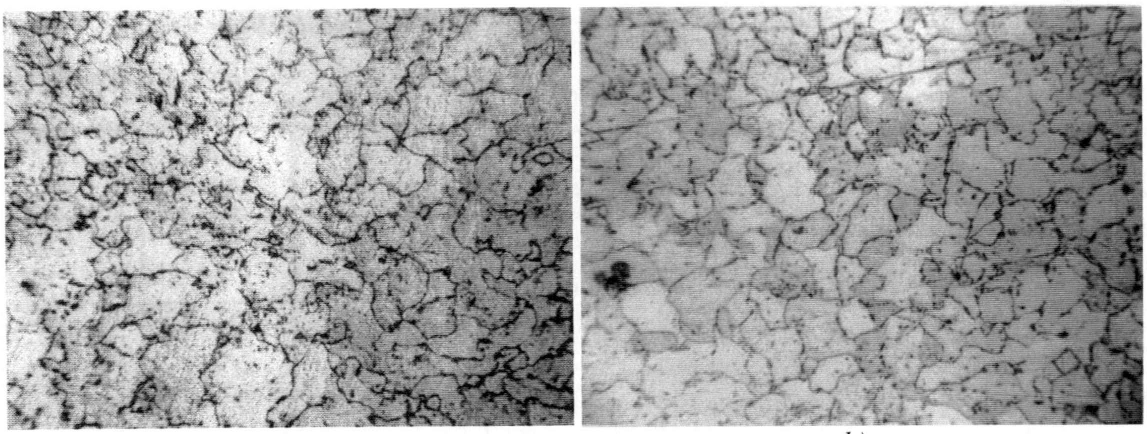

a) *b)*

Fig. 2. Structure of steel 12H1MF with different microdamages, × 500:
a – point In according to Appendix A of the OST 34-70-960-96 (microdamage by optical methods are not detected); *b* – point 6-7 by Appendix G OST 34-70-960-96

To detect defects, various test methods are used that fall into two large groups [2]: destructive and non-destructive research methods.

Destructive testing methods include tensile tests, impact tests, carbide analysis and metallographic tests [9, 11]. The main drawback of the above methods is that to conduct the experiment it is necessary to disable the equipment, and to stop, therefore, the production.

Therefore, in order to detect defects in production conditions and without failure of equipment, various methods of nondestructive testing are used.

At present, the sensitivity of devices of traditional methods of nondestructive testing makes it possible to detect microdamage only from 4 points. To detect microdamages of 1, 2 and 3 points, it is necessary to use instruments with greater sensitivity.

One of the most promising methods for monitoring the state of the metal of welded joints, the base metal and the detection of discontinuities is currently considered an acoustic method [10, 12, 13]. Earlier, the authors of [5-8] established that the spectral-acoustic control method is a sensitive method for estimating local fields of internal stresses and microstructure parameters (dislocation density, curvature-torsion of the crystal lattice, etc.). In addition, the acoustic characteristics are related to the mechanical properties of the materials. This allows us to judge the mechanical properties of the material. The authors of [5-8] proposed complex criteria for the degree of metal damage in relative units.

Method of experimental research

The authors carried out studies of the state of metal samples, straight sections with welded joints of the main steam pipelines, bends of the steam-through pipes from the output collectors of the convective superheater of the IV stage to the steam-collecting chamber of the boiler, made of 12H1MF steel and which were operated at different power stations and the Kemerovo region heat and power plant (table).

Table

General characteristics of the test samples

№ i/o	Object	Parameters of Exploitation	Grain value	Point of microdamage	Defects
1.	Straight section with welded joint No. 23 of thread "A" of the main steam pipeline of Block 1; Ø325.0 × 43.0	T = 545 °C, P = 140 kgf /cm^2 Working time: 342 820 h.	№8	Ip (not detected)	None
2.	Bent portion No. 16 of the steam-through pipe from the output collectors of the convective superheater of the IV stage into the steam-collecting chamber of the boiler No. 5; Ø133.0 × 17.0	T = 555 °C, P = 140 kgf /cm^2 Working time: 182 621 h.	№8	6-7	Crack on the outer surface of 50 mm long, blind, depth 6.5-7.0 mm
3.	Bent and straight section of bend with welded joint of the boiler steam duct st. №10; Ø325.0 × 24.0	T = 540 °C, P = 140 kgf /cm^2 Working time: 175 354 h.	№8	Ip (not detected)	None
4.	Bent portion of the steam-through pipe No. 6 from the output collectors of the convective superheater of the IV stage into the steam chamber of the boiler; 133.0 × 17.0	T = 555 °C, P = 140 kgf /cm^2 Working time: 230 000 h.	№7	6-7	Brack on the outer surface of 140 mm
5.	Bend of the steam-through-pipe from the output collectors of the convective superheater of the IV stage into the steam chamber of the boiler; 133.0×17.0	T = 555 °C, P = 140 kgf /cm^2 Without work	№1	Ip (not detected)	None

The study was carried out using the ASTRON measuring and computing complex and the INTROSCAN metal stress and structure analyzer. Precise measurements of the propagation time (delays) and the ratio of the amplitude of ultrasonic pulses (attenuation coefficient) propagating in the material of the object under study (ASTRON) with the help of a 4 MHz surface acoustic wave sensor, as well as the intensity of magnetic noise (INTROSCAN) were made. To do this, control zones were prepared on the surface of the samples. After that, measurements were made of the above parameters, with the location of the sensors along and along the rivers relative to the central axis of the bends and steam lines.

Results and its discussion

The results of measurements of the acoustic and magnetic characteristics, as well as their dependences on the time taken and the microdamage score, are shown in Fig. 3 and 4. Analysis of the obtained graphs (Fig. 3) makes it possible to judge the absence of any dependencies between acoustic and magnetic characteristics and operating time. At the same time, there are clear relationships between the acoustic and magnetic characteristics and the point of microdamage.

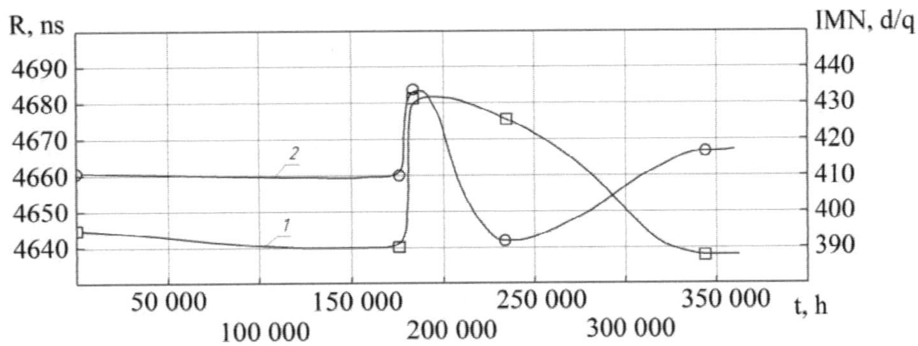

Fig. 3. Dependences of the values of the intensity of magnetic noise – IMN, b / in (*1*) and the delay time of surface acoustic waves – *R*, ns (2)

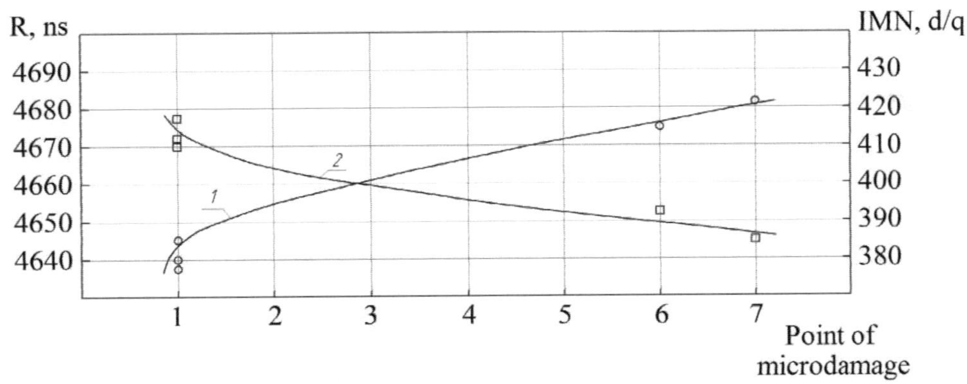

Fig. 4. Dependences of the values of the delay time of surface acoustic waves – R, ns (*1*) and the intensity of magnetic noise – IMN, d/q in (*2*) from the point of microdamage

Thus, the spectral-acoustic and magnetic noise control methods can be used for evaluation of micro metal continuously operating FEC objects.

Conclusions

1. A study of metal samples of damaged steam pipelines by acoustic and magnetic methods was carried out. The sample is not damaged, samples with creep cracks and samples from the detected microdamage.

2. Established values of acoustic and magnetic characteristics corresponding to the maximum, the defective point of microdamages steampipes metal specimens made of steel 12H1MF. Spectral-acoustic

IOP Conf. Series: Materials Science and Engineering **253** (2017) 012015 doi:10.1088/1757-899X/253/1/012015

and magneto-noise control methods can be used to assess the microdamage of metal of long-term operating facilities of the fuel and energy complex.

The work was supported by the RNF grant agreement №14-09-00724 and Russian President's grant for young candidates, MK-1341.2017.8.

References

[1] Criteria for the evaluation of the technical state of the long-lived metal of HPP equipment based on acoustic structuroscopy / Smirnov A. N., Ababkov N. V., Muravev V. V., Folmer S. V. // Russian Journal of Nondestructive Testing. 2015. T. 51. 2. p. 94-100.

[2] Aleshin, N. P. Fizicheskiye metody nerazrushayushchego kontrolya svarnykh soyedineniy: Uchebnoye posobiye. M.: Mashinostroyeniye, 2006. 368 p.; In Russian.

[3] Makhutov N.A., Gadenin M.M. Tekhnicheskaya diagnostika ostatochnogo resursa i bezopasnosti. M.: Spektr, 2011. 185 p.; In Russian.

[4] Bugai, N. V. Working capacity and durability of metal power equipment [Text] / N. V. Bugai, T. G. Berezina, I. I. Trunin. M., Energoatomizdat, 1994. 272 p.; In Russian.

[5] Substructure, grain boundaries, and microcracks in long-operating metal / Smirnov A. N., Kozlov E. V., Koneva N. A., Popova N. A. // Metal Science and Heat Treatment. 2005. T. 47. 3-4. p. 155-161.

[6] Dolzhanskij, P. R. Kontrol' nadezhnosti metalla objektov kotlonadzora: sprav. posobie. – M.: Nedra, 1985. – 263 p.

[7] Kontrol' metalla v jenergetike / pod. red. V. G. Borisova [i dr.]. Kiev : Tehnika, 1980. 134 p.

[8] Mitenkov, F.M. O novom metode kontrolja povrezhdaemosti materiala oborudovanija JAJEU i apparatno-programmnyh sredstvah dlja ee realizacii / Mitenkov F.M., Uglov A.L., Pichkov S.N., Popcov V.M. // Problemy mashinovedenija i nadezhnosti mashin, 1998. 3. p.3–9.

[9] Ivanov, JU. F. Gradientnye strukturno-fazovye sostojanija, formirujushhiesja v stali 08H18N10T pri mnogociklovoj ustalosti do razrushenija / JU.F. Ivanov, V.E. Gromov, S.V. Gorbunov i dr. // Fizika metallov i metallovedenie. 2011. T. 112. 1. p. 85–93.

[10] Analysis of acoustic and physical-mechanical parameters Of Cr-Mo-Va steel in the vicinity of the crack / Smirnov A., Logov A., Ababkov N. // AIP Conference Proceedings 2. "Advanced Materials in Technology and Construction, AMTC 2015: Proceedings of the II All-Russian Scientific Conference of Young Scientists "Advanced Materials in Technology and Construction"" 2016. Ch. 020009.

[11] V.P. Gagauz, et al., Structure, Phase Composition and Mechanical properties of Thick Welded Joints [in Russian]. Novokuznetsk: SibGIU, 2008. 150 p.

[12] Miroshin I. V. Otsenka vzaimosvyazey parametrov mekhanicheskogo sostoyaniya metalla s signalami akusticheskoy emissii / I. V. Miroshin, O. A. Ostanin // Uprochnyayushchiye tekhnologii i pokrytiya. 2006. 2. p. 44-50.

[13] Blyumenshteyn, V. Yu. Issledovaniye vliyaniya istorii nagruzheniya na signaly akusticheskoy emissii (AE) / V. Yu. Blyumenshteyn, A. A. Krechetov, I. V. Miroshin i dr. // Fundamental'nyye i prikladnyye problemy tekhniki i tekhnologiy. 2005. 4. p. 54-57.

IOP Conf. Series: Materials Science and Engineering **253** (2017) 012016 doi:10.1088/1757-899X/253/1/012016

Acoustic Emission Control of Strain State in Simple Loading Condition

V Yu Blumenstein [a], **I V Miroshin** [b], **K P Petrenko** [c]

28 Vesennyaya str., Kemerovo, 650000, Russia T.F. Gorbachev Kuzbass State Technical University

E-mail: [a] blumenstein.vu@gmail.ru, [b] imiroshin@rambler.ru, [c] konst.petr@mail.ru

Abstract. In this paper, review, technique and experimental research results of strain parameters correlation with acoustic emission signal properties are presented. Researches are carried out in condition of simple one- and two-step compression. Results obtained were used for assessing accumulation of properties in terms of technological inheritance in cutting and burnishing processes.

Introduction

When engineering and optimization of machine parts hardening technology by surface plastic deformation with technological inheritance (TI) taken into account phenomenological theory is used. Mentioned theory is based on end-to-end description of physical state of part surface layer on all its productive life stages in uniform terms and categories of continuum and fracture mechanics. The fundamentals of this approach is built on the conception of continuous surface layer formation and plasticity reserve depletion during surface plastic deformation. Metal condition evaluation is performed using mechanical parameters such as degree of shear deformation Λ, degree of plasticity reserve depletion Ψ (DPRD) etc.

However, the complexity of mechanical parameters identification restrain possibility of using this approach in actual practice. On the other hand, physical nature of inherited phenomenological theory makes it possible to employ physical research methods for obtaining high quality results [1].

One of such methods that accounts for nature of plastic deformation on micro-level is acoustic emission method. Acoustic emission (AE) is defined as transient mechanical waves generated due to localized physical changes in a solid material under mechanical or thermal stresses and fracture. AE wave is an acoustic impulse with wide frequency spectrum. The fundamental features of the AE signals include count, cumulative event count, signal energy E_C, amplitude, duration, signal power, e. t. c. Some problems require pulse-height distribution and frequency spectrum researches. These parameters have different information capability and should be selected according to research type and equipment available.

AE method is widely used during mechanical tests for revealing correlation between AE parameters and number of accumulated defects in metal loading process. AE parameters reproduce nature of strain accumulation and enable to apply phenomenological approach more objectively. AE method is commonly used for fracture prediction because of its capability of revealing real damage and assessing probability of defects, which is crucial for durability estimation of machine parts.

As has been mentioned above, there is a problem in establishing relationship between phenomenological parameters of surface layer and AE parameters since TI mechanics is expressed in macro-parameters (Λ , Ψ – macro-level) while AE reproduces micro-level (dislocation density, slip lines developing). However, there is a correlation between dislocation arrangement alteration and mechanical characteristics of surface layer [2-4].

AE method is in common use for control and examination of parts and assembly units for various machinery.

I. Baran, M. Nowak and W. Darski [5] investigated the possibility of using AE method to determine the technical state of the slide bearings in engines with self-ignition. The tests were carried out on three sliding bearing: without defects, with an opening and with longitudinal and circumferential scratches. Results obtained have shown considerable difference in AE signals for each type of bearing, especially in root mean square (RMS) and frequency distribution of signals. The sensitivity of the AE method allows one to record the signals, which indicate the transition from the fluid friction into the mixed friction.

J. Yan et al. [6] have applied AE method to diagnose valve leakage, which is one of most common defect. It was revealed that in the process of fluid leakage high frequency (100 kHz – 1 MHz) AE signals are generated and that they can be easily identified by environmental noise from the factory. The results showed stable positive correlation between disk wear and RMS and between foreign objects size and AE parameters. Error of actual leakage measurement with leakage calculated by AE signals is less than 8 %.

A. Keshtgar and M. Modarres [7] investigated relationship between particular features of AE signals and fatigue crack initiation on wrought aluminum specimens. Fatigue experiments were implemented under cyclic loading with frequency of 20 Hz and loading ratio of 0.5. The minimum and maximum applied loads were 4.5 kN and 9 kN, respectively. The fatigue cracks were monitored by direct measurement using an optical microscope. Signals from the AE sensor were filtered using a band pass filter (200 kHz- 3 MHz) to eliminate emissions from outside sources. It was discovered that the first detected jump (more than 50% sudden increase) in intensity of AE signals having a relatively fast rise time and high amplitude corresponds to the crack initiation.

S. A. Niknam et al. [8] investigated the usage of AE method for unbalanced analysis in rotary systems. The parameter of interest in this study was PAC-energy parameter that covered the significance of count, duration and peak amplitude in AE signals. As a test object self-aligning ball bearings or three kinds were used: new, used and faulty. To make a faulty bearing, a groove was made by using electrical discharge machine on the inner race of a bearing. To impose the unbalance force two uniform disks with the off-center holes were used. As a result, the categorical data analysis based on generalized linear models was presented. It was observed that the faulty bearing was the most significant in producing more AC-energy and the effect of used bearing was more significant than new bearings in this respect. The results demonstrated the possibility of using AE method for detecting and estimating unbalance in rotary systems.

L. Gao et al. studied the possibility of AE method application for fault diagnosis of low-speed heavy-duty gears [9]. The authors developed special de-noising technique based on redundant second-generation wavelet. It was revealed, that with the wear process developing of gears the amplitude of mesh frequency increased and so did the ultra-harmonics amplitudes. The examination of gears confirmed detected regularities. Developed technique allows one to effectively process AE signals for on-spot fault diagnosis of fears.

K. Shrama et al. [10] investigated crack initiation and growth on mild steel specimens using AE method and Digital Image Correlation (DIC). The specimens were subjected to fatigue load ranging from 41 % to 85 % of the ultimate tensile strength. Part of the specimens were exposed to alternating spraying of 5 % NaCl solution. A combination of AE and DIC was used to monitor the strains values and fatigue crack growth during the fatigue tests. Cumulative counts and cumulative absolute energy are two parameters that were used to develop plots that correlate to the fatigue crack growth process with time. The results obtained showed the capabilities of AE method for detecting fatigue fractures.

V. Greshnov [4] examined the alteration of surface layer condition depending on dislocation arrangement changing while plastic deforming. That research work is of great interest since it makes possible to monitor the metal damage from micro-level (dislocation) to macro-level (visible crack). The author considers plastic deformation and metal fracture as a united kinetic, multi-stage, stochastic and activated (mechanically and thermally) process driven by fault motion and transformation of various kinds. For non-monotonous processes with multiple loading micro-crack density increment was assessed as follows:

$$dN_{m(i)} = [\frac{10^{-5}\sigma_i^2(\varepsilon_i)}{(\alpha\overline{m}Gb)^2} - N_{m0(i-1)}\overline{k}_{md(i)}]d\overline{\varepsilon}_i \tag{1}$$

where $dN_{m(i)}$ is micro-crack density increment on i loading stage; $N_{m_0(i-1)}$ is accumulated micro-crack density after $(i-1)$ loading stage; $\overline{k}_{md}$ is factor of proportionality. The author used addition principle of metal damage and generalized equation $\sigma_i(\varepsilon_i)$ obtained in [11].

By metal damage (fissuring, faultiness), V. Greshnov means degree of plasticity reserve depletion since he accepted known equation derived by A. Bogatov [12]:

$$\omega = \sum_{i=1}^{n}\int_{0}^{\Lambda_i}\frac{a\Lambda^{a-1}}{\Lambda^a}d\Lambda \tag{2}$$

where ω represents elementary volume metal damage after n deforming stages (monotonous, quasi-monotonous); $a = a(\kappa, \nu_\sigma, H, T)$ – parameters dependent on material nature, thermo-mechanical conditions and the kind of deforming.

We use the approach as in expression (1) on transformation from micro-level to macro-level. It can be seen that metal damage accumulation depends on strain and strain rate with stress state taken into account, while micro-level in the end is considered as faultiness of initial microstructure state.

Considering that, initial metal state has no micro-damage and $\Psi = 0$, we can assume that at surface layer loading metal micro-damage results in certain plasticity reserve depletion, which in turn involves AE signal generation. We will take a closer look as to how metal damage occurs while surface layer loading and its connection with AE signal generation.

All known models can be expressed as follows:

$$(\dot{N},\ A,\ E,\ W) = f(\Delta\rho), \tag{3}$$

where $\Delta\rho$ is dislocation density alteration.

It has been noted by many authors, that there is a relationship between AE parameters and strain rate $\dot{\varepsilon}$ since plastic flow velocity depends on dislocations motion rate (density alteration) in sliding direction [13-14]:

$$\dot{\varepsilon} = b\frac{d\rho}{dt}, \tag{4}$$

where b is Burgers vector.

D. Merson suggested model that establishes a link with not only dislocation arrangement but stress-strain state as well [13]:

$$W = D\frac{\nu_0^2 S\dot{\varepsilon}}{\sigma^2} \tag{5}$$

In expression given the author employs such analogous concepts as dislocation movement velocity and strain rate, that were used, for example, in Eq. 4. D-parameter (constant for sample given) represents its physical and mechanical properties. The author claims that AE signal power depends on surface area on which deformation occurs. However, such assertion is right only if constant volume of material is engaged in deformation, otherwise AE signal should be attributed to material volume.

Analysis of this model allows one to suggest that AE signal power while metal deforming is function of direct components: physical and mechanical material properties, geometry of specimen (part) and stress-strain indicators.

Such assumption does not contradict A. Kibalchenko's researches in which the author distinguishes AE signal power from deformation in metal cutting as part of deformation power [15]:

$$W = k \cdot W_d = k\tau_s \dot{\varepsilon} S, \tag{6}$$

where τ_s is material shear yield stress, $\dot{\varepsilon}$ is strain rate, S is volume of metal being deformed, k is factor which takes into account portion of process energy emitted in form of AE.

Studies conducted show that AE signal power in this case too depends on same constants and strain rate, without stress-strain indicator taking into account though.

Thus, AE signal is connected with dislocation density alteration on the one hand and with strain rate on the other hand (in plane strain condition with strain rate intensity ξ_i) [4].

Models examined are of a great research value, but using of dislocation density in practice especially in engineering calculations when manufacturing processes designing presents a challenge for research workers.

In the context of present work TI theory has been developed in course of estimation of surface layer physical inherited condition in terms of AE signals. Relationship between mechanical condition parameters and AE parameters has been used later on for identification inherited mechanical condition and surface layer condition control at any loading stage. In doing so the results of various sources on AE signal parameters control has been taken into consideration which showed that in simple loading condition count rate $\dot{N}$ and signal amplitude A held an ample information capability.

Experimental researches technique

Due to some complications of integrated criteria Ψ and Λ estimation at machine cutting and surface plastic deformation, which occur in deformation sign changing conditions, at first relationships between mechanical parameters and AE signal have been investigated in more simple condition.

The aim of experimental researches was to obtain steady correlation between surface layer mechanical parameters and AE signals. Press compression with simultaneous AE signal recording was performed at metal test lab on "AZOT" stock company (Kemerovo city).

Researches were carried out on two batches of specimens. Barrel-type annealed state specimens made of 1045 carbon steel were subjected to AE investigations. Table 1 shows initial data for specimens of the first batch, where D_0, h_0; S_0; V ; P ; HV are dimensions before loading, sectional area, volume, loading force and original hardness respectively. Through preliminary hardness test four groups of specimens were sorted out. Specimen's data of second batch (AE signals included) is given below. Specimens of that batch were also made of 1045 steel but from other delivery lot. Tests were carried out in uniaxial compression condition of uniform and monotonous character; maximum load was 500 *kN*.

The compression was performed by installing specimen 3 in special device 4, which was mounted onto lower press crossbar 5 (fig. 1). AE information recording was carried out during all the loading time. At the same time, loading condition control was implemented by annotation recording when reaching of certain value (with interval of 50 *kN*).

Figure 1. Experimental set-up: 1 – pre-amplifier; 2 – AE sensor; 3 – specimen; 4 – special device; 5 – press lower crossbar

Table 1

Initial Data for Experimental Specimens

Specimen number	h_0, mm	D_0, mm	S_0, mm^2	V, mm^3	*HV*	Group number	*P, kN*
1	19.8	19.9	311	6158.305	174-177	1	250
2	19.9	19.9	311	6189.408	177-183	1	350
3	20	19.8	308	6158.15	174-177	1	450
4	20.3	20	314	6377.433	171-177	2	350+250
5	20.1	20.1	317	6377.905	166-171	3	350+250
6	20.1	19.9	311	6251.613	162-166	3	250+350
10	20.1	20	314	6314.601	171-174	2	250+350
11	20.1	20	314	6314.601	152-154	4	250
12	20.4	19.9	311	6344.921	162-166	3	350
13	20.2	19.9	311	6282.716	171-174	2	200
14	20.4	19.9	311	6344.921	166-171	2	250
15	20.4	19.6	302	6155.059	171-177	2	300
16	20.1	19.9	311	6251.613	162-166	3	350
17	20.2	20	314	6346.017	158-162	4	400
18	20.7	19.9	311	6438.228	158-162	4	150
19	20.5	19.9	311	6376.023	166-171	2	250
20	20.3	19.9	311	6313.818	166-171	3	350
21	19.9	20	314	6251.769	158-162	4	450

AE specimens control was performed on AE monitor “Vulcan 8SM”. The system represents an eight channel signal-processing unit interfaced with PC. The digital signal from monitor was received by PC serial port. Further signal processing was carried out with monitor support software package. For the signal recording the resonance sensors NS2002 (Acoustic Emission Sensor) with 200 kHz working

IOP Conf. Series: Materials Science and Engineering **253** (2017) 012016 doi:10.1088/1757-899X/253/1/012016

frequency were used. As a pre-amplifier model NA200 (Acoustic Emission Pre-amplifier) was used with 190B filter type.

AE parameters investigation at mechanical tests was implemented by using two signal detectors mounted onto special "tray" device due to specimen small size and its deformation. For better acoustic contact, wave-conducting liquid poured beforehand in device was used. The sensors were mounted on specially machined surfaces (optimal roughness – *Ra* 6.3). Before sensors installation, the surface was lubricated with "LITOL-24" grease.

AE monitor allows one to obtain the following information from loading area: count rate – $\dot{N}$ and signal amplitude A , dB. Monitor software package produces this information in graph forms: signals amplitude distribution (fig. 2) and "cumulative count rate versus time" (fig. 3).

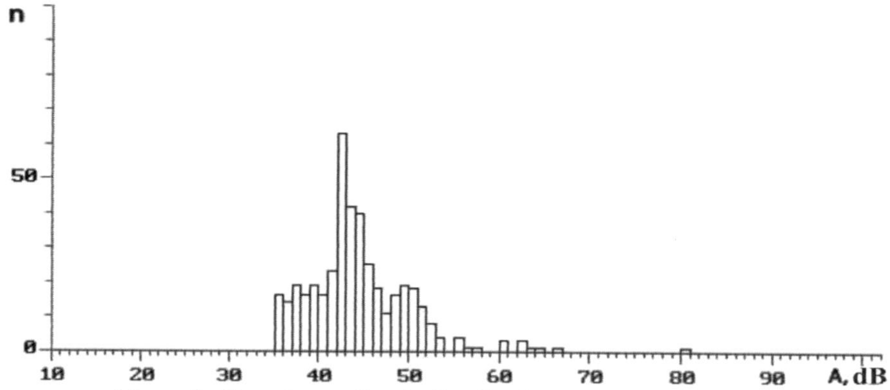

Figure 2. Signals amplitude distribution: n – counts

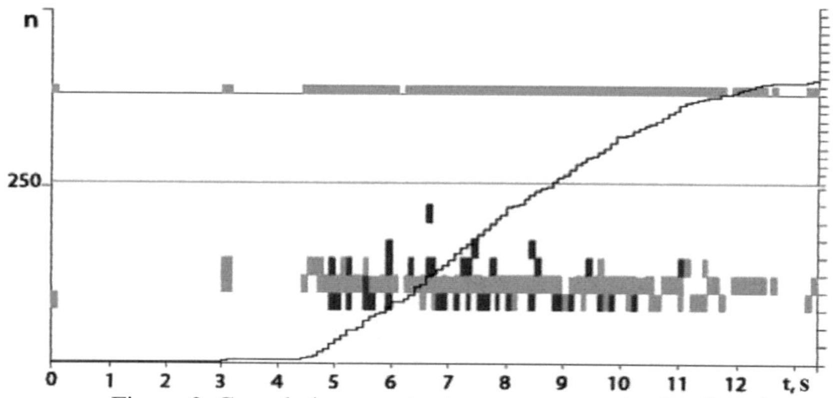

Figure 3. Cumulative count rate: n – counts, t – loading time

For obtaining the information in a text form, the processing procedure was performed using monitor software package. For further processing the data was converted into tabular processor files by tabulating (fig. 4). To make the processing more convenient data for each specimen was presented as shown in fig. 5.

Date	Time	Frame	Sensor Number	First sensor amplitude	Second sensor amplitude
17-Jul-11	11:25:28	OBR-14-2			
17-Jul-11	11:25:37	START	35 T		
17-Jul-11	11:26:23	0	10	65	58
17-Jul-11	11:26:23	64	10	46	41

Figure 4. Initial data table fragment

Time	Amplitude, *dB*		Mean amplitude, *dB*	Cumulative amplitude, *dB*	Time, *N*	Cumulative time, s	$\dot{N}$	$\sum \dot{N}_i$
	from 1st sensor	from 2nd sensor						
11:26:17	46	42	44	44	0:00:00	0:00:01	1	1
11:26:18	35	44	39.5	83.5	0:00:01	0:00:02	1	2
11:26:23	65	58	61.5	145	0:00:05	0:00:07	1	3
11:26:23	46	41	43.5	188.5	0:00:00	0:00:07	1	4
11:26:23	39	35	37	225.5	0:00:00	0:00:07	1	5

Figure 5. Information obtained processing sequence

Degree of shear deformation Λ for monotonous deformation and degree of plasticity reserve depletion were calculated using technique reported in [17]. Results are presented in table 2.

To reveal the relationship between mechanical parameters of metal condition and AE parameters cumulative count rate $\dot{N}$ and signal cumulative amplitude A were used. Results obtained are presented in Table 3.

Table 2

Degree of shear deformation estimation results

t, s	Δh_i	ΔR_i	ε_Y	ε_X	ξ_Y	ξ_X	H	Λ
1	0.01052	0.0042	0.000687	0.000365	0.000687	0.000365	0.000687	0.000687
2	0.02104	0.0084	0.001373	0.000731	0.001373	0.000731	0.001374	0.001374
3	0.03156	0.0127	0.00206	0.001097	0.00206	0.001097	0.002061	0.002061
4	0.04208	0.0169	0.002747	0.001463	0.002747	0.001463	0.002749	0.002749
5	0.0526	0.0212	0.003433	0.001829	0.003433	0.001829	0.003436	0.003436

Table 3

AE Signal cumulative amplitude

Number	Group	*HV*	*P, kN*	$\sum \dot{N}_i$	Mean amplitude	Cumulative amplitude
1	1	174-177	250	984	46.1	56112
2	1	177-183	350	574	46.4	42838
3	1	174-177	450	680	45.4	46554
4	2	171-177	350	672	46.83	31593
15	2	171-177	250	447	44.2	19815
5	3	166-171	200	233	44.1	10305
6	3	162-166	250	676	45.5	30410
12	3	162-166	300	728	46.25	33780
16	3	162-166	350	637	46.8	29720
20	3	166-171	400	1138	43.1	49310
11	4	152-154	150	335	45.68	15349
17	4	158-162	250	563	47.1	26659
18	4	158-162	350	737	45.8	34016
21	4	158-162	450	1000	45.2	45435

Since during AE investigation the issue reliability of results received constantly arises, the following versions of their processing were verified:

- signals registered in null frame were considered as significant ones;
- signals registered in null frame as well as signals in frames next to null were considered as significant ones given they had large amplitude;
- signals having amplitude more than 40 dB were considered as significant ones (this enables eliminating probability of noises, random and reflected signals registration;
- all received signals were considered as significant ones.

The drops of number of registered impulses from sequence for number one specimen are attributed to the fact that at the earliest loading not only specimen is deformed but special devise as well. During next loadings, the devise does not "sound" anymore and thus valid data are registered.

Assessment of cumulative AE parameters was carried out in the following way: the amount of events was registered every second and placed in the column $\dot{N}$. Cumulative amplitude $\sum A$ was registered in a similar manner (Table 4).

Table 4

The results of cumulative AE signals registration

t, s	ε_Y	ε_X	ξ_Y	ξ_X	H	Λ	$\dot{N}_i$	$\sum \dot{N}_i$	$\sum A$
1	0.00109	0.00058	0.00109	0.00058	0.00109	0.00109	1	1	55
2	0.00218	0.00117	0.00218	0.00117	0.00218	0.00218		1	
3	0.00327	0.00175	0.00327	0.00175	0.00327	0.00327	6	7	381
4	0.00436	0.00234	0.00436	0.00234	0.00436	0.00436	1	8	429
5	0.00545	0.00293	0.00545	0.00293	0.00545	0.00545		8	

As strain rate was variable, the extra factor was introduced to consider this point at a degree of shear deformation calculation for specimens of third and fourth groups:

$$k = 1/U \tag{7}$$

where U represents straining rate.

To make sure there is a consistency of hereditary mechanisms, extra experimental run was carried out. Analysis of second batch specimens was conducted in the same way. Since the loading rates of two bathes were closely matched, additional data processing was not performed (Table 5).

Table 5

Results of second batch specimens

Specimen number	h_0, mm	D_0, mm	S, mm^2	V, mm^3	HV	P, kN	Degree of shear deformation Λ		$\sum \dot{N}_i$
							On current stage	Total	
32-1	20.7	19.7	305	6309.46	179-183	220	0.054	0.054	372
32-2	19.52	20.3	323	6309.46	179-183	280	0.078	0.132	496
33-1	20.15	20.1	317	6393.77	174-183	200	0.212	0.212	710
33-2	19.5	20.43	328	6697.22	174-183	300	0.058	0.058	869
34-1	20.1	20.12	318	6390.60	174-187	225	0.073	0.073	473
34-2	19.35	20.5	330	6386.73	174-187	275	0.041	0.041	389
38-1	20.75	19.85	309	6421.38	170-183	240	0.047	0.047	583
38-2	19.55	20.45	328	6421.30	170-183	260	0.058	0.058	270

Results and discussion

Fig. 6 shows relationships between degree of shear deformation and cumulative count rate for compression test.

Statistical treatment of experimental data made it possible to obtain the following equation:

$$\dot{N} = 1218 \cdot \exp\left(-\frac{(\Lambda - 0.6)^2}{0.17}\right) \quad (8)$$

It was revealed that loading history has a significant effect on AE signals generation (fig. 7). At the one-stage loading of number 14 specimen by 250 *kN* force the degree of shear deformation Λ amounts to 0.34, of number 16 specimen by 350 *kN* Λ amounts to 0.42. It is clearly seen that at one-stage loading the curves actually coincide.

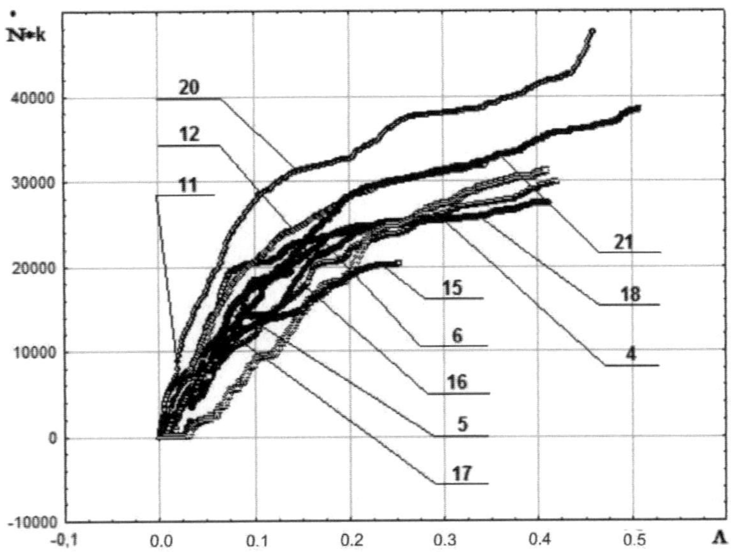

Figure 6. Cumulative count rate versus degree of shear deformation for specimens of second, third and fourth groups (numbers in figure correspond with specimens' numbers)

At the second stage of number 14 specimen loading by 350 *kN* the degree of shear deformation keeps growing but there is no increase of cumulative count rate until point reached at first loading. Only at the certain threshold crossing there is a leap of AE signal as distinctive kink of the curve demonstrates. The degree of shear deformation Λ amounts to 0.2 in doing so.

The feature of second batch experiments lies in the fact that after first loading stage and following specimens unloading second loading stage was performed by greater force. It is should be noted that total force magnitude for all specimens was 500 *kN*.

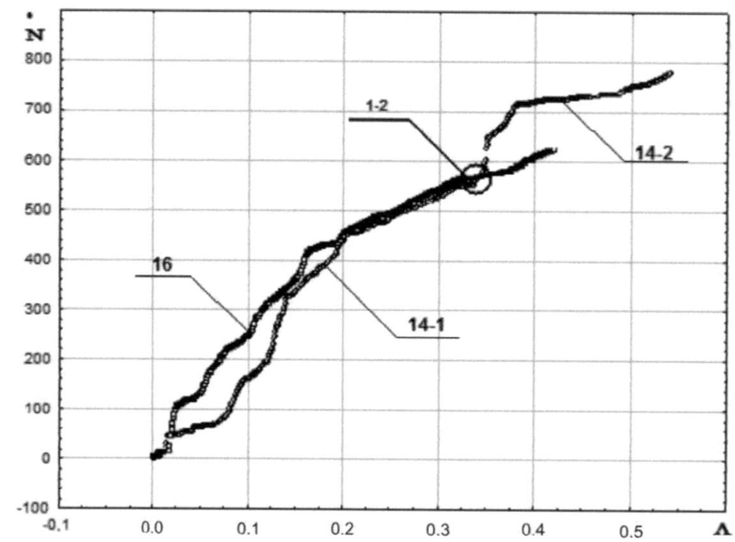

Figure 7. Cumulative count rate versus degree of shear deformation for two-stage loading: 14-1 and 14-2 – 1st and 2nd loading stage for number 14 specimen; 16 – 1st stage for number 16 specimen

Fig. 8 shows relationships between cumulative count rate and degree of shear deformation during compression test for second batch of specimens.

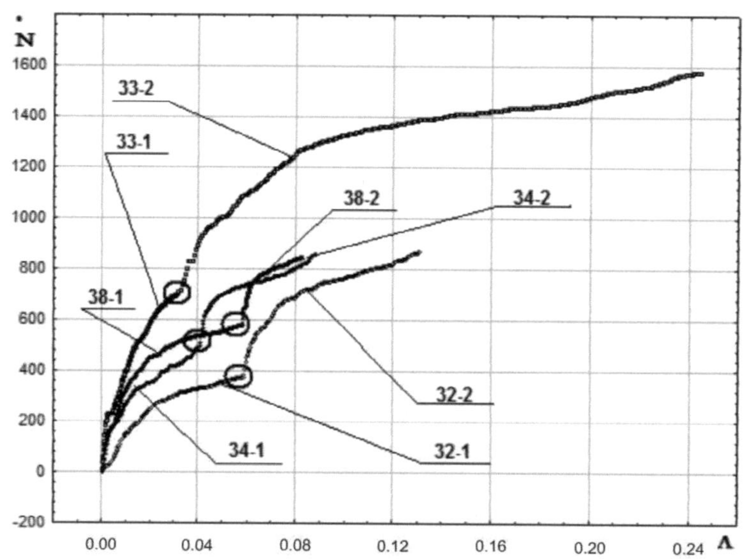

Figure 8. Cumulative count rate versus degree of shear deformation for total loading of number 32-38 specimens (numbers in figure correspond with specimens' numbers)

Conclusions

Analysis of two-stage loading of specimens showed that:

- the loading history has a significant effect on AE distribution pattern and AE signal value;
- greater value of cumulative AE signal on second loading stage at the same degree of shear deformation value is more for number 33 specimen, which on the first stage was loaded by lesser force and on the second stage by greater force. This rule applies for all specimens.

- greater total degree of shear deformation value (greater faultiness) leads to greater cumulative count rate.

References

[1] A. M. Dalskiy, Reliability technological support for high-precision machine parts, Moscow, Mechanical Engineering, 1975, 223 pp.
[2] V. K. Starkov, Dislocation conceptions of metal cutting, Moscow: Metallurgy, 1979, 160 pp.
[3] V. M. Smelyanskiy, Machine Parts Surface Plastic Deformation Hardening Mechanics, Mechanical engineering, Moscow, 2002.
[4] V. M. Greshnov, Yu. A. Lavrinenko, A. V. Napalkov, Engineering physical model of metal deformability // Forging and Stamping Production, 1998, № 7, pp. 5–9.
[5] I. Baran, M. Nowak, W. Darski Application Of Acoustic Emission In Monitoring Of Failure In Slide Bearings, Journal Of Acoustic Emission Volume 25, 2007, pp. 341-347.
[6] J. Yan, Y. Heng-hu, Y. Hong, Z. Feng, L. Zhen, W. Ping, Y. Yan Nondestructive Detection of Valves Using Acoustic Emission Technique, Hindawi Publishing Corporation Advances in Materials Science and Engineering, Volume 2015, pp. 1-9.
[7] Keshtgar A., Modarres M., 2013, Detecting crack initiation based on acoustic emission, Chemical Engineering Transactions, 33, 547-552 DOI: 10.3303/CET1333092
[8] S. A. Niknam, T. Thomas, J. W. Hines, R. Sawhney Analysis of Acoustic Emission Data for Bearings subject to Unbalance, International Journal of Prognostics and Health Management, ISSN 2153-2648, 2013 015, pp. 1-10.
[9] L. Gao, F. Zai, S. Su, H. Wang, P. Chen, L. Liu Study and Application of Acoustic Emission Testing in Fault Diagnosis of Low-Speed Heavy-Duty Gears, Sensors 2011, 11, pp. 599-611; doi:10.3390/s110100599.
[10] Kadhum Shrama, Alastair Clarke, Rhys Pullin, Sam L. Evans Detection of Cracking in Mild Steel Fatigue Specimens Using Acoustic Emission and Digital Image Correlation, 31st Conference of the European Working Group on Acoustic Emission (EWGAE) – We.4.B.2, pp. 1-8.
[11] V. M. Greshnov, Yu. A. Lavrinenko, A. V. Napalkov, Bauschinger's engineering physical model and constitutive equation of isotropic material with anisotropic hardening (tensor correlation) // Forging and Stamping Production, 1998, № 6, pp. 3–6.
[12] A. A. Bogatov, O. I. Mizhiritskiy, S. V. Smirnov, Metal plasticity reserve at forming operations. Moscow: Metallurgy, 1984, 144 pp.
[13] D. L. Meerson Acoustic emission physical nature during deforming processes of metals and alloys, thesis: 01.04.07, Tol'yatti, 2001. 327 pp.
[14] V. E. Vainberg Acoustic Emission during specimens deformation at various rates with sign changing // Defectoscopy, 1975. № 5, pp. 133–135.
[15] A. V. Kibalchenko, N. N. Rasskazov, Acoustic emission during metal cutting processes // Institutes of higher education Bulletin. Mechanical Engineering, 1991. № 6, pp. 34–37.
[16] V. Yu. Blumenstein, V. M. Smelyanskiy, Technological Inheritance Mechanics in the Machine Parts Treatment and Operation Stages, Mechanical engineering-1, Moscow, 2007.
[17] V. L. Kolmogorov, Mechanics of metal forming processes, Moscow: Metallurgy, 1986, 688 pp.

Finite Element Modeling of Strengthening Process by Means of Surface Plastic Deformation Using a Multiradius Tool

Valeriy Yu Blumenstein, Maksim S Mahalov, Anastasia G Shirokolobova

T. F. Gorbachev Kuzbass State Technical University,
28, Vesennyaya St. 650000, Kemerovo, the Russian Federation

E-mail: blumenstein@rambler.ru

Abstract: New designs of deforming tools with a complex working profile, based on the mechanics of technological inheritance, have been developed. The finite element method modeling of the surface plastic deformation process by a multiradius roller was performed and possibility to accumulate large values of deformation without destroying the metal of the surface layer was shown.

Introduction

In modern machine building, constant attention is paid to improving the schemes and methods of machining. Serious attention is given to increasing the accuracy of processing and the quality of the surface layer of machine parts. Today, there is no doubt that the quality of the product is laid in at the stage of design. This involves transferring of the product creation from full-scale testing of prototypes or batches to mathematical modeling of product properties, as well as modeling of product manufacturing process, which allows detecting and eliminating design and technological faults before the production stage begins.

The development of new tools and technologies is based, first of all, on the modeling of machining processes. Modern finite-element program products allow the development of high-quality models and their comparison with experimental tests proves it.

Thus, new achievements in hard metals mechanical processing based on physics of modeling are presented in the paper [1]. The authors show that the reduction in the cycle time and providing quality assurance depends on the physical model activity for hard metal processing.

Knowing physical laws of metal plastic flow allowed to develop and successfully apply the cutting processes with advanced plastic deformation (APD) [2]. The effectiveness of the APD method is achieved by the purposeful change in physico-mechanical properties of material of the cut-off layer by its preliminary deformation, which is carried out in the process of cutting by additional mechanical energy source.

Calculations of the stress-strain state and evaluation of ductility in the deformation zone allowed to improve the deforming-cutting pulling (DCP) scheme with advanced plastic deformation [3]. To eliminate the elastic shrinkage of the product (bush) a number of deforming-cutting and pulling ways were developed with the combination of cutting and deformation zones with the cutting edges in the zone of contact deformation. This allowed to reduce the energy consumption of the cutting process by 20% compared with the deforming-cutting pulling with the APD, as well as the total energy costs of cutting and deformation by 10-14%.

The development of methods and technologies of SPD is associated with the refinement of FEM models, complication of deforming tools geometry, imposition of thermal and vibration effects, consideration of metal plasticity and so on.

So, for example, Yu.I. Sidyakin and his co-authors carry out scientific researches, engineering calculations and design the SPD processes by modeling the contact interaction of the indenter with the processing material and with evaluation the elastoplastic deformation [4].

In the research area of S.A. Zaides the processes of cross-sectional SPD are investigated and developed, allowing to ensure the accuracy and high quality of the surface layer of the parts [5]. In the elastoplastic putting the FEM model of the process was developed, calculations of the stress-strain state of the deformation center were performed. Mathematical modeling and subsequent engineering calculations made it possible to create a theory of the process, to develop technological recommendations and a set of technological equipment that provides specified values of accuracy, hardening and residual stresses based on the assigned operating conditions of the slightly hard axel.

T. Altan carried out a study of the mechanics of the SPD process, creating 2D and 3D FEM models of rolling-off by a roller [6]. The results of modeling, including surface deformations and residual stresses, have showed high convergence with experimental data, obtained from literature sources.

V.M. Smelyansky developed a mechanical theory of SPD process, the stress-deformation state modeling of the deformation zone was carried out and the rules of formation of the surface layer of machine parts were established. It was established that the asymmetric deformation zone (DZ) arises under the SPD processing, the shape and size of which depends on technological factors [7]. The motion of a metal particle in the DZ is carried out along the current flow line, taken at a certain depth and determined according to the tasks solution of the mechanics of rigid body. Moving along the current flow line, equidistant to the deformation zone, the particle passes through three states: initial, current and final. The model analysis made it possible to obtain the distribution of stress tensor components, deformations and rates and allowed to evaluate the effect of hydrostatic pressure on the accumulation of deformation and the exhaustion of the reserve of plasticity of metal.

One of the important trends in the development of machining technologies, including SPD, is the creation of nanostructured state in a thin surface layer of the metal.

A.V. Kirichek and D.L. Soloviev developed a method of gradient static-impulse processing (SIP) of SPD on the basis of the physical model, using shock waves for plastic deformation, creating high pressure in the deformation zone and forming great depth of the hardened surface layer [8]. Crushed metal particles measuring up to 100 ... 300 nm have been revealed in the hardened sample; great concentration of nanoparticles was found in the surface layer at the depth of 3 ... 8 mm. The authors showed that the method of deformation hardening by shock waves makes it possible to create heterogeneous ultrafine-grained gradient-hardened structures, alternating solid and plastic zones according to the given law.

In dynamic FEM modeling V.P. Kuznetsov and co-authors performed the process of smoothing out a thin surface layer of a steel sample under the indenter, introduced with constant force and with constant speed [9]. Under the conditions of plane deformation, the regularities of changes in the stress-strain state (SSS) of the material and the mechanisms of the nanostructured layer formation have been studied. The results of the studies are in good agreement with the experimental data.

One of the priority areas is the development and implementation of intensive plastic deformation (IPD) methods, the feature of which is high hydrostatic pressure in the deformation zone [10-11]. This allows to achieve high plastic deformations, to provide a nanostructured state without destroying the metal, and, as a result, to achieve a unique combination of high strength and ductility at room temperature.

It is known that the action of the cutting and / or deforming tool on the nanostructured metal of the surface layer leads to plastic and thermal deformations, which lead to grain growth and loss of most useful properties. Therefore, the efforts of researchers are also aimed at preserving these properties, including the determination of rational regimes of mechanical processing [12].

The analysis showed that increase in the efficiency (intensification) of the processes of "cold" mechanical hardening treatment is possible by creating complex stress-strain state with predominantly high hydrostatic pressure in deformation zone. In turn, such schemes are realized by tools with complex geometry and which provide a certain kinematics of the metal flow.

Theoretical investigations

The author has developed the theory of formation and transformation of the inherited state of the surface layer in the processing and exploitation processes – the mechanic of technological inheritance (TI) [13]. The description is based on the idea of continuous accumulation of deformations and depletion of the metal plasticity reserve in the surface layer of the part under the influence of loading programs. The life cycle is represented by the stages of cutting, surface plastic deformation and fatigue loading, which in turn consists of two stages – cyclic durability and cyclic crack resistance.

To solve the problems of the TI mechanics, the parameters known from the mechanics of a deformable solid are used:

- stress condition scheme index:

$$\Pi = \frac{\sigma}{\mathrm{T}} = \frac{1/3(\sigma_1 + \sigma_2 + \sigma_3)}{1/\sqrt{6}\sqrt{(\sigma_1 - \sigma_2)^2 + (\sigma_2 - \sigma_3)^2 + (\sigma_3 - \sigma_1)^2}}; \tag{1}$$

- accumulative shear strain level:

$$\Lambda = \langle \frac{2}{\sqrt{3}} \int_0^t \sqrt{\frac{1}{2}\left[\left(\xi_x - \xi_y\right)^2 + \left(\xi_y - \xi_z\right)^2 + (\xi_z - \xi_x)^2\right] + + \frac{3}{4}\left(\eta_{xy}^2 + \eta_{yz}^2 + \eta_{zx}^2\right)} \rangle \, dt; \tag{2}$$

- residual stresses tensor:

$$[\mathrm{T}\sigma_{residual}]_{ij} = \left[\mathrm{T}\sigma_{def}\right]_{ij} + [\mathrm{T}\sigma_{unload}]_{ij} + [\mathrm{T}\sigma_t]_{ij}; \tag{3}$$

- plasticity reserve exhaustion level [14]:

$$\Psi = \Psi_1 + \Psi_2 = \Psi_1 + (\Psi_{21} + \Psi_{22}) = \\ = n\varphi_0 \int_0^{\Lambda_k} \Lambda_i^{n-1} d\Lambda + \left(\int_0^{\Lambda_k} \frac{d\Lambda}{\Lambda_p} - \varphi_0 \int_0^{\Lambda_k} \Lambda_p^{n-1} d\Lambda \right), \tag{4}$$

where σ – average normal stress; T – intensity of shear stress; $\sigma_1, \sigma_2, \sigma_3$ – the main components of stresses tensor; $\xi_x, \xi_y, \xi_z, \eta_{xy}, \eta_{yz}, \eta_{zx}$ – components of strain rate tensor; $\left[\mathrm{T}\sigma_{def}\right]_{ij}$ – load stress tensor; $[\mathrm{T}\sigma_{unload}]_{ij}$ – unload stress tensor; $[\mathrm{T}\sigma_t]_{ij}$ – heat stress tensor; Ψ_1 – component, depending on the yield stress or on the accumulated deformation; Ψ_2 – component, depending on metal plasticity in conditions $\Pi = const$; Λ and Λ_p – accumulated and limited shear strain level at a given index of the stress state Π; n – strain-hardening coefficient; φ_0 – coefficient, determined on the basis of the ductility tests. In non-reinforced metal $\Psi = 0$, and with a complete depletion of ductility resource $\Psi = 1$.

The plasticity curve is used as initial hardening characteristics $\sigma_s = \sigma_s(\Lambda)$, the curve of limited plasticity $\Lambda_p = \Lambda_p(\Pi)$ and the diagram of cyclic crack resistance V=V(K) in the coordinates "stress intensity factor" K – fatigue crack growth rate V".

In the same terms and categories the solution of the problems of mechanics at the investigated stages is fulfilled and it is shown that technological inheritance manifests itself in the formation of hereditary loading programs depending on the geometric parameters of the deformation centers (DZs), acting as a set of initial and boundary conditions for solving the problems of deformation mechanics. The loading program was presented in the coordinates "stress state index Π – accumulative shear strain level Λ".

The regularities of the metal plastic flow, the accumulation of deformation, the depletion of the metal plasticity reserve, the formation and transformation of the surface layer according to the studied loading stages are revealed, the rules of technological inheritance are formulated. The results of the investigations made it possible to reveal in the processes of free orthogonal cutting, and in the SPD processes to confirm the presence in the deformation zone of three sections of quasimonotonic deformation, at the edges of which the deformation changes sign.

In particular, when processing by a SDP toroidal roller the force P=2 500 N it was found that most of the material in deformation zone is under compression, while the largest value of the average normal stress is observed in the zone of the tool contact with a part. The highest intensity values of shear stresses occur in the top zone of the wave front of a deforming tool. When moving deep into the surface layer absolute values decrease, but the distribution pattern of these components is practically unchanged. Such character of the loading scheme leads to the fact that the most intense accumulation of strain stress occurs in the front area of the deformation zone. As metal is reinforced, every subsequent operating cycle leads to a more "rigid" loading schemes, resulting in the accumulation of the limited deformations and complete depletion of metal ductility.

The results of fatigue tests of samples hardened by SPD showed that depletion of the plasticity reserve to a certain limit is useful from the point of view of increasing cyclic durability [13]. At the same time, this leads to the opposite effect at the stage of cyclic crack resistance, increasing the crack growth rate and shortening the duration of this stage. It has been established, that for each material and for each load program there is a certain level of depletion of the plasticity reserve, starting from which there is an increase in the intensity of the cracks growth rate.

Thus, the available range of variation, reached by SPD processing of the parameters of the mechanical state of the metal surface layer is limited both by the initial properties of the metal of the part and by the permissible range of changes in the technological parameters of the processing regime, which include the shape of the profile of the base circle roller.

At the same time, the expansion of the technological capabilities of SPD is possible due to the creation of processing schemes with large hydrostatic pressure and the number of sections of quasi-monotonic deformation. This will lead to the accumulation of large deformations with relatively low exhaustion of the ductility reserve without destroying the surface layer of metal.

Results and discussions

The results of theoretical and experimental studies allowed to develop new designs of deforming tools, having complex working profile, that allows [15-17]:

• to vary widely the geometry of the working profiles, both in radial and axial direction;

• to create large interference of deforming elements in deformation zones;

• to create a complex stress state with predominantly large hydrostatic pressure;

• to obtain large deformations with a relatively low degree of exhaustion of the metal plasticity reserve.

The multiradius base circle roller (MR-roller) has the form of the working surface profile in the form of a combination of successively located deforming elements (DE) with radii of constant magnitude, disposed relative to each other with some displacement in the radial and axial directions [17].

Evaluation of the mechanical state of the deformation zone by the MP roller processing was carried out by the FE modeling method; the modeling material was assumed to be isotropic with mechanical parameters corresponding to the parameters of steel 45 (GOST 1050-88) in the state of delivery. A bilinear approximation of the flow curve was adopted, taking into account the hardening of metal. A flat-deformed setting was used; while the process of deformation accumulation and depletion of the plasticity reserve was simulated in the plane of feed that passes through the axis of rotation of the processed part [18].

As a processing surface a plane part of the main deformations of a cylindrical part with length L = 70 mm and height h = 30 mm was modeled (Fig. 1). As it is known, the plane of main deformations is located in the meridional section of the cylindrical work piece – the feed plane [7]. The lower and side boundaries of the work piece were rigidly fixed along both axes. The dimensions of the deformation zone appeared during the processing, are much less than the modeled fragment, so the edge effects, which occur in this situation, introduce a very insignificant error in the simulation results.

The indenter was modeled as an absolutely rigid body and was in the form of a circle modeling roller of a given profile radius. The surface material of the part, with the properties given in Table 1, was affected by a multiradius roller with profile radii: $R_{pr1} = 1\ mm$ and interference $h_{d1} = 0.05\ mm$; $R_{pr2} = 1\ mm$ and interference$h_{d2} = 0.1\ mm$; $R_{pr3} = 1\ mm$ and interference $h_{d3} = 0.15\ mm$; $R_{pr4} = 3\ mm$ and interference$h_{d4} = 0.05\ mm$.

In the plane of the main deformations, the displacement of the deformation zone in the direction of feed has discrete nature: a new DZ occurs with a displacement of previous one by the feed quantity.

In the initial position, the indenter was disposed to the surface with a certain gap. At the first and any subsequent odd step of modeling the loading was performed – the displacement of the indenter in the direction of the surface by a certain amount, assuming the introduction and creation of the interference specified for modeling. At the second and any subsequent even step of the simulation the unloading was performed – the withdrawal of the indenter from the surface to the initial distance with its simultaneous displacement along the surface by the feed quantity.

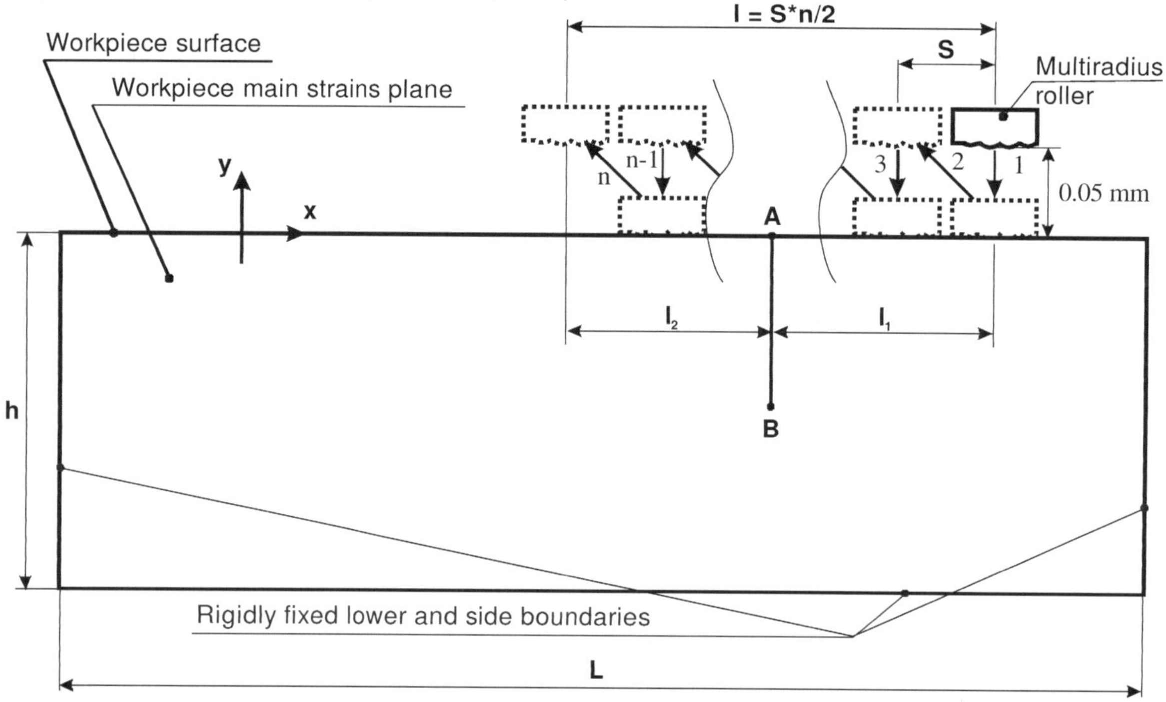

Fig. 1. Scheme for the formulation of the FEM modeling problem

In total, 340 steps were simulated – 170 loading steps and 170 unloading steps. The length of the processing surface in the presented model was $l = S \cdot n/2 = 17\ mm$.

Thus, in the process of realization of all simulation steps, this cross-section passes through the deformation zone space and is the very cross-section along the depth of the processing surface layer in which deformation accumulation took place, partial exhaustion of the plasticity reserve, and the formation of the residual stress tensor.

Table 1

Physical and mechanical properties of metal and parameters of the flow curve (steel 45, GOST 1050-88, 160-180*HV*)

Young's modulus	E, MPa	2×10^{11}
Density	$\rho, kg/m^3$	7800

Poisson ratio	ν	0.3
Friction ratio	η	0.21
Extrapolated range of fluidity	σ_{ext}, MPa	3.66×10^8
Tangent modulus	T_{mod}, MPa	2.596×10^6

After solving the model in the selected section, the following values were fixed for each simulation step: the units' coordinates, the displacement components of the units, the stress tensor components, the tensor components of the elastic, plastic, and total elastic-plastic deformation.

For further calculations of the accumulated parameters of the mechanical state, the angle of deformation in the circumferential direction was assumed equal to 10^0, the rotation frequency of the part being equal to 50 s^{-1} [7]. Thus, time of one rotation of the part was 0.2 s. For the assumed deformation angle in the circumferential direction, time of one cycle, during which the surface layer is loaded and loaded, was 0.0054 s. At the same time, half of this time (0.0027 s) was spent on loading and the same time on unloading.

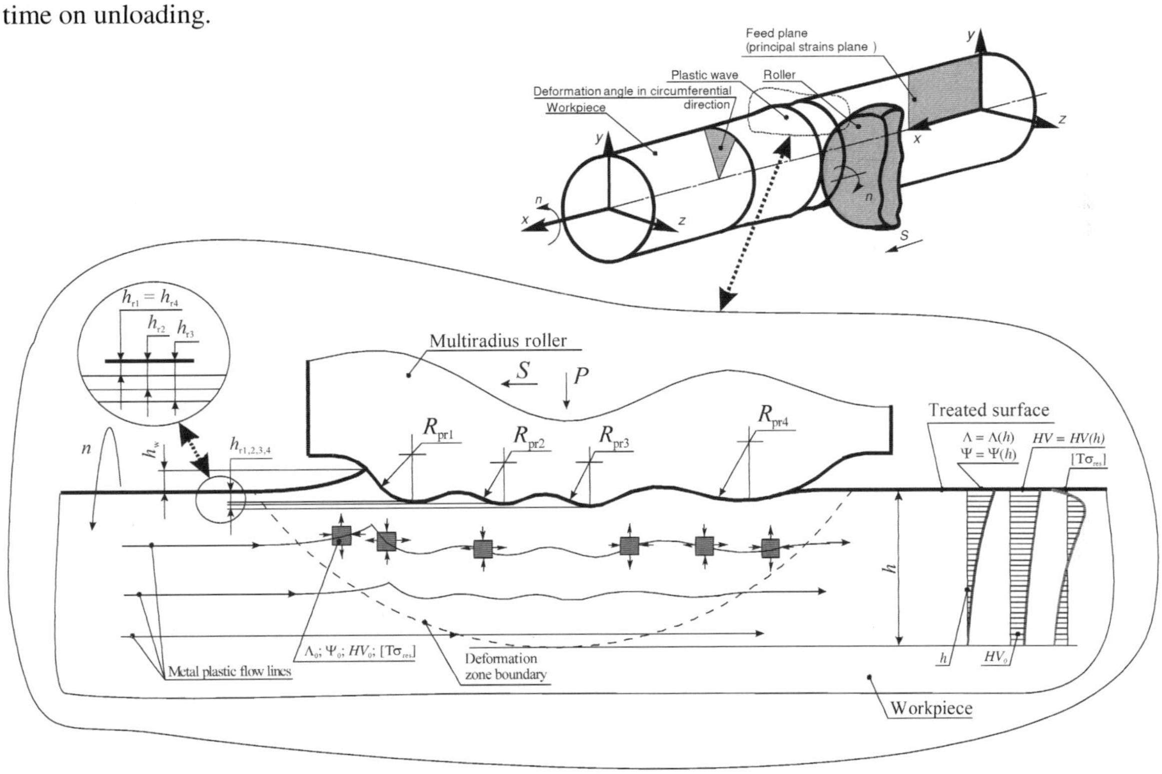

Fig. 2. The scheme of SPD processing by a multiradius roller

Subsequently, the components of the stress and strain tensors were recalculated into the coordinates of current flow line points in the deformation zone [19].

Figure 3 shows the profile of the deformation zone, reconstructed according to the results of FEM modeling. Figures 4-7 show the components distribution of the stress-strain state along the profile line of the deformation zone; while the abscissa points on the diagrams coincide with the abscissa of the profile.

The deforming element, which first comes in contact with the original (unprocessed) surface layer of the part, has a profile radius $R_{pr1} = 1\ mm$ and moves with interference $h_{d1} = 0.03\ mm$. As a result, the DZ arises that is identical in shape and size to the zone of deformation in the SPD processing toroidal roller.

The second deforming element also has a profile radius $R_{pr2} = 1\ mm$ and moves with interference $h_{d2} = 0.08\ mm$ relatively to the initial surface and interference of 0.05 mm with respect to the first deforming element. In this case, the zones of plastic flow are overlapped because of the action of the first and second deforming elements. Based on the ideas about the mechanics of SPD in the zone of this overlap, a change in the stress state scheme occurs, which leads to a change in the sign of plastic deformation.

The same effect is observed in all zones located between adjacent deforming elements.

The third deforming element with a profile radius $R_{pr3} = 1\ mm$ also moves with interference $h_{d3} = 0.13\ mm$ relative to the initial surface and interference of 0.05 mm towards the second deforming element.

The fourth deforming element with a profile radius $R_{pr4} = 3\ mm$ moves with interference $h_{d4} = 0.03\ mm$ relatively to the initial surface and interference of -0.1 mm towards the third deforming element.

Such construction of a multiradius roller leads to complex stress state of the metal, under which a repeated change in the sign of plastic deformation occurs.

Analysis of the results of FE modeling showed that the profile of the deformation zone with a confidence greater than 85% coincides with the experimental profile, obtained by the method of profilography after rapid removal / shooting of the MR roller from the surface of the work piece (Fig. 3). It indicates the formulation correctness of the FEM model and the solution of the problem of determining the components of the stress and strain tensors.

The second important conclusion is that metal is under compression in the upper deformation zones (Figure 4).

The components of the stress tensor change in accordance with the profile of the deforming elements of the MR roller. The component σ_x in the zone of the front non-contact surface before the first deforming element is firstly reduced to the level of -380 MPa, then increases to -230 MPa in the top zone of the plastic wave. The component σ_y decreases to the level of -850 MPa, and the σ_{xy} component first increases to +75 MPa, then decreases to -85 MPa in the top zone of the plastic wave. In the zone of the front contact surface of the first deforming element, growth occurs firstly, the numerical value of the component σ_x decreases to -630 MPa, then again grows to -550 MPa at the top point of the profile of the first deforming element. In the same zone, the component σ_y continuously increases to a value of -190 MPa, and the σ_{xy} component reaches positive values of +175 MPa.

In the transition zone between the first and second deforming elements at points with abscissas equal to 4.2-5.2 mm, the component σ_x increases, the component σ_y decreases to -1300 MPa, and the σ_{xy} component first decreases to -180 MPa, then grows.

In the front contact zone of the second deforming element, the σ_x component decreases sharply to -980 MPa, the component σ_y begins to increase, and the σ_{xy} component increases to +200 MPa.

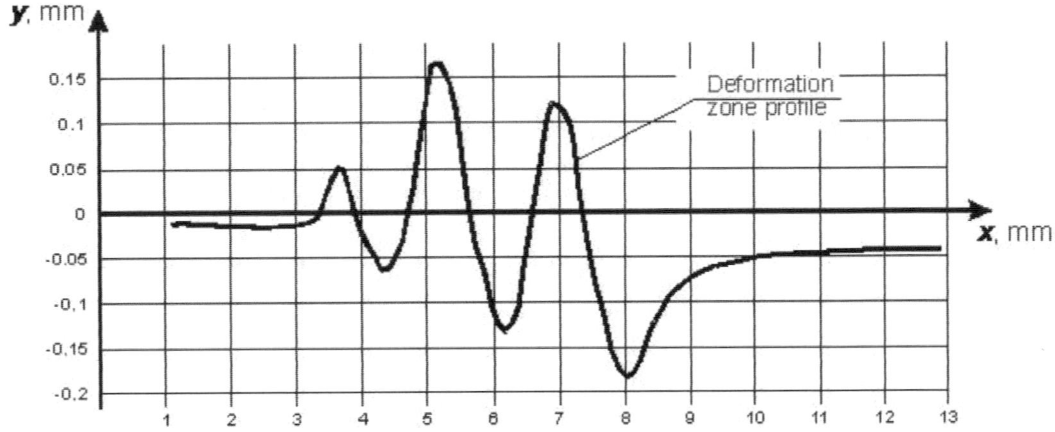

Fig. 3. Profile of deformation zone reconstructed from the results of FEM modeling

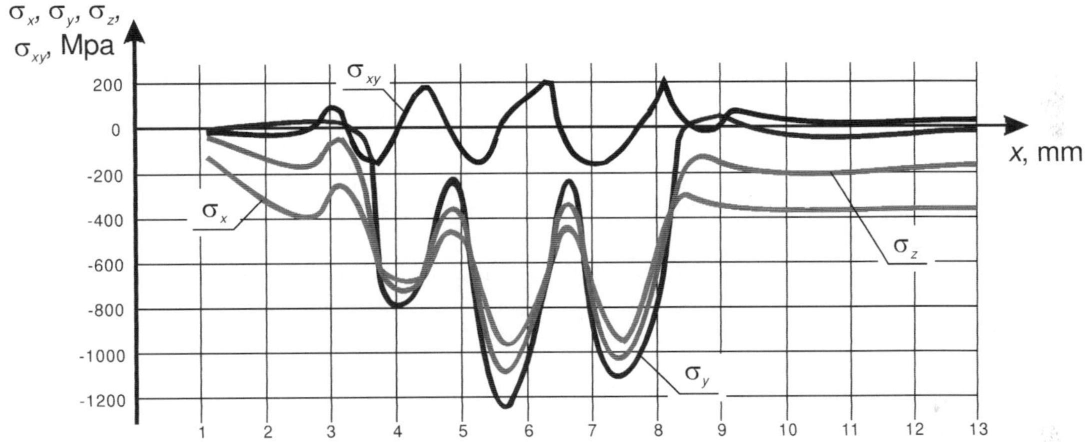

Fig. 4. Distribution of stress tensor components

In the transition zone between the second and the third deforming elements, as well as in the contact zone of the third deforming element, there is the same example of changes in the components of the stressed state.

In the transition zone between the third and the fourth deforming elements, the components σ_x and σ_y grow, but the σ_{xy} component decreases.

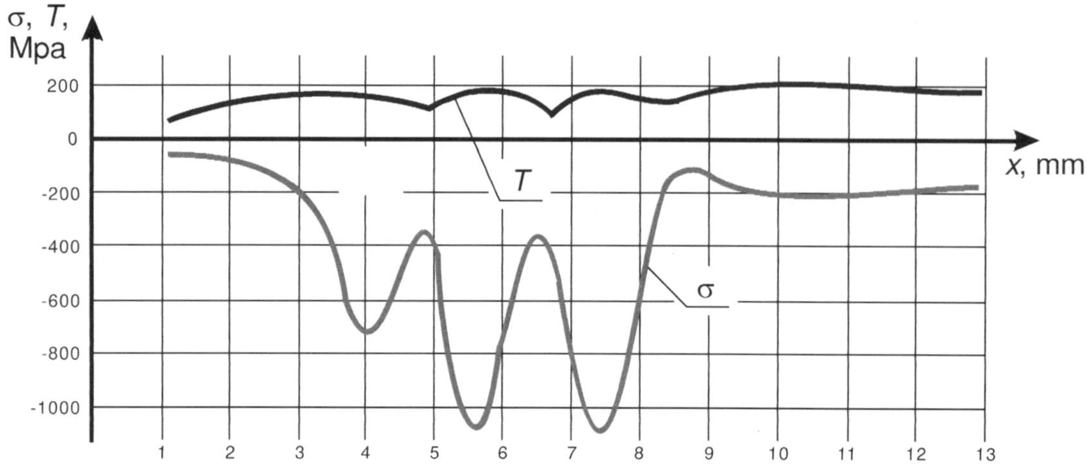

Fig. 5. Distribution of average normal stress and shear stresses intensity

The average normal stress repeats the tendencies of the changes in the components of the stressed state and shows that the deformation zone is under compression (Fig. 5). In this case, under the second and the third deforming elements, special conditions are attained at which maximum hydrostatic pressure reaches 1150 MPa.

The third important conclusion is that, in difficult situation of metal flow in the same characteristic zones, the values and sign are changed repeatedly according to the same tendencies of deformation rates (Fig. 6).

Thus, the accumulation of the deformation degree of the shear and the depletion of metal ductility reserve occurs continuously with the index negative value of the stress state diagram (Figure 7). Moreover, even with a significant accumulated deformation of $\Lambda \approx 7$ due to a favorable deformation scheme, the degree of depletion of ductility reserve does not exceed $\Psi \leq 0.62$.

The results of the research showed that when a traditional construction is processed by a toroidal roller:

- the accumulation of the limiting deformation $\Lambda \approx 1.24$ would lead to the limited metal state $\Psi = 1$ at the profile radius $R_{pr} = 1\ mm$ and interference $h_d = 0.03\ mm$;
- the accumulation of the limiting deformation $\Lambda \approx 1.24$ would lead to the limited metal state $\Psi = 1$ for a profile radius $R_{pr} = 3\ mm$ and interference $h_d = 0.05\ mm$;
- the interference $h_d = 0.15$ mm at the profile radius $R_{pr} = 1\ mm$ is not permissible, because when the tool is inserted for several rotations of the part results to a nonpermanent deformation zone. This is the catastrophic growth of the plastic wave ahead of the deforming tool and the destruction in its top zone;
- SPD processing with the specified strains and profile radii successively by second, third and fourth operating cycle leads to accumulation of limited deformations and complete exhaustion of metal plasticity reserve after the first operating cycle.

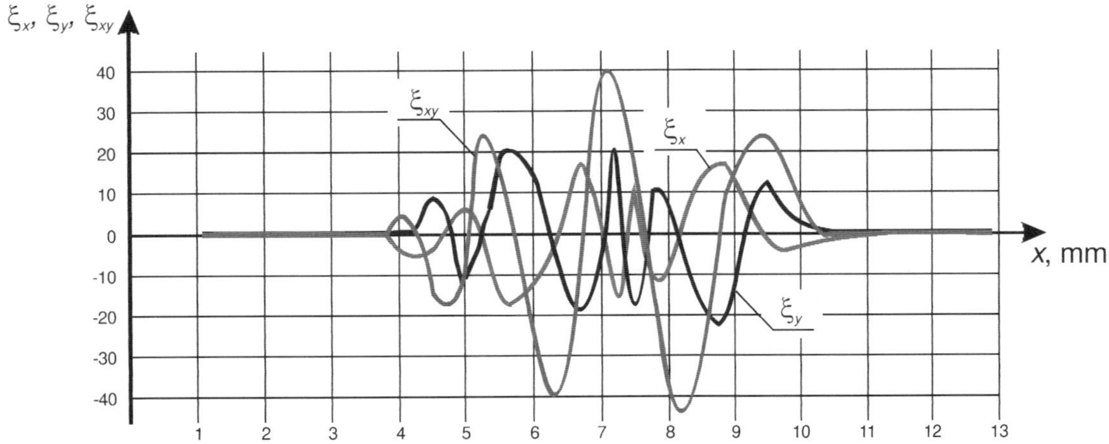

Fig. 6. Distribution of the tensor components of the strain rate

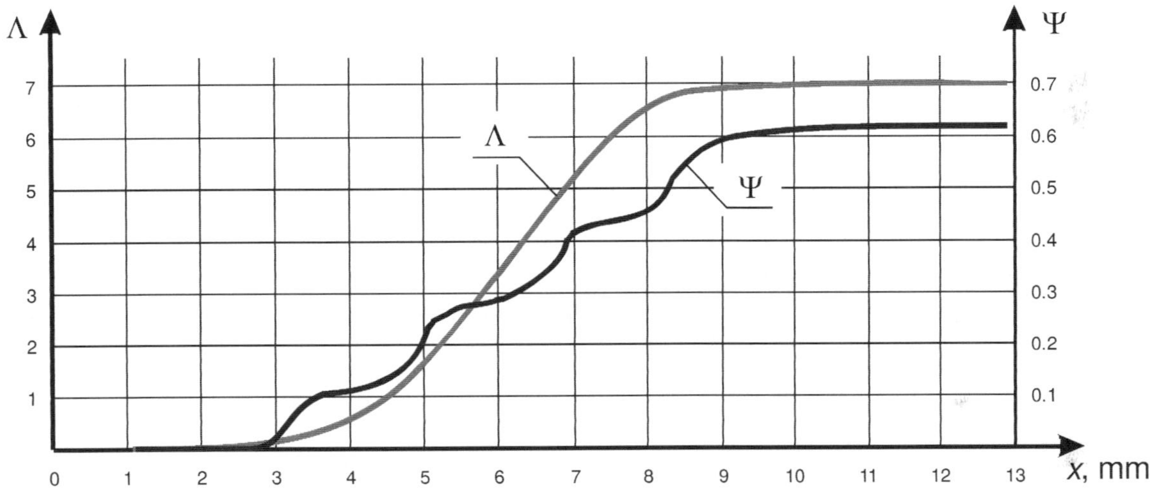

Fig. 7. Accumulation of shear deformation degree and change in the depletion degree of metal plasticity reserve

The conducted researches showed that the considered profile design of the working piece of the roller allows to accumulate large deformation values without destroying the metal of the surface layer, which provides additional possibilities for assurance both cyclic durability and cyclic crack resistance.

Metal micro-hardness investigations of the deformation zone and the surface layer hardening, carried out on the device mod. DuraScan 20, confirmed the received laws.

Conclusions

1. Schemes and methods improvement, development of new tools and technologies is based, first of all, on modeling of machining processing. Modern finite-element software products allow to develop high-quality models, as their comparison with experimental tests testify. At the same time, the quality of calculations is determined by the quality of the physical model, based on modern ideas about the mechanics of a deformed solid body.

2. Knowing physical regularities of the metal plastic flow, calculation of the stress-strain state and evaluation of the accumulated deformations and plasticity in the process and after the processing has

allowed to develop a number of unique methods and processes that ensure high quality of the surface layer and, accordingly, high durability of the processed machines parts.

3. The efficiency (intensification) increase of the processes of "cold" mechanical hardening processing, is possible by creating complex stress-strain state with predominant schemes with high hydrostatic pressure in the deformation zone. In turn, such schemes are implemented by tools that have complex geometry and provide a certain kinematics of metal flow.

4. The results of the theoretical and experimental studies have allowed to develop new designs of deforming tools having the complex working profile. The proposed designs make it possible to create large tension and complex stress state in the deformation zone with predominantly large hydrostatic pressure, which leads to the accumulation of large deformations at relatively low degree of metal plasticity reserve depletion.

5. There was the FE-modeling by SPD processing with a multiradius roller, which has the form of a working surface profile in the form of a combination of successively located deforming elements with radii of constant magnitude, located relative to each other with some displacement in the radial and axial directions. The situation of the metal plastic flow in the deformation zone under the conditions of a complex stressed state is revealed and is shown that the considered profile of the working part of the roller allows to accumulate large values of strains without destroying the metal of the surface layer. In turn, this provides additional opportunities to ensure both cyclic durability and cyclic crack resistance of products.

References

[1] **Marusich, T.D., Usui, S., Zamorano, L., Marusich, K., Leopold, J.** New advances in the machining of hard metals using physics-based modeling (2013) Engineering Transactions, 61 (1), pp. 3-13. http://www.scopus.com/inward/record.url?eid=2-s2.0-4882622323&partnerID=40&md5=9eab415aedec1f30497cdc3dced9ca57

[2] **Yaroslavtsev, V.M.** Cutting with advanced plastic deformation in the technology of metal chip recycling. .http://technomag.bmstu.ru/doc/567548.html # 07, July 2013 DOI: 0.7463/0713.0567548.

[3] **Ambrosimov, S.K.** Determination of technological parameters of the deforming-cutting stretching process with advanced plastic deformation and elastic and plastic loading of the cutting zone // Strengthening technologies and coatings. – 2008. – №8. – P. 3-7.

[4] **Sidyakin, U.I., Osipenko, A.P., Bocharov, D.A.** Perfection of technology of finishing-hardening processing of shafts by surface plastic deformation // Hardening technologies and coatings.– 2007. – №8. – P. 23-26.

[5] **Zaides, S.A.** Covering surface plastic deformation. – Irkutsk: Publishing House of Irkutsk State Technical University, 2001. – 309 p.

[6] **T. Altan.** Finite Element Modeling of Roller Burnishing Process. http://www.ercnsm.org

[7] https://www.researchgate.net/publication/222072381_Finite_Element_Modeling_of_Roller_Burnishing_Process [accessed Apr 18, 2017].

[8] **Smelianskyi V.M.** Mechanics of parts hardening by plastic surface deformation / V. M. Smelianskyi. – M.: Engineering , 2002. – 300 p.

[9] **Kirichek, A.V., Soloviyev, D.L.** Nanostructure changes in iron-carbon alloys as a result of impulse deformation wave action (2013) Journal of Nano- and Electronic Physics, 5 (4), art. no. 04009. http://www.scopus.com/inward/record.url?eid=2-s2.0-84891092852&partnerID=40&md5=56582ee23a5485f83b372e015129248f

[10] **Kuznetsov, V.P., Smolin, I.Y., Dmitriev, A.I., Konovalov, D.A., Makarov, A.V., Kiryakov, A.E., Yurovskikh, A.S.** Finite element simulation of nanostructuring burnishing (2013) Physical Mesomechanics, 16 (1), pp. 62-72. http://www.scopus.com/inward/record.url?eid=2-s2.0-84876512923&partnerID=40&md5=00ee115f8291c87727443f4baab07b3e

[11] **Valiev, R.Z., Aleksandrov, I.V.** Nanostructured materials obtained by intensive plastic deformation. M .: Logos, 2000. 272 p.

[12] **Valiev, R.Z.** Nanostructured alloys: Large tensile elongation (2013) Nature Materials, 12 (4), pp. 289-291. http://www.scopus.com/inward/record.url?eid=2-s2.0-84875424964&partnerID=40&md5=085556eae51045bb3f2d4dede01acfa2

[13] Cutting of metals with a bulk nano- and submicrocrystalline structure: monograph / A.I. Grabchenko, J. Kaptay, A.A.Simonova, A.P. Tarasyuk, V.V. Dragobetsky, N.V. Verezub. - Kharkov: Publishing house "Tochka", 2012. – 217p. - Bibliography: p.194-217 (239 names)

[14] **Blumenstein V.Yu.** Mechanics of technological inheritance at the stages of processing and machine parts operation / V. Yu. Blumenstein, V.M. Smelyansky. – Moscow: Mashinostroenie-1, 2007. 400 p.

[15] **Filippov, Yu.K.** Criterion for assessing the quality of parts obtained by cold forging // Blacksmithing and Forging. 1999. 2. P. 3-9.

[16] **Patent 2529335 The Russian Federation, IPC V24V 39/04 (2006.01)** Reeling combined roller / Blumenstein V.Yu., Krechetov A.A., Mahalov M.S., Ostanin O.A.; Applicant and patent owner is KuzSTU. - №2013135796 / 02; Application 07/30/2013; Published 27.09.2014, Bul. №27.

[17] **Patent 2530600 The Russian Federation, IPC V24V 39/04 (2006.01)** Reeling double radius roller / Blumenstein V.Yu., Krechetov A.A., Mahalov M.S., Ostanin O.A.; Applicant and patent owner is KuzSTU. – №2013135794/02; Application 30.07.2013; Published 10.10.2014, Bul. № 28.

[18] **Patent 2557377 The Russian Federation, IPC V24V 39/04 (2006.01)** Reeling multiradius roller / Blumenstein V.Yu., Krechetov A.A., Mahalov M.S., Ostanin O.A.; Applicant and patent owner is KuzSTU.. – №2013135795/02; Application 30.07.2013 ; Published 20.07.2015, Bul. № 20.

[19] **Makhalov, M.S.** Modeling of residual stresses at the stage of the product life cycle / M.S. Makhalov, V.Yu. Blumenstein, // Machine Building Bulletin. 2014. 12. P. 21-25.

[20] **Krechetov, A.A.** The technique to calculate the mechanical state parameters of the surface layer of machine parts // KuzSTU Bulletin. 2001. 5. P. 27-31.

IOP Conf. Series: Materials Science and Engineering **253** (2017) 012018 doi:10.1088/1757-899X/253/1/012018

Structure and Phase Composition of 09G2S Steel Modified by Different Types of Welding

Aleksandr N Smirnov [1,2,a], Natal'ya A Popova [3,4], Elena L Nikonenko [3,5], Evgenii A Ozhiganov [1,2], Nikolai V Ababkov [1,2], Nina A Koneva [3]

[1] OOO 'Kuzbass Center for Welding and Control', 33/2, Lenin Str., Kemerovo, 650055, Russia
[2] T.F. Gorbachev Kuzbass State Technical University, 28, Vesennyaya Str., Kemerovo, 650000, Russia
[3] Tomsk State University of Architecture and Building, 2, Solyanaya Sq., Tomsk, 634003, Russia
[4] Institute of Strength Physics and Materials Science SB RAS, 8/2, Akademicheskii Ave., Tomsk, 634021, Russia
[5] National Research Tomsk Polytechnic University, 30, Lenin Ave., Tomsk, 634050, Russia

E-mail: [a] agalvas.kem@gmail.ru

Abstract. The paper presents the transmission electron microscopy (TEM) investigations of the structure and phase composition of the type 09G2S weld steel modified by four types of welding, namely: electrode welding and electropercussive welding both with and without the introduction of artificial flaws. Artificial flows are aluminum pieces. TEM investigations are carried out within the heat-affected zone, i.e. between the deposited and base metal, at 0.5 mm distance to the former. Welding electrode of the type E50A is used for welding 09G2S steel specimens. It is shown how the type of welding affects the steel morphology, phase composition, defect structure and its parameters. After each type of welding, the dislocation structure is polarized. This, however, does not cause internal stresses which can destroy the specimen.

Introduction

At present, the technological development in all industries is impossible without welding applications. Welded joints are part of the majority of structures. Modern technology utilizes many types of welding [1-10]. However, each type has its own strengths and weaknesses. The structure and phase composition of metal which form during welding affect physical-mechanical properties of final products [11-14]. Changes in the structure and phase composition of welded items caused by a long-term operation, and particularly deformations they are often subjected to, might lead to negative consequences [11, 13, 14]. Therefore, the quality improvement and the reliability of welded joints are being considered as a high priority until now. The solution to these problems is based on knowledge of processes that determine the formation and properties of welded joints. This is the case of the heat-affected zone (HAZ) between the deposited and base metal. It is HAZ that provides dangerous stress concentrators which lead to the formation of cracks and various defects [14], thereby decreasing the strength and safety of welded items [11-13]. Knowledge of the structure and phase composition of

IOP Conf. Series: Materials Science and Engineering **253** (2017) 012018 doi:10.1088/1757-899X/253/1/012018

material, and particularly HAZ, could be useful not only for the assessment of strength properties of welded item, but also for predicting the behavior of welded joint during its operation.

The main objective of the paper is to explore HAZ (between the deposited and base metal) structure and phase composition depending on the type of welding.

Materials and methods

Four types of welding were used to investigate the welded joint, namely: 1) electrode welding without the introduction of artificial flaws; 2) electrode welding with the introduction of artificial flaws; 3) electropercussive welding without the introduction of artificial flaws; 4) electropercussive welding with the introduction of artificial flaws. Artificial flaws represented pieces of aluminum. The type 09G2S steel was used as a weld material the chemical composition of which is summarized in the Table below. Electrode of the UONI-13 brand (type E50A) was used for welding the type 09G2S steel specimens. Its chemical composition is also given in the Table. The steel specimens were plates having the test portion of 200×15×4 mm^3 in size. The welded joint occupied the central part of the test portion of the specimen. Specimens were preliminary macroetched with a 4 % aqueous HNO_3 solution to visualize the deposited metal and heat-affected zones. The width of the welded joint did not exceed 12 mm, while that of HAZ was not over 5 mm.

Table. Chemical composition of steel and welding electrode (wt.%)

	09G2S	E50A
C	0.09	0.09
Si	0.5–0.8	0.42
Mn	1.3–1.7	0.83
Ni	~0.3	–
S	~0.04	0.022
P	~0.035	0.024
Cr	~0.3	–
N	~0.008	–
Cu	~0.3	–
As	~0.08	–
Fe	~96–97	

A method of electroerosion cutting was used to obtain the test foils. They were cut parallel to the welded joint. The transmission electron microscopy (TEM) investigations were performed nearby the HAZ at 0.5 mm distance to the deposited metal. Observations of thin foils were carried out on an EM-125 transmission electron microscope with 125 kV accelerating voltage and 25000× magnification. TEM images are used to measure the volume fractions of the steel morphology within the HAZ; the size of α-phase fragments; size, density distribution and volume fractions of carbide particles; scalar and excess dislocation densities; bending-torsion amplitude of the crystal lattice; amplitudes of internal stresses.

Phase identification was carried out using the known methods. With this view, microdiffraction patterns calculated by the reference values of the crystal lattice parameters were used. The size and volume ratios of observed phases were measured using the TEM images confirmed by the microdiffraction patterns and dark-field images obtained in reflections of the respective phases. Linear sizes and scalar dislocation density of the structural elements were detected on micrographs using the secant method in accordance with the standard procedures. Two types of internal stresses are determined: 1) shear stress (stress fields created by the dislocation structure) and 2) long-range or local stress [15].

Results and discussions

It has been found that the steel matrix in the weld metal is α-phase after each type of welding, *i.e.* a solid solution of carbon and alloying elements in α-Fe with the body-centered cubic (BCC) crystal system. Depending on the welding type, the structural components of α-phase are lamellar perlite, ferrite and martensite as illustrated in Fig. 1.

Ferrite is mostly fragmented after using each type of welding (Fig. 1a). A fragmented substructure consists of dislocation subboundaries (fragment's walls) and the inner space with or without dislocations [16]. The fragments are mainly isotropic [11], *i.e.* $L/D \approx 1$, where L and D are respectively longitudinal and transverse sizes of the fragment. These fragments contain a reticulated dislocation structure. Rounded carbide particles – cementite – are observed on these dislocations. After using various types of welding, the average size of these particles ranges between 10-20 nm, and their volume fraction in fragmented ferrite is not over 0.5%.

After each type of welding, martensite is observed either in the form of parallel laths which are not practically misoriented relative to each other (Fig. 1b) or separate lamellae (Fig. 1c). Unlike lamellar martensite which forms during the electropercussive welding, lath martensite forms at electropercussive welding with the introduction of artificial flaws. At the interface between martensite laths, lamellar cementite particles are present. They are ~15×40 nm in size and the volume fraction of them is not over 1%. Inside martensite laths austenite (γ-phase) is observed in the form of separate islands. Its volume fraction is ~3%. Cementite particles are not found in lamellar mertensite, but there are austenite (γ-phase) layers. The average size and the volume fraction of these layers is 0.3×0.07 μm and ~3% respectively. After using the electrode welding without the introduction of artificial flaws, martensite is not found in the weld metal.

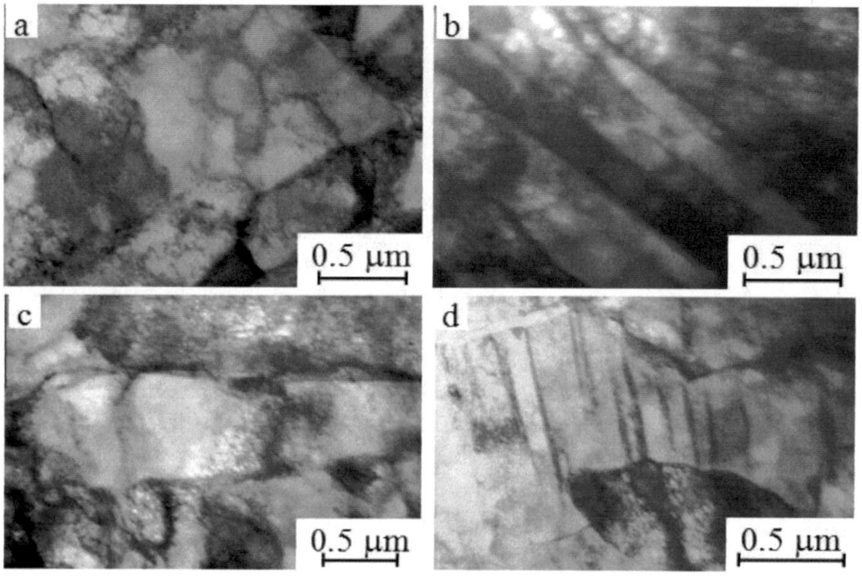

Fig.1. TEM images of structural elements of weld metal: *a* – fragmented ferrite (electrode welding); *b* – lath martensite (electrode welding with introduction of artificial flaws); *c* – lamellar martensite (electropercussive welding); *d* – perlite (electropercussive welding with introduction of artificial flaws) Fig.1. TEM images of structural elements of weld metal: *a* – fragmented ferrite (electrode welding); *b* – lath martensite (electrode welding with introduction of artificial flaws); *c* – lamellar martensite (electropercussive welding); *d* – perlite (electropercussive welding with introduction of artificial flaws)

Perlite is lamellar regardless of the type of welding. As Tushinskii *et al.* report in [17], lamellar perlite is a conglomerate of alternating parallel (or almost parallel) layers of ferrite and cementite. In other words, lamellar perlite is a layered aggregate with clearly marked anisotropy conditioned by the crystalline structure and contrasted properties of consisting phases. Cementite is a solid and brittle phase, while ferrite is soft and plastic. The contrast observed in mechanical properties of cementite and ferrite is stipulated by their chemical composition and the type the crystal lattice. Thus, ferrite has a BCC crystal system, while cementite being a chemical compound of carbon and iron (Fe_3C), has an orthorhombic lattice. Perlitic grains usually concentrate inside the structure of fragmented ferrite (Fig. 1 d) and have the size of not over 0.5 μm.

Figure 2 demonstrates the dependence of the quantitative parameters of the weld metal structure on the type of welding. One can see that the welding process affects, first of all, the structural components of the weld metal. Thus, the introduction of artificial flaws during the electrode and electropercussive welding results in the reduction of the volume fraction of ferrite and the increase in that of perlite and particularly martensite.

We already mentioned above that the dislocation structure is reticular in each structural component. The measurements show that the scalar dislocation density depends on the welding type (see Fig 2). The dislocation structure of steel is polarized after welding that is indicated by bend extinction contours observed in each structural steel component. Measurements of bend extinction contours allow detecting the bending-torsion amplitude of the crystal lattice [15]. In general, the bending-torsion amplitude χ can be found from $\chi = \chi pl + \chi el$, where χpl is its plastic component and χel is its elastic component. Experiments show that during the electrode welding with and without the introduction of artificial flaws, the bending-torsion amplitude χ in each structural component and the material as a whole, is described only by the plastic component, *i.e.* $\chi = \chi pl$. This indicates to its plasticity and $\rho > \rho\pm$ and $\sigma sh > \sigma l$ are true [15], *i.e.* the scalar dislocation density ρ is higher than the excess dislocation density $\rho\pm$, and the shear stress amplitude σsh exceeds the long-range stress amplitude σl. The electropercussive welding with and without the introduction of artificial flaws leads to the plastic-elastic bend of the crystal lattice, *i.e.* $\chi = \chi pl + \chi el$. The contributions of χpl and χel to different structural components are different. It is observed that $\rho < \rho\pm$ and $\sigma sh < \sigma l$ in the whole material (Fig. 2).

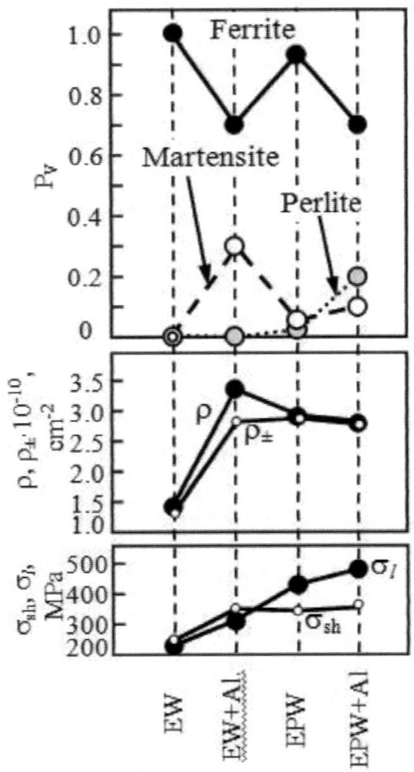

Fig.2. Qualitative parameters of welded joint depending on the type of welding.

According to Fig. 2, the weld metal structure subjected to the electrode welding is characterized by the lowest values of ρ, $\rho\pm$, σsh and σl. This means that the hardening of weld metal in this case will be the lowest as compared to other types of welding. The introduction of artificial flows in electrode welding enhances the quantitative parameters (ρ, $\rho\pm$, σsh and σl) of the structure, thereby leading to a stronger hardening of material [15]. The electrode welding with the introduction of artificial flows modifies the weld metal structure more significantly than the electropercussive welding. At last, at the electropercussive welding with the introduction of artificial flows, the amplitude of long-range internal stresses is practically two times as high as the amplitude of shear stress (see Fig. 2) on the average in the material. On the one hand, this hardens the material, and on the other facilitates microcracking in the weld metal.

Conclusions

TEM investigations were conducted for the structure and phase composition of the type 09G2S weld steel modified by four types of welding, namely: electrode welding and electropercussive welding both with and without the introduction of artificial flaws. The following results were obtained:

- Regardless of the type of welding, lamellar perlite, ferrite and martensite are the structural components of the weld metal. Their volume fraction depends on the type of welding;
- The introduction of artificial flaws during any type of welding leads to the destruction and elimination of perlitic component and the increase in the amount of fragmented structure in ferrite;
- The steel structure after the electrode welding has the lowest values of ρ, $\rho_{\pm}$, σ_{sh}, and σ_l as compared to other types of welding;
- The introduction of artificial flaws during any type of welding leads to the decrease in the quantitative parameters (ρ, $\rho_{\pm}$, σ_{sh}, and σ_l) of the defect structure in each structural component;
- At the electropercussive welding with the introduction of artificial flows, the amplitude of long-range internal stresses is practically two times as high as the amplitude of shear stress on the average in the material. On the one hand, this hardens the material, and on the other facilitates microcracking in the weld metal.

References

[1] Ju. Hu, *et al.* Structure and performance of welding joint of Q235 steel welded by SHS welding, Front. Mech. Eng. China. 2010. No. 5 (2). Pp. 189-193.

[2] B.A. Grinbreg, Elkina O.A., Antonova O.V., *et al.* Peculiarities of Formation of Structure in the Transition Zone of the Cu-Ta Joint Made by Explosion Welding. The Paton Welding Journal, 2011. No. 7. Pp. 20-25.

[3] V.E. Rubtsov, S.Yu. Tarasov, A.V. Kolubaev One-Dimensional Model of Inhomogeneous Shear in Sliding. Physical Mesomechanics . 2012. V.15. No. 5-6. Pp. 337-341.

[4] S.F. Gnyusov, V.A. Klimenov, Yu.V. Alkhimov, *et al.* Formation of Ti Structure and Corrosion Resistant Steel at Laser Welding. Svarochnoe proizvodstvo [in Russian]. 2012. No. 1. pp. 17-22.

[5] N.V. Boiko, *et al.*, Structure of titanium alloy austenitic steel welds formed by pressure welding with intermediate coatings, Met. Sci. Heat Treat. 2013. V.54. pp. 9-10.

[6] F. Foadian, M. Soltanieh, M. Adeli, M. Etminanbakhsh, A study on the formation of intermetallics during the heat treatment of explosively welded Al-Ti multilayets, Metal. Mater. Trans. A. 2014. V. 45A. No. 4. pp.1823–1832.

[7] M.Yu. Kollerov, S.D. Shlyapin, Gusev D.E *et al.*, Metally [in Russian]. 2015. No. 6. pp. 32-36.

[8] E.S. Konovalenko, *et al.*, Izvestiya vuzov. Fizika [in Russian]. 2015. V.58. No. 6-2. pp. 137-141.

[9] N.M. Rusin, A.L. Skorentsev, E.A. Kolubaev Dry Friction of Pure Aluminum Against Steel, Journal of Friction And Wear 2016. V. 37. No. 1. pp. 86-93.

[10] N.A. Popova, *et al.*, Structure and Phase Composition of Deformed Heat-Affected Zone of the Weld Steel St3, AIP Conf. Proc. 2016. V.1772. P. 030006 (1-6).

[11] A.N. Smirnov, E.V. Kozlov, Substructure, Internal Stress Fields and Destruction of Steam Lines in Steel 12Kh1MF [in Russian]. Kemerovo: Kuzbassvuzizdat, 2004. 163 p.

[12] V.P. Gagauz, *et al.*, Structure, Phase Composition and Mechanical properties of Thick Welded Joints [in Russian]. Novokuznetsk: SibGIU, 2008. 150 p.

[13] A.N. Smirnov, *et al.*, Damageability of Welded Joints, Spectral-Acoustic Control [in Russian]. Moscow: Mashinostroenie, 2009. 240 p.

IOP Conf. Series: Materials Science and Engineering **253** (2017) 012018 doi:10.1088/1757-899X/253/1/012018

[14] A.N. Smirnov, *et al.*, Gradient Structures at Metal Cutting [in Russian]. Kemerovo: Sibirskaya izdatel'skaya gruppa, 2013. 179 p.
[15] N.A. Koneva, E.V. Kozlov, Regularities of Substructural Hardening Rus. Phys. J. 1991. V. 34. No. 3. pp. 224-236.
[16] E.V. Kozlov, N.A. Popova, N.A. Koneva, Fragmented Substructure in BCC Steels at Deformations, Izv. RAN. Seriya fizicheskaya [in Russian]. 2004. V.68. No. 10. pp. 1419-1427.
[17] L.I. Tushinskii, A.A. Bataev, L.B. Tikhomirova, Steel Structure and Strength. Novosibirsk: Nauka, 1993. 280 p.

The Increase in Wear Resistance of Low Carbon Steel by Flux-Cored Wire Surfacing Followed Be Electron Beam Processing

Yuriy F Ivanov [1, 2, a], Vasiliy E Kormyshev [3], Victor E Gromov [3], Sergey V Konovalov [3], Elena V Martusevich [3], Andrey D Teresov [2], Alexander P Semin [3]

[1] Institute of High-Current Electronics of Siberian Branch of Russian Academy of Sciences, Tomsk, 634021, Russia
[2] Tomsk National Research State University, Tomsk, 634050, Russia
[3] Siberian State Industrial University, Novokuznetsk, 654007, Russia

E-mail: [a] yufi55@mail.ru

Abstract. Structure-phase state and tribological properties of the coating deposited on Hardox 450 martensitic low-carbon steel by flux-cored wire Fe-C-Cr-Nb-W and modified by subsequent electron-beam processing were studied. It is shown that the electron beam processing of ~ 5 mm thick deposited layer leads to the formation of ~ 20 μm thick modified surface layer with the main phases of α-Fe and NbC, Fe_3C and $M_6C(Fe_3W_3C)$ carbides. Wear resistance of the weld layer is 70 times higher than the one of the original steel.

Introduction

The production of coatings with high service properties making possible the increase in service durability of the articles in the extreme conditions of high wear, corrosion, mechanical loads and temperatures is the important fundamental problem. The thorough analysis of the relation «wear parameters – hardness - microstructure» is necessary in scientific and practical use of various kinds of surfacing in the critical articles. Only in this case it is possible to produce the articles with high service parameters [1 – 8].

The purpose of the research is the analysis of structure and tribological properties of the layer formed on Hardox 450 steel by electrocontact surfacing of powder wire of system Fe-C-Cr-Nb-W and modified by high intensity pulsed electron beam irradiation.

Hardox 450 steel (weight %): (0.19-0.26 C; 0.70 Si; 1.6 Mn; 0.025 P; 0.010 S; 0.25 Cr; 0.25 Mo; 0.004 B; balance – Fe) was used as a test material. The surfaced electrode 1.6 mm in diameter had the following chemical composition (weight %): 1.3 C; 7.0 Cr; 8.5 Nb; 1.4 W; 0.9 Mn; 1.1 Si, balance – Fe.

The testing techniques of the surfacing by methods of modern physical material science are presented in [1].

The results and their discussion

The formation of the surfaced layer results in the increase in wear resistance of steel. The performed tests revealed the increase in wear resistance of the surfaced layer modified by intense pulsed electron beam by more than 70 times in relation to wear resistance of steel.

Fig.1 shows the variation of friction coefficient in tribological tests of the surfaced layer modified by electron beam. Attention is drawn to the two-stage variation in friction coefficient. At the first stage the friction coefficient value is ~ 0.17, at the second stage it is ~ 0.5. The friction coefficient of steel without surfacing is ~ 0.26. When analyzing the variation in friction coefficient in tribological tests (Fig.1) it can be supposed that the modification of the surfaced layer by the intense pulsed electron beam results in the significant (~ 3-fold) decrease in friction coefficient of the surfaced layer.

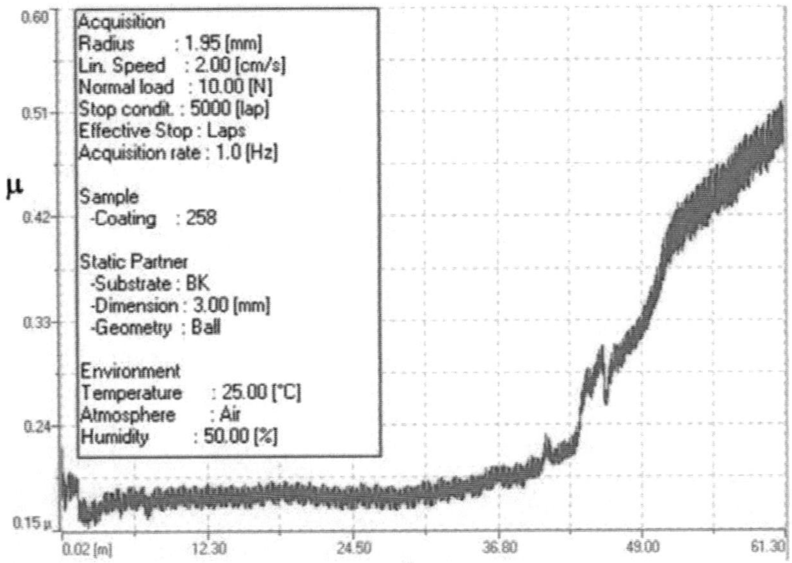

Fig.1 Dependence of friction coefficient on track length of tribological tests. The insert shows the tribological test conditions.

The electron beam processing (EBP) of the surfaced layer results in the formation of the modified surface layer up to 20 μm in thickness. The modified layer differs from the main volume of the surfaced material by the degree of structure dispersion detected on ion etching of the transverse metallographic section. If in the volume of the surface unirradiated by electron beam the sizes of etched structural elements amount to 1.5 μm then after EBP the sizes of high-melting compounds having the low level of ion beam etching vary within 150 to 750 nm (Fig.2)

The phase composition analysis of the surfacing modified layer was performed by methods of microdiffraction analysis using dark field technique. The results of the analysis are presented in Fig.3.

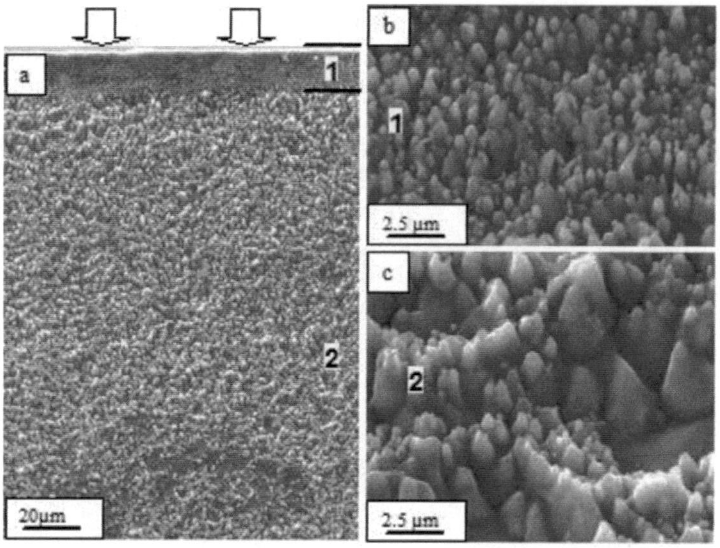

Fig.2 Structure of the surface layer.The transverse metallographic section. The arrows designate the irradiation surface of the surfaced layer by pulsed electron beams. The figures designate: 1 – the layer modified by electron beam processing; 2 – the main volume of the surfacing.

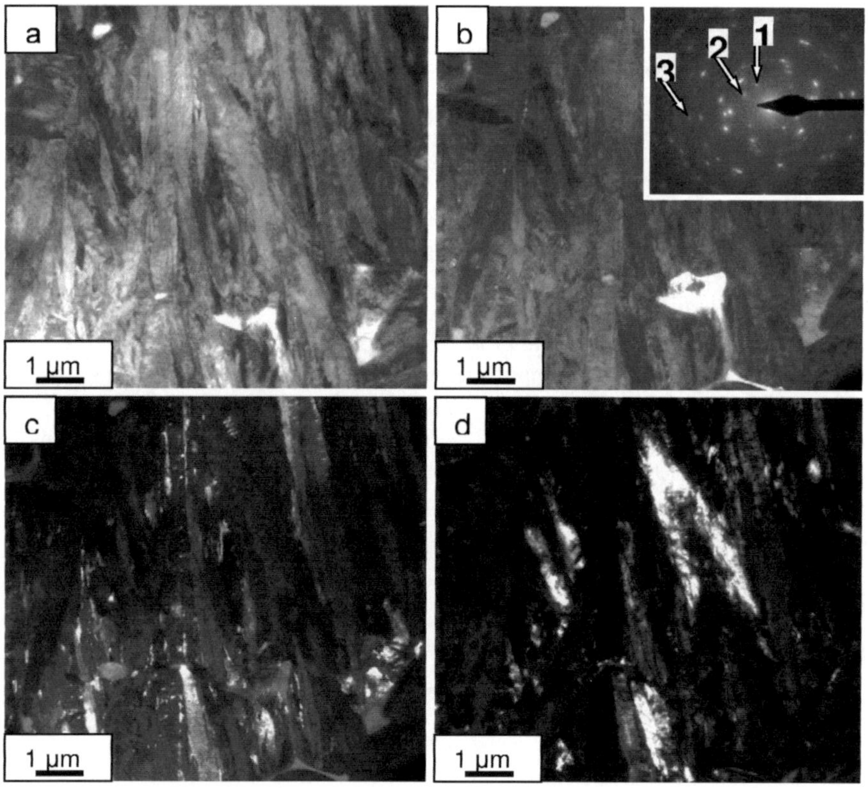

Fig.3.Electron microscope image of surfaced layer structure irradiated by intense pulsed electron beam; a – light field; b-d – dark field obtained in reflections [511] $M_6C(Fe_3W_3C)$ (b); [110] α-Fe+[102] Fe_3C (c); [024] α-Fe (d). The microelectron diffraction pattern (insert in "b") shows the reflections of obtaining the dark field images 1(b), 2(c), 3 (d).

It is established that the second phase inclusions located at grain boundaries as interlayers are the carbide $M_6C(Fe_3W_3C)$ (Fig. 3, b). In the volume and along the boundaries of martensite crystals the particles of iron carbide Fe_3C (Fig. 3, c) are revealed.

The surface layer structure of the surfacing irradiated by intense pulsed electron beams is characterized by the presence of the facet inclusions located chaotically in the volume of grains (Fig.4). The sizes of these inclusions amount to 2 μm and they are niobium carbide NbC.

Thus, by methods of electron diffraction microscopy it is shown that the surface layer of the surfacing modified by intense pulsed electron beam is a multi-phase aggregate with the main phases of α-iron-base solid solution and carbides M_6C, NbC and Fe_3C.

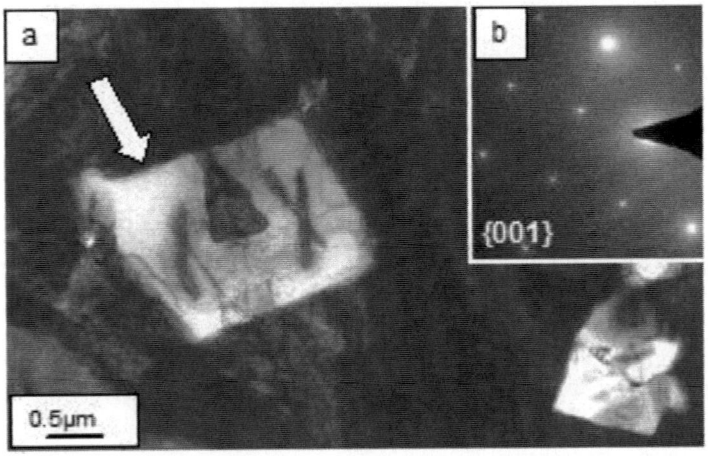

Fig. 4.The structure of the surfaced layer modified by electron beam (a). Microelectron diffraction pattern (b) is obtained from the particle designated by the arrow.

Conclusions

By methods of modern physical material science the investigation of structure and tribological properties of the layer formed on Hardox 450 steel by flux-cored wire Fe-C-Cr-Nb-W modified by EBP were carried out. It is shown that electron-beam processing of the surfacing top layer results in the refinement of structure and variation in the morphology of carbide phase of layer and the significant decrease in its friction coefficient and increase in wear resistance.

The research was financially supported by RNF grant (project № 25-19-00065). Electron beam processing was supported by RFFI grant (project № 16-43-790659 p-a).

References

[1] Kapralov E.V., Raikov S.V., Budovskikh E.A., Gromov V.E., Vaschuk E.S., Ivanov Yu.F. Structural phase states and properties of coatings welded onto steel surfaces using powder // Bulletin of the Russian academy of sciences. Physics. – 2014. – Vol. 78. – No. 10. – P. 1015–1021.

[2] Yüksel N., Şahin S. Wear behavior–hardness–microstructure relation of Fe-Cr-C and Fe-Cr-C-B based hardfacing alloys // Materials & Design – 2014. – Vol. 58. – P. 491-498.

[3] Venkatesh B., Sriker K., Prabhakar V.S.V. Wear characteristics of hardfacing alloys: state-of-the-art // Procedia Materials Science – 2015. – Vol. 10. – P. 527-532.

[4] Gualco A., Marini C., Svoboda H., Surian E. Wear resistance of Fe-based nanostructured hardfacing // Procedia Materials Science – 2015. – Vol. 8. – P. 934-943.

[5] Teker T., Karatas S., Osman Yilmaz S. Microstructure and wear properties of AISI 1020 steel surface modified by HARDOX 450 and FeB powder mixture // Protection of Metals and Physical Chemistry of Surfaces. – 2014. – Vol. 50. – №1. – P. 94-103.
[6] Zahiri R., Sundaramoorthy R., Lysz P., Subramanian C. Hardfacing using ferro-alloy powder mixtures by submerged arc welding // Surface and Coatings Technology. – 2014. – Vol. 260. – P. 220–229.
[7] Chang C.-M., Chen Ye.-C., Wu W. Microstructural and abrasive characteristics of high carbon Fe–Cr–C hardfacing alloy // Tribology International. 2010. Vol. 43. Issues 5-6. P. 929-934.
[8] Fan C., Chen M.-C., Chang C.-M., Wu W. Microstructure change caused by (Cr, Fe)$_{23}$C$_6$ carbides in high chromium Fe–Cr–C hardfacing alloys // Surface and Coatings Technology. 2006. Vol. 201. Issues 3-4. P. 908-912.

IOP Conf. Series: Materials Science and Engineering **253** (2017) 012020 doi:10.1088/1757-899X/253/1/012020

Assembling of Steel Angle Headframe of Multifunctional Purpose

E G Kassikhina [a], V V Pershin, V M Volkov

Mining Institute, FSBEIHE (Federal State Budgetary Educational Institution of Higher Education) T.F. Gorbachev Kuzbass State Technical University,
28 Vesennya Str., Kemerovo, Russia, 650000

E-mail: [a] kalena-07@mail.ru

Abstract. The article reviews new technical solution on equipment provision of vertical shaft utilizing steel headframe of multifunctional purpose. Practice of construction of coal and ore mines provides application of various designs for steel angle headframes which are divided into separate large assembly blocks and constructive elements during assembling operations. Design of these blocks and elements, their weight and dimensions effect the chose of the method of assembling on which economic and technological indicators, as well as duration of down-time, depend on during performance of construction operations in shaft. The technical solution on equipment provision for mine vertical shaft using headframe of multifunctional purpose will allow changing the management construction of vertical shaft. The proposed headframe combines the functions of sinking and operation that eliminates costs for assembling/dissembling of temporary headgear. The constructive design of the headgear allows application of the effective method of assembly and thus to provide improvement of the technical and economic indexes, and high calendar time rate of the shaft construction due to reduction of duration of works on equipment provision for the shaft and to refurbishment of the shaft in order to carry out horizontal mining.

Introduction

The construction project of the vertical mine shaft consists of the following main stages: provision of equipment for mine surface and sinking of the technological part of the shaft (including shaft collar); provision of equipment for surface for a shaft sinking, including assembling of the sinking headgear; shaft work; shaft walling; refurbishment of the shaft to carrying out horizontal and inclined developments, including dissembling of headgear unit during sinking operations and, construction of the permanent headframe [1 –6].

Object of survey. The method of assemble of headframe significantly impacts the entire construction schedule of a mine as the duration of the shut-down in a shaft depends exactly on it.

The practice of construction of the coal and ore mines provides various designs for the steel angle headframes, both sinking headgear and permanent headframe: 4-post headframe, 4-post headframe with two stay legs, tent headframe and two-post headframe which are divided into separate large assembly blocks and constructive elements during assembly operations [1 –10].

IOP Conf. Series: Materials Science and Engineering **253** (2017) 012020 doi:10.1088/1757-899X/253/1/012020

Method

Depending on the constructive design of these blocks and elements, their weight and dimensions they choose a method of assembly on which economic and technological indicators, as well as duration of shut-downs during construction works in shaft depend.

Aim of survey. 'Construction of Underground Structures and Mines' department at KuzSTU developed technical solution regarding furnishing of the vertical shaft with equipment utilizing steel headframe of multifunctional purpose [11, 12, 13] what will allow to change mining management on some of the hereinabove mentioned stages.

Results

As the proposed headframe (fig. 1) due to assembly/disassembly of the replaceable functional blocks combines the functions of both sinking and operation, then such stages as disassembly of gear complex during sinking operations and construction of the permanent headframe will be eliminated from the schedule.

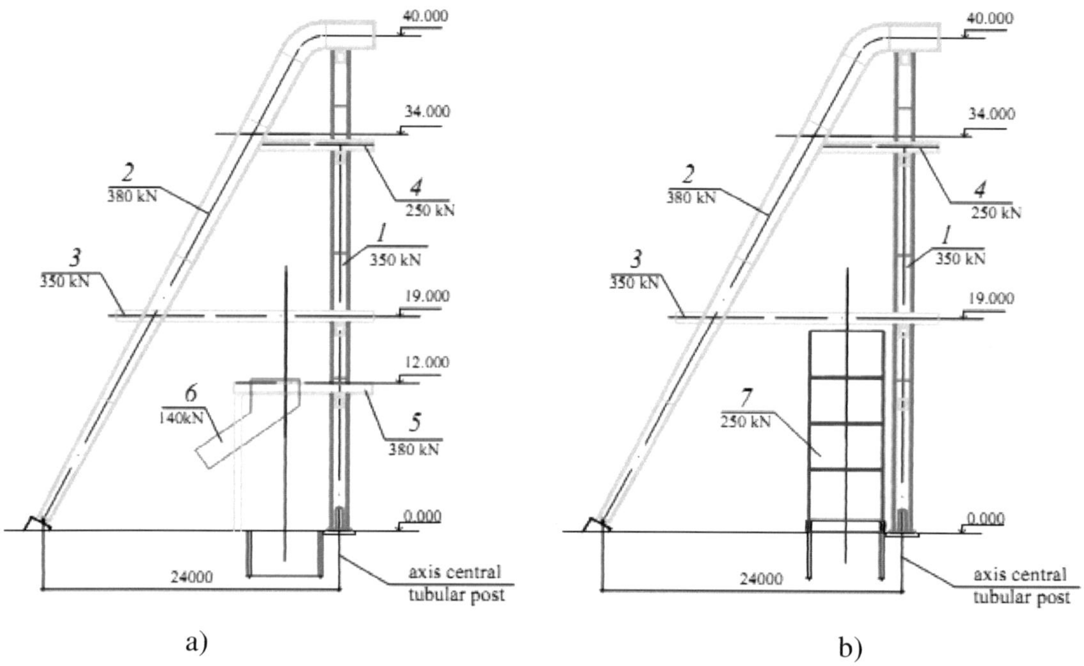

Fig. 1. Breakdown of the headgear on blocks: a –scheme for sinking period; b – scheme for operation; *1* – central tubular post; *2* –stay leg; *3* – sheave wheels platform; *4* – permanent sheave wheels platform; *5* – unloading machine; *6* – reception bunker; *7* – rig

The assembly of the headframe is carried out in the following consequence. On the assembling stand (fig. 2) located in the close proximity from the constructing shaft, the assembly is performed in horizontal position of the central tubular post (block #1). The assembled post is pulled to the shaft by sliding using two valves working as masts, erected to the vertical position and installed on the foundation as per the project position (fig. 3, a).

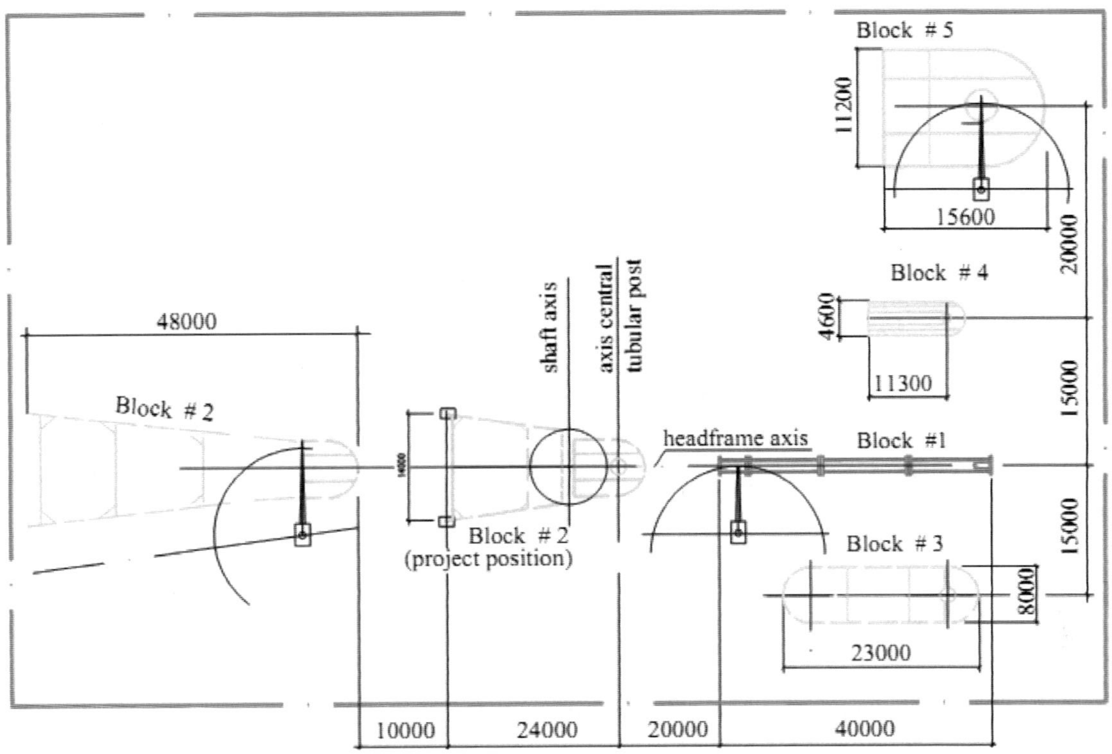

Fig. 2. Assembly schedule for blocks of the headframe

On the assembly stand (see fig. 2) they perform pre-assembly of stay leg (block #2) in horizontal position with the subsequent movement by sliding to the shaft by two cranes, there it is erected to the project position and fixed on the central tubular post *1* (fig. 3, b).

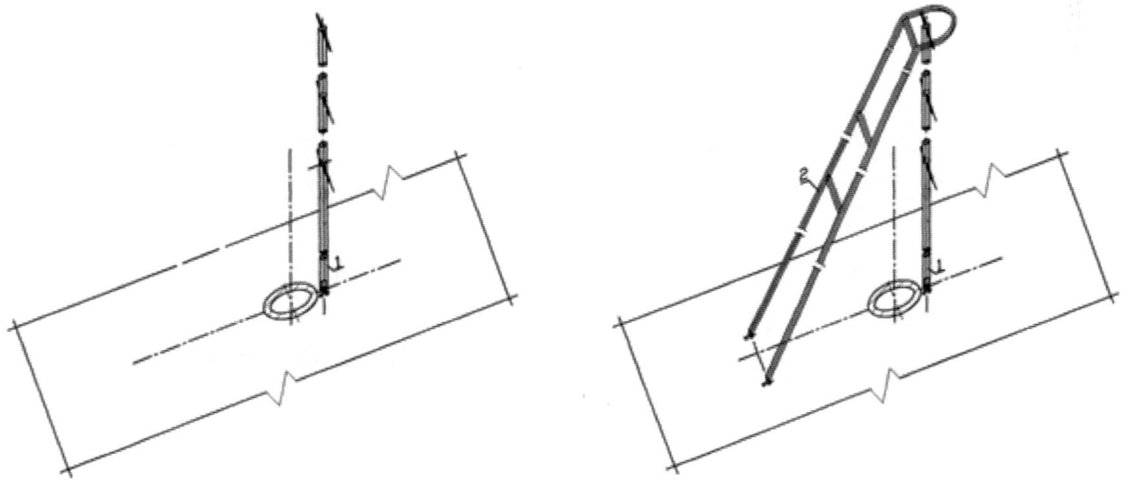

a) b)

Fig. 3. Scheme of assembly of pulley device: a – erection of central tubular post; b – assembly of stay leg; *1* – central tubular post; *2* – stay leg

The driving sheave wheels platform (block #3) is put together on the assembly stand in the form of two blocks, each of them is separately moved to the position 'on the ground' on the level ±0.00 m where they are connected to each other (fig. 4, a). By two cranes using the central tubular post *1* as a guide, the driving sheave wheels platform *3* is erected to the project position (fig. 4, b).

a) b)

Fig. 4. Scheme of assembly of the driving sheave wheels platform: a – assembly on the ground; b – lifting to the design position; *1* – central tubular post; *2* – stay leg, *3* – the driving sheave wheels platform

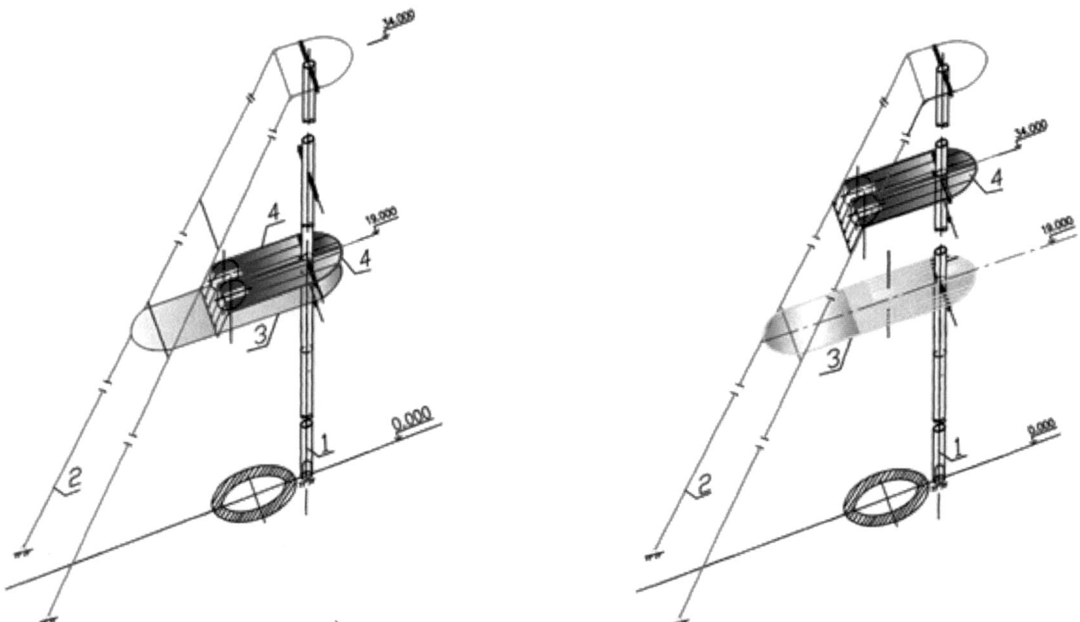

a) b)

Fig. 5. Scheme of assembly of permanent sheave wheels platform: a – assembly on driving sheave wheels platform; b – erecting to project position; *1* – central tubular post; *2* – stay leg, *3* – driving sheave wheels platform, *4* – permanent sheave wheels platform

IOP Conf. Series: Materials Science and Engineering **253** (2017) 012020 doi:10.1088/1757-899X/253/1/012020

The permanent sheave wheels platform (block #4) is put together on the assembly stand also as of two blocks, each block is moved and placed on the driving sheave wheels platform *3* by two cranes (fig. 5, a). On the driving sheave wheels platform *3* they perform the final assembly of the permanent sheave wheels platform *4* from the two enlarged blocks and erect it up to the project position (fig. 5, b). The lifting is carried out by two cranes using the central tubular post *1* as a guide.
The assembly of the enlarged block #5 (the driving unloading platform) is performed as per analogy with the assembly of the block #3 (the driving sheave wheels platform) with the subsequent erection of the bunker (block #6). At this stage of assembling of the winch on the level 40.000 m is already possible for the elevator moving inside of the tubular post *1* which can be in turn used for needs of assemblers.
After a long break when mining and walling of the shaft are complete, it is necessary to dissemble the unloading machine and the receiving bunker, part of the driving sheave wheels platform and the mining equipment to assemble the unit (block #7) (see fig. 1).
The unit *5* [15] consists of ring elements *6*; each element is separately moved and stack on the driving sheave wheels platform *3* for the subsequent assembly by the 'from up to down' method over the shaft collar *7* (fig. 6). Thus, the driving sheave wheels platform is used again as assembly and installation platform already during transition to the operation mode.

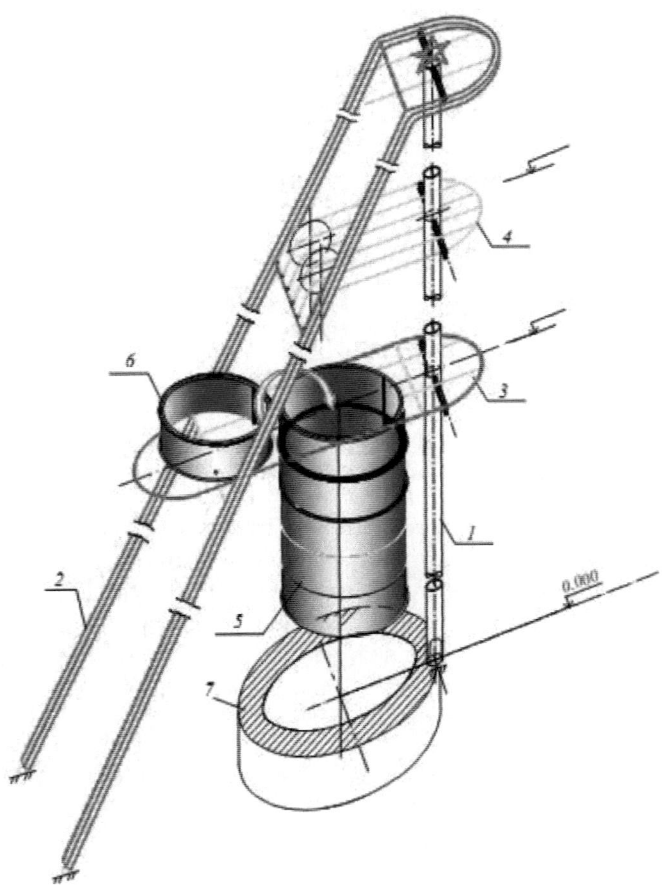

Fig. 6. Scheme of assembly of driving sheave wheels platform: *1* – central tubular post; *2* – stay leg, *3* – driving sheave wheels platform, *4* – permanent sheave wheels platform, *5* – rig, *6* – ring element, *7* – shaft collar

IOP Conf. Series: Materials Science and Engineering **253** (2017) 012020 doi:10.1088/1757-899X/253/1/012020

The efficiency of the suggested method of assembly of the headframe of multifunctional purpose is provided by:

– Reduction of the total time for assembling of the headframe and, in this regard, considerable reduction of shutdown during operations in shaft, considerable reduction of the overall labor costs and the cost of headframe assembly;

– Reduction of labor input and cost of installation due to breakdown of the headframe on the assembly blocks of small weight (250 kH - 350 kH);

– Application of the sheave wheels platforms for assembly and erection;

– Use of the central tubular post as a guide what considerably reduces the number of operations on aligning and adjustment of structures during assembling;

– Reductions of labor costs on rigging up of the unit which is put together separately from the sheave wheels platform and other constructive elements of the headframe;

– Lack of labor costs on arrangement of the rolling track to deliver ring elements of the rig to the place of installation due to their small weight;

– Minimizing of works at height.

The suggested option of the vertical shaft construction utilizing headframe of multifunctional purpose versus the traditional technical solution [1-7] allows to reduce labor input of assembly operations in 1.5 times, duration of works in 1.22 times, steel consumption in 1.5 times (table 1).

Table 1

Index	Sinking headgear 'Sever-1' (excluding dissemble)	Permanent angle headframe	Headframe of multifunctional purpose
Labor input of assembly MK, man/hr.	95.9	153	174
Assembly duration as per time-table, days	20	25	15
Steel consumption, kN	1200	1890.5	2060

Conclusion

The technical solution on equipment provision for the mine vertical shaft utilizing headframe of multifunctional purpose will allow changing of the arrangement of construction operations in vertical trunk. The proposed headframe combines the functions of sinking and operation that eliminated costs for assembling/dissembling of the temporary headgear. The constructive design of the headgear allows application of the effective method of assembly and thus to provide improvement of the technical and economic indexes, and high calendar time rate of the shaft construction due to reduction of duration of works on equipment provision for the shaft and to refurbishment of the shaft in order to carry out horizontal mining.

References

[1] A.P. Maksimov, Gornotexnicheskie Zdaniya i Sooruzheniya, Nedra, Moskva, 1984.

[2] Ya.V. Brovman, Nadshaxtnye Kopry, Gosgortexizdat, Moskva, 1961.

[3] G.L. Rozenblit, Stalnye Konstrukcii Zdanij i Sooruzhenij Ugolnoj Promyshlennosti, Ugletexizdat, Moskva, Leningrad, 1953.

[4] Yu.N. Kulikov, A.P. Maksimov, Proektirovanie i Stroitelstvo Gornotexnicheskix Zdanij i Sooruzhenij. Texnologiya Stroitelstva Zdanij i Sooruzhenij, Nedra, Moskva, 1991.

[5] A.P. Yakimenko, S.F. Zolotarev, Proektirovanie ukosnyx shaxtnyx koprov novogo tipa, Shaxtnoe stroitelstvo. 9 (1981) 20-21.

[6] D.I. German, K.Yu. Xablo, D.A. Fedorovskij, Issledovanie i usovershenstvovanie metodov proektirovaniya nad-shaxtnyx koprov krivbassa, Visnik KTU. 27 (2011) 3-6.
[7] F. Wei, I. F. R. Wang, 3D Parametric design for steel headframe of coal mine based on Solidworks, Key Engineering Materials. 455, Issue 1 (2010) 340-344.
[8] Ian Ross, Goldcorp's Cochenour gold project, Northern Ontario business. 33, Issue 12 (2013) 15.
[9] L. Kelly, Historical end, Northern Ontario business. 34 Issue 3 (2014) 31-32.
[10] Elena. G. Kassikhina, Weiguo Qiao, Vladimir V. Pershin, Nikita O. Butrim, Selection of a rational form for the steel winding tower as a preventive measure to increase its industrial safety, Taishan Academic Forum – Project on Mine Disaster Prevention and Control. (2014) 1-4.
[11] E.G. Kassikhina, V.V. Pershin, Patent № ru 2120013 s1. (1998)
[12] E.G. Kassikhina, Obosnovanie konstruktivnyx reshenij stalnyx koprov mnogofunk-cionalnogo naznacheniya, Gornyj informacionnyj analiticheskij byulleten. 4 (2014) 28-32.
[13] E.G. Kassikhina, V.V. Pershin, N. O. Butrim, Patent 154286 u1 (ru), e 04 n 12/26. Stanok metallicheskogo shaxtnogo kopra. № 2015103084/03. (2015)

IOP Conf. Series: Materials Science and Engineering **253** (2017) 012021 doi:10.1088/1757-899X/253/1/012021

New Design Heaters Using Tubes Finned by Deforming Cutting Method

N N Zubkov [1], S M Nikitenko [2], M S Nikitenko [2]

[1] Bauman Moscow state technical university, 105005, Russia, Moscow, ul. Baumanskaya 2-ya, 5
[2] Federal State Budget Scientific Institution «Federal Research Center for Coal and Coal Chemistry of the Russian Academy of Sciences Siberian Branch», 650000, Russia, Kemerovo, Sovetskiy avenue, 18

E-mail: zoubkovn@bmstu.ru

Abstract: The article describes the results of research aimed at selecting and assigning technological processing parameters for obtaining outer fins of heat-exchange tubes by the deformational cutting method, for use in a new design of industrial water-air heaters. The thermohydraulic results of comparative engineering tests of new and standard design air-heaters are presented.

1. Introduction

Heat exchangers (HE) for heat discharge into the atmosphere from the cooled devices or for buildings and premises' air heating are widespread in practically all branches of Russia's economic complex. HE, using hot water as heat carrier, are one of the most efficient and economical heat exchangers for heating large premises. The design of HE (water heaters) should provide maximum efficiency in heat transfer of heat from hot water to heated air with minimum dimensions, metal consumption and price. The heat transfer conditions at the interface "hot surface - heated air" are limiting for the above factors due to the low heat transfer coefficient and the air low heat flux.

Therefore, the main focus is concentrated on the heat transfer coefficient increasing at the solid body - air interface. The low coefficient of heat transfer from the outer surface of the heat exchange elements to air conditions a large mass, size and cost of the HE, which can be reduced by the air contacting surface development. To increase the efficiency of heat transfer, innovative technical solutions are also used to change the organization of the heat carrier flow within a single module HE, changing its hydrodynamic parameters [1, 2], an example of such a HE is a new design heater (NDH) [3] (fig. 1).

For the development of an air contacting surface, pipe-plate and pipe HEs were most widely used. Pipe-type HE have advantages in terms of their manufacturability, which has a significant impact on the price of heat exchangers. Plain-pipe HE have a restricted use. At present in pipe-type HE systems for heating, cooling and air conditioning they use pipes with packed-on plate, spiral-wound, rolled-on, wire and other types of fins, as well as pipes with hole intensifiers [4]. Prefabricated structures of finned tubes with packed-on, wound and wire fins have a high thermal contact resistance and are not manufacturable. Rolled-on finning on monometallic pipes has low surface area increase factor after treatment. Bimetallic pipes with rolled finning are of high cost and low compactness due to the impossibility of obtaining small finning pitch.

Studies of convective heat transfer on surfaces with applied hole intensifiers have shown an increase in heat transfer by 35-40% with lagging behind in growth and even a decrease in aerodynamic resistance [5]. This method of heat transfer intensification is very promising, but the problems of technological order restrain its use.

Fig. 1. NDH Heater module

An alternative option that minimizes the shortcomings of existing technologies for the development of the outer surface of heat-exchanging pipes is the technology of deforming cutting (DC), based on the pipe billet surface layers undercutting and deforming them with the formation of spiral fins without forming chips (Fig.2).

Fig. 2. Finning by DC

The finning done by the DC have such advantages as: absence of thermal contact resistance, large compactness of HE due to the possibility of obtaining small fining pitches, high manufacturability and productivity of the pipe finning process.

2. Development and methodology

At the Federal Research Center for Coal and Coal Chemistry of the Russian Academy of Sciences Siberian Branch, a bench installation for comparative hydrodynamic and thermal engineering studies of air-water air heaters (water heaters) of various types was developed and manufactured. In order to study the hydrodynamic and thermotechnical parameters, a methodology was developed and comparative tests were carried out on this stand of the NDH module and the factory module of the water air heater (WAH). The heater modules had identical overall dimensions, the number and type of heat exchange pipes, and the heat transfer surface. The difference consisted in the hydraulic piping (connection of heat-exchange pipes) and, accordingly, the organization of the heat carrier flow inside the single module of the NDH.

Ventilation air supply was provided by two independent axial fans with the same parameters. The parameters monitoring during stand operation: flow rate, temperature, pressure were provided by measuring instrumentation systems, including heat meter "Vzlet" TSRV-020 with data recording on a computer.

The thermo-technical characteristics obtained as a result of comparative tests showed that the NDH module more efficiently transfers heat to the air environment in comparison with the WAH module, as seen from the lower temperature of the return water and the higher air temperature at the exit of the bench installation. This trend also persists with the flow rate change of the heat carrier and is graphically presented in Figure 3.

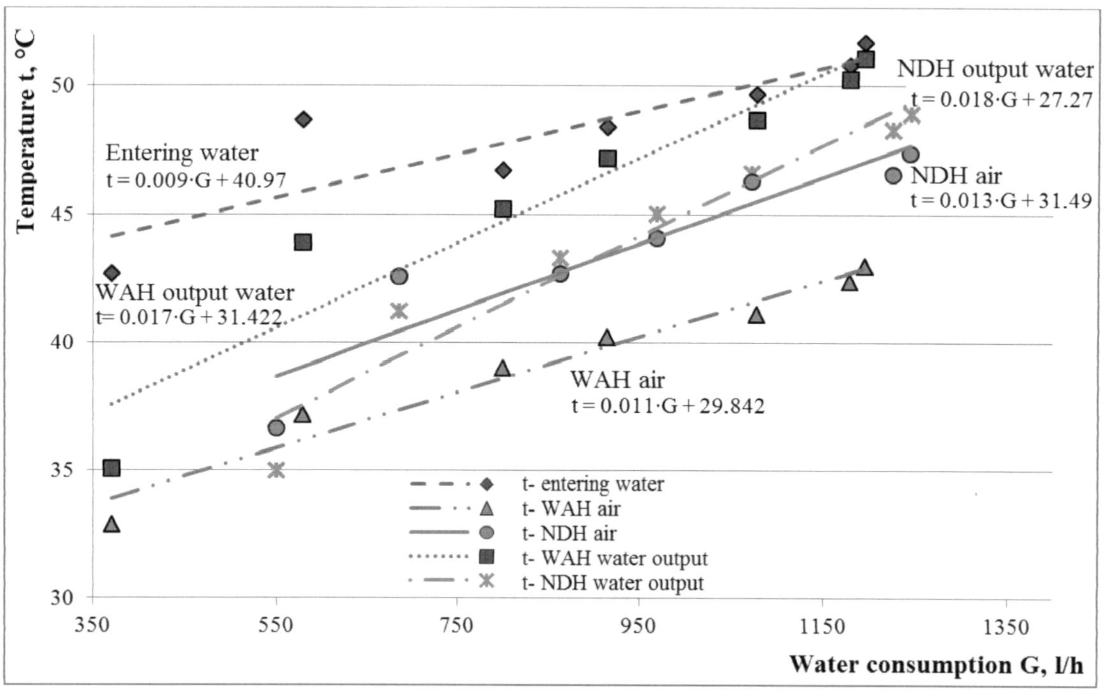

Fig. 3. Comparative thermohydraulic characteristics of the NDH and WAH modules

According to the authors' estimates, the application of the DC method for the NDH heating pipes' surface development [3] in conjunction with the technical solutions for performing the connection of heat-exchange finned pipes in a row and rows among themselves in a module, one by one in a single branch and branches between each other by interpipe transitions in the form of steeply bent branches or tangential-slit transitions will additionally allow:

- to reduce the metal consumption of HE (air heater) by an average of 10-12% compared to domestic and imported air heaters;
- to increase the heat carrier velocity by 30%;
- to reduce the heat carrier and energy consumption for its pumping;
- to achieve the HE heat transfer efficiency increase by 30-40% [7].

Using DC method and lathes, finning can be obtained with pitches of 0.2 to 2.5 mm, with a width of inter-fin clearance in the range 0.05-0.65 (usually 0.5) of the fin pitch [8].

The maximum fin height when processing copper for large fining pitches is 4 mm. In addition to the fin pitch, the fin height also depends on the thickness of the initial workpiece wall and can not be more than twice of its value. When processing titanium, copper-nickel alloys, low-carbon and corrosion-resistant steels, the maximum fin height is 3-5 pitches of finning, but not more than 3 mm. Due to the spread in the pipe workpiece wall thickness, which is 10% of the normal accuracy pipe nominal value [9], the residual thickness of the pipe wall after treatment with DC is not recommended to be less than 0.5 mm. Fine-pitch finning with a narrow inter-fin gap is effective for heat exchange processes based on phase transitions [10]. For air heat exchangers in conditions of convective heat exchange of air blown at low velocities (up to 0.5 m/s), it is necessary to provide a width of inter-fin gap of at least 0.3 mm with the maximum increase of the pipe area after the finning operation [11].

The DC method allows processing both annealed and non-annealed workpieces. The more plastic the pipe workpiece is, the wider are the method possibilities, however, work on a conventional lathe with long annealed (soft) workpieces is inconvenient, since long pipe workpieces (especially copper ones) are easily bent and additional measures are required to eliminate pipe bending. For this reason, non-annealed (solid or semi-solid) copper and aluminum pipe workpieces are recommended. When finning steel, titanium or copper-nickel pipes, it is advisable to use annealed workpieces, which in their properties have an increased plasticity with sufficient technological rigidity.

Pipe workpieces intended for processing by DC method must not have nicks, upset forgings and dents. The deflection of the pipe workpiece should not exceed the deflection, regulated by GOST 617-2006 (for diameters up to 60 mm this tolerance is 2...3 mm per running meter of the pipe, depending on the class of its accuracy). To implement the DC process, lathes can be used with both manual and programmed control. When an automatic chuck is available, computer controlled machines provide the ability to rearrange a long pipe in an automatic cycle.

3. Results and discussion

As the initial workpiece, a copper pipe of grade M2 with 20 mm outer diameter and 2 mm wall thickness DKRNP was used, according to GOST 617-2006 (cold-deformed, round, normal precision, semi-solid). The total length of the samples was 320 mm, the finned section length was 300 mm, the smooth endings length was 10 mm each.

As a tool For making samples, high-speed steel P6M5 was used as tool maretial. The DC tool was manufactured by way of sharpening of the high-speed workpiece 120 mm long and 30x20 mm cross section in three planes. The pipes' finning was done by DC method using a 1E61MT screw-cutting lathe. The tool for DC 2 was fixed in the tool holding device 1 of the machine (Figure 4). The machine was equipped with a moving lunette 4, mounted on the support of the machine. To reduce the friction between the rotating pipe and the movable work rest 4, a rolling bearing 5 was used, into which a bushing 6 having an inner diameter of 20 mm was pressed. The pipe workpiece was clamped in a three-jaw chuck of the machine through a split bushing (the pipe fixing in the chuck in Fig. 2 is not shown).

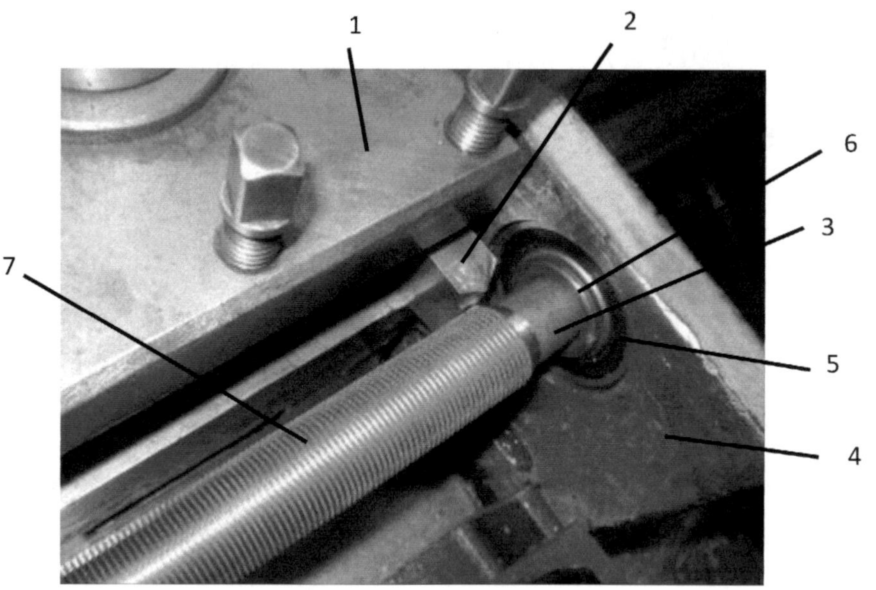

Fig. 4. - Adjustment for pipe finning by DC method:
1 - cutting tool, 2 - DC cutter, 3 - pipe workpiece, 4 - moving work rest, 5 - rolling bearing, 6 - bushing, pressed into the rolling bearing, 7 – the obtained finning

The feed was carried out from the running roller and was 1.5 mm/rev, cutting depth t = 1.2 mm. The fins were produced at the spindle speed n = 800 rpm. The fins' cutting time at 300 mm section was 15 seconds. The photo of the obtained samples is shown in Fig. 5a. Based on the measuring results of the finned pipes sections (Figure 5b) on an instrumental microscope UIM23, the following geometric parameters of the finned structure were obtained, which are presented in Table 1.

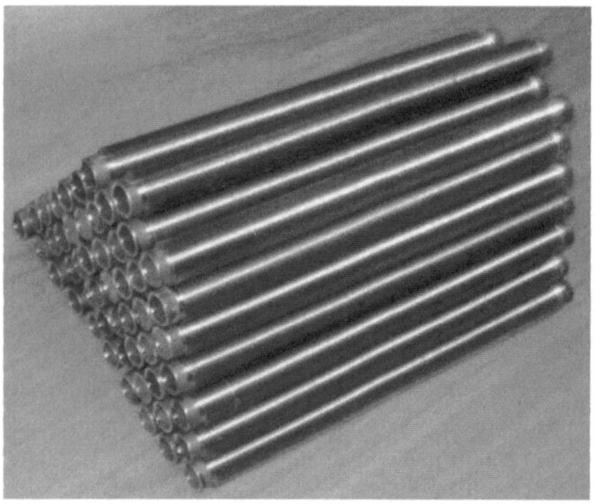

a

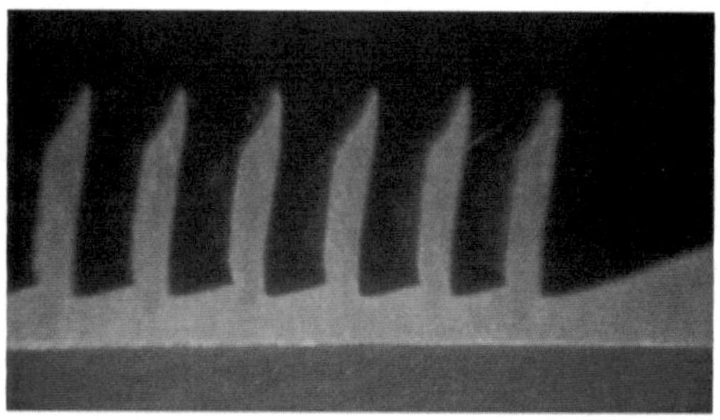

b

Fig. 5 – Pipes finned by DC:
a – photo of finished pipe samples for NDH module, b – section of the finning obtained

Table 1. Geometric parameters of samples' finning

Finning pitch S, mm	1.5
Fin gap width, b,mm	0.95
Fin thickness, a, mm	0.55
Fin height, h, mm	3.3…3.4*
Residual pipe wall thickness , mm	0.8…0.85*
Diameter at the fin tops, Dp, mm	24.0

* - Measurements for diametrically located sections

4. Conclusion

As a result of the performed works, it is shown that the deforming cutting method is promising for the NDH heaters pipes outer surface development, since it allows to obtain highly efficient small pitch finning, high-efficiency, waste-free, does not require the use of lubricating-cooling technological media, is realized with a relatively simple cutting-deforming tool, does not require any specialized equipment.

The generalizing thermotechnical characteristics of the NDH air-heater module and the factory-made WAH air-heater module obtained confirmed the NDH heating systems constructive solution in the form of series-connected heat exchange pipes of the same cross section in rows and next to each other advantages.

References

[1] Paleev D.Yu., Kiselev Yu.Ye., Kozlov V.I., Slivnoi V.N. Supply ventilation - problems and prospects. Materials of the 10th International Scientific and Practical Conference "Safety of Enterprises Activity in Industrially Developed Regions", Kemerovo, 2013, pp.166-169.

[2] Kiselev Yu.Ye., Getsman A.Ye., Slivnoi V.N. Water (steam) heater of new design, advantages and experience of application. Collected Works of the XX All-Russian Scientific and Technical Conference with International Participation "ENERGY: EFFICIENCY, RELIABILITY, SAFETY", Tomsk, 2013, pp. 306-310.

[3] Patent RU 2567884 the Russian Federation, IPC F 24 H 3/6. Water heater/Kiselev Yu.Ye., Nikitenko M.S., Getsman A Ye. (Russia). 2014147411 / 06; Claimed. 11/25/2014; Publ. 10.11.2015, Bul. 31.

[4] Dreitser G.A. On some problems of high-efficiency pipe heat exchangers creation. "Heat supply news". 5, 2004.

[5] Numerical simulation of the turbulent air flow in the narrow channel with a heated wall and a spherical dimple placed on it for vortex heat transfer enhancement depending on the dimple depth. Isaev S.A., Schelchkov A.V., Baranov P.A., Gulcova M.E., Leontiev A.I./ International Journal of Heat and Mass Transfer. 2016. V. 94. P. 426-448.

[6] Pat. RU 2341732 Russian Federation, IPC F 24 H 1/12. Water heating boiler / Kiselev Yu.E., Getsman A.E., Nikitenko M.S. (Russia). 2006132157 / 06; Claimed. 06.09.2006; Publ. 20.12.2008, Bul. No. 35.

[7] Nikitenko S.M., Nikitenko M.S., Kiselev Yu.E. Development of heat exchangers of a new generation for the fuel and energy sector enterprises. // Innovations in the Fuel and Energy Complex and Mechanical Engineering (TEK-2017), a collection of works of the International Scientific and Practical Conference. 2017. pp. 29-34.

[8] Zubkov N.N. A multifunctional technology of the surface area increasing for improving the parts heat-exchange and technological properties / Poliot. All-Russian Scientific and Technical Journal. 2003. 3. pp. 41-46.

[9] GOST 617-2006. General purpose copper and brass pipes.

[10] Water boiling on microstructured surfaces. Gortyshov Yu.F., Popov I.A., Zubkov N.N., Kaskov S.I., Shchelchkov A.V. Proceedings of Akademenergo. 2012. 1. pp. 14-31.

[11] Putilin V.Yu. Estimation of heat transfer efficiency of finned pipes made by deforming cutting method/ Bulletin of the International Academy of Refrigeration. 2003. 2. pp. 31-33.

IOP Conf. Series: Materials Science and Engineering **253** (2017) 012022 doi:10.1088/1757-899X/253/1/012022

The Morphology of Intermediate Structures Formed During Bainite Transformation in HSLA Steels

Mikhail N Seidurov, Sergey V Kovalev, Alexander S Zubkov

Altai State Technical University after I.I. Polzunov
Lenin av., 46, Barnaul, 656038, Russia

E-mail: seidurov@mail.ru

Abstract: The paper deals with the structure of bainite formed under the influence of thermal deformation cycles of welding in low-carbon bainitic class steels. Morphology features associated with the formation of mesoferrite and granular bainite determines the high cold resistance of welded joints.

Introduction

A complex of mechanical properties of steel depends on the state of austenite before the accelerated cooling, speed of cooling, temperature of the $\gamma \rightarrow \alpha$ – transformation. These mechanical properties are associated with the formation of a particular morphology of bainite, largely depending on the level of alloying of steel [1].

Currently there is no unambiguous interpretation of intermediate structures that are observed in low-carbon low-alloy steels. According to the classification of the intermediate structures shown in [2], it is proposed to distinguish needle bainite, rack and pinion, and a globular (granular) bainite by the form of crystals.

The development of bainitic transformation in low-carbon low-alloy steels at high temperatures for a certain time, when redistribution of not only carbon but also other alloying elements in the chemical composition of steels is possible, allows to predict that even under the isothermal $\gamma \rightarrow \alpha$ transformation a variety of morphological forms of bainite may exist [1, 3].

The study of the composition of intermediate structures in low-carbon low-alloy steels carried out in [4, 5] showed that at the isothermal decay bainite structures of different morphology are formed. In the upper temperature range, comprising 580-450 °C, decomposition of the austenite occurs by diffusion mechanism with the formation of mesoferrite (the term was proposed in [3]) and globular shape carbide $(Fe, Cr)_{23}C_6$. At temperatures below 450 °C, the austenite decomposes at a shear mechanism at bainitic α-phase with the selection of the carbide $(Fe, Cr, Mn)_3C$.

The aim of this work was to determine the characteristics of bainite morphology, formed in the heat-affected zone (HAZ) of high-strength low-alloy (HSLA) steels.

Material and methods of research

Material for research served as the industrial bottoms of steel 20Cr2Ni and 24Cr2Ni. The samples were exposed to both thermal and deformation cycles of welding (transverse welding stresses and elastic-plastic deformations) using the experimental setup according to the method described in [6]. The heating rate in the temperature range of phase transformations was 150 °C/s. The range of investigated cooling rates of $\omega_{6/5}$ from 0.35 to 17.0 °C/s spanned the full range of structures from ferrite-pearlite to martensite. The temperature of the sample was raised from 850 to 1350 °C in increments of 50 °C. The anisothermal decomposition diagram of austenite charts were supplemented by structural diagrams. Data on the development of stresses in the sample under test were obtained by means of a system of load cells resistance with data transfer to computer.

The phase composition, total and fine structure were studied by means of optical microscopy (NEOPHOT-32), x-ray analysis (DRON-2,0), transmission electron diffraction microscopy of thin foils (EM-125 and EM-125K). Chemical etching of the sections was performed on 4% solution of nitric acid in ethyl alcohol.

Grain size was determined by the method of secant with GOST 5639-82 using digital processing [7]. To identify the boundaries of former grain of austenite in the case of the collapse of the supercooled austenite in the shear-diffusion mechanism of the micro-sections were subjected to etching in a hot saturated solution of picric acid.

The results of the study and their discussion

Structures, formed in the HAZ of HSLA steels were investigated based on the results of the research presented in works [8-10].

Top bainite, having a plumose structure, consists of a supersaturated carbon α-phase and carbide particles located at the boundaries of ferritic plates (fig. 1, a). Lower bainite has a needle (martensite) structure. At its formation, due to the lower diffusion mobility of carbon, bainitic α-phase to a greater extent is supersaturated with this element, and the carbides are distinguished mainly inside the plates of ferrite immediately after their formation (fig. 1, b).

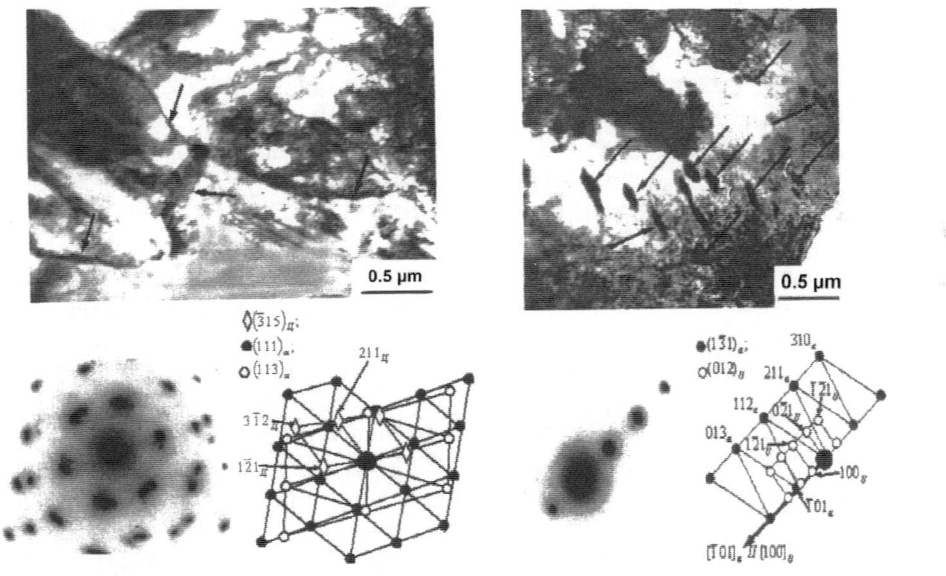

a) b)

Fig. 1 – Bainitic structure, steel 24Cr2Ni (electronic microscopy with diffraction patterns): a) upper bainite; b) lower bainite, cementite particles are marked by arrows

Intermediate structures of granular morphology represent a multiphase mixture of mesoferrite, granular bainite, residual austenite, cementite and globular carbides. It is established that the chromium carbides, which do not have a specific orientation in the layout and differ from each other in particle size (fig. 2), together with α-phase compose the basis of granular bainite [9].

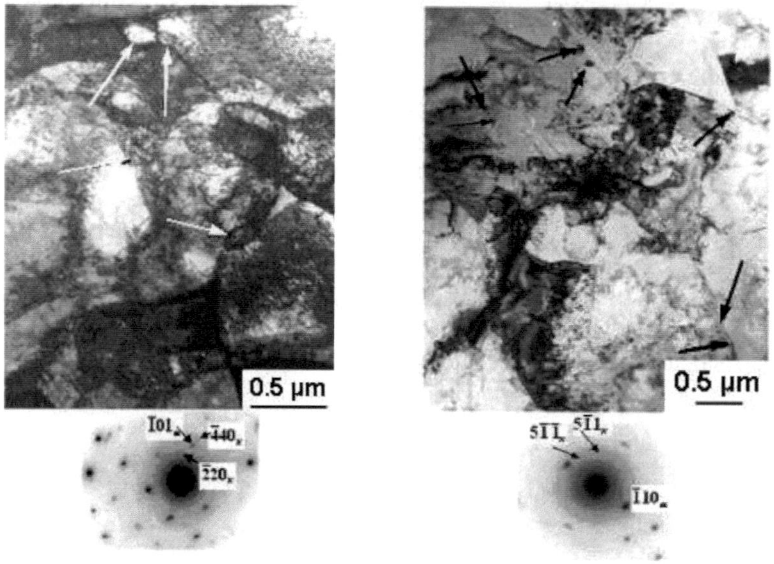

a) b)

Fig. 2 – Bainitic structure, steel 20Cr2Ni (electronic microscopy with diffraction patterns): granular bainite, large (a) and small (b) globular chromium carbides $(Fe, Cr)_{23}C_6$ marked by arrows

Large particles having a diameter of approximately 200 nm are located mainly in the joints and on the borders of fragments of intermediate structures of granular morphology (fig. 2a). Their volume fraction in the individual local areas rises up to 10 %. Small carbides of globular shape with a diameter of about 10-20 nm are located mainly on the dislocations inside of all structural components of the α-phase (fig. 2b). Despite the fact that the volume fraction of small carbides is not more than 0.2 %, their presence has a positive effect on the deformation capacity of the metal and it is the main factor preventing the birth of pockets of slow destruction [10].

In the first stage, the transformation is preceded by the fluctuation redistribution of carbon, resulting in individual local areas, which are depleted in carbon and become $\gamma \rightarrow \alpha$ – transformation diffusion centers with the formation of mesoferrite with winding borders. Next comes the increase in the number of misoperate due to the removal of carbon from adjacent to the front of recrystallization sections of austenite with simultaneous appearance of new centers of crystallization of the α-phase. High temperature bainitic transformation and a greater degree of hypothermia contribute to the diffusion mobility both carbon atoms and the atoms of carbide-forming elements.

Because the rate of redistribution of carbon is high, large amount of α-phase both bainite and mesoferrite origin is quickly produced, which covers and blocks the unconverted austenite and gives it either a sort of islet inclusions on the background of fragmented ferrite and mesoferrite, or interleaving inclusions on the background of the rack and pinion and needle bainite. The emerging volume changes cause not only bainite α-phase hardening, but the carbon-enriched austenite inclusions as well. As a result, there is a large number of crystal structure defects in the latter, serving as centers of crystallization of the carbides of globular shape in the allocation of carbon.

The formation of globular carbides contributes to the heterogeneity of the austenite, as in the carbon content and in the content of carbide-forming elements, which is inevitable in high speed at its collapse. In the enriched carbon austenite during bainitic transformation the formation of globular products of decay is possible, as α-phase and carbides, which does not contradict modern ideas about the nature of bainitic transformations in steels [11-13].

Advanced hot plastic deformation in the austenitic region has a significant influence on phase transformation during continuous cooling and, consequently, to the formation of the final structure.

IOP Conf. Series: Materials Science and Engineering **253** (2017) 012022 doi:10.1088/1757-899X/253/1/012022

Plastic deformation in the austenitic region, above 50 %, leads to a partial growth of grain and increasing the stability of austenite, the upper bainite and structurally free ferrite are formed. The presence of the structural imperfections in austenite, crystal structure defects and concentration gradients caused by hot plastic deformation lesser extent causes the formation of intermediate structures of granular morphology, and shifts the region of their formation towards higher cooling rates and temperatures of transformation.

Microplastic deformation develops along the grain boundaries of austenite (the locations of the large carbides of the type $Me_{23}C_6$) and inside them (the location of the fine carbides of the type $Me_{23}C_6$), which greatly facilitates the process of accommodation of grains in the total plastic deformation, reduces the level of internal stresses at slow speeds relaxation, and contributes to the increased dispersion of the emerging structures. After hot plastic deformation and continuous cooling in a typical the HAZ interval velocity, volume fraction of lath bainite is minor, mostly there are mesoferrite and granular bainite. The increase in cooling rates leads to changes in morphology of a resulting bainite from granular to needle. The formation of needle structures in the whole volume of the original austenite grains leads to a significant heterogeneity. Ferrite is allocated along the former grain boundaries of austenite.

Conclusions

1. In high-strength low-alloy steel carbon significantly affects the intensity and temperature range of decomposition of austenite to mesoferrite and granular bainite. In addition to carbon, bainitic structure in the heat-affected zone is determined by the grain size and homogeneity of austenite, dependent on heat input of welding.
2. The crystal structure of bainite in high-strength low-alloy steels subjected to thermal cycle of welding is characterized by high density of dislocations and subgrain structure. It is connected not only with the development of phase $\gamma \rightarrow \alpha$ – transformation in the formation of intermediate structures of granular morphology, but also with the defects of the crystal structure acquired by the austenite at microplates deformation, inherited by bainitic structure. Peculiarities of morphology associated with the formation of mesoferrite and granular bainite determine the high cold resistance of welded joints.

References

[1] V.M. Schastlivtsev, T.I. Tabatchikova, I.L. Yakovleva et al., Effect of austenite- decomposition temperature on bainite morphology and properties of low-carbon steel after thermomechanical treatment, The physics of metals and metallography. vol. 114, 5 (2013) 419-429.

[2] M.A. Smirnov, I.Y. Pyshmintsev, A.N. Boryakova, Classification of low-carbon pipe steels microstructures, Metallurgist. vol. 54, 7-8 (2010) 444-454.

[3] V.V. Svishchenko, D.P. Cheprasov, O.V. Antonyuk, Formation of mesoferrite and granular bainite in low-carbon low-alloy steel, Metal science and heat treatment. vol. 46, 7-8 (2004) 324-328.

[4] D.P. Cheprasov, Constitution and conditions of formation of intermediate structures with granular morphology in low-carbon low-alloy steels of bainitic class, Metal science and heat treatment. vol. 52, 1-2 (2010) 7-12.

[5] D.P. Cheprasov, Yu.A. Filatov, Structure and mechanism of formation of granular bainite in the HAZ of high-strength steels, Svarka i diagnostika [Welding and diagnostics]. 6 (2014) 14-18. (in Russian)

[6] M.V. Radchenko, D.P. Cheprasov, Yu.O. Shevtsov et al., Study of the processes of welding and surfacing using modern methods of collecting and processing experimental data, Obrabotka metallov [Metal finishing]. 1 (2008) 7-10. (in Russian)

[7] E.I. Khlusova, A.A. Kruglova, V.V. Orlov, Effect of chemical composition and heat and strain treatment on austenite grain size in low-carbon steel, Metal science and heat treatment. vol. 49, 11-12 (2007) 549-555.

[8] D.P. Cheprasov, M.N. Seidurov, Ensuring quality and properties of welded joints of bainitic steels, Svarka i diagnostika [Welding and diagnostics]. 3 (2012), 30-33. (in Russian)

IOP Conf. Series: Materials Science and Engineering **253** (2017) 012022 doi:10.1088/1757-899X/253/1/012022

[9] M.N. Seidurov, Investigation of mesoferrite structure in high strength low alloy steels, Polzunovskii vestnik [Polzunovskii Bulletin]. 4, T.2 (2016) 53-57. (in Russian)
[10] M.N. Seidurov, Influence of the formation of mesoferrite and granular bainite in the HAZ on the formation of cold cracks, Polzunovskii vestnic [Polzunovskii Bulletin]. 1 (2017), 99-104. (in Russian)
[11] H.K.D.H. Bhadeshia, Bainite in steels, second ed., IOM Communications Ltd., London, 2001.
[12] V.M. Schastlivtsev, New concepts of the nature of bainitic transformations in steels, Metal science and heat treatment. vol. 47, 7-8 (2005) 276-281.
[13] V.M. Schastlivtsev, Y.V. Kaletina, E.A. Fokina, A.Y. Kaletin, Effect of cooling rate on the amount of retained austenite upon bainitic transformations, The physics of metals and metallography. vol. 115, 10 (2014) 990-1000.

IOP Conf. Series: Materials Science and Engineering **253** (2017) 012023 doi:10.1088/1757-899X/253/1/012023

Machine-Building for Fuel and Energy Complex: Perspective Forms of Interaction

S M Nikitenko [1, 3, a], E V Goosen [1,2], E A Pakhomova [4], O V Rozhkova [4], M A Mesyats [3]

[1] Federal Research Centre of Coal and Coal Chemistry of Siberian Branch of the Russian Academy of Sciences, 10 Leningradsky Ave, Kemerovo, Russia, 650065
[2] Kemerovo State University, 6 Krasnaya Street, Kemerovo, Russia, 650065
[3] Kemerovo Institute (branch) of Plekhanov Russian University of Economics, 39 Kuznetskiy Ave, Kemerovo, Russia, 650992
[4] Dubna State University, 19 Universitetskaya Street, Dubna, Russia, 141980

E-mail: [a] nsm.nis@mail.ru

Abstract. The article is devoted to the study of the existing forms of cooperation between the authorities, business and science in the fuel and energy complex and the machine-building industry at the regional level. The possibilities of applying the concept of the "triple helix" and its multi-helix modifications for the implementation of the import substitution program for high-tech products have been considered.

Introduction

International approaches to the solution of territorial formations problems caused by raising regions' role and empowerment due to institutional and resource self-sufficiency are based on the sustainable development paradigm, aimed at population living standards improving under the positive dynamics of interrelated economic, social, environmental, etc. indicators.

The spread of information technologies has complicated the spatial organization of economic systems and brought to the foreground, horizontal and network type of cooperation that is more flexible than hierarchical interaction with the traditional administrative approach, and more open and focused on collaboration, cooperation and integration than the "individualized" competitive market economy [1]. Unlike the competition, inevitably leading to concurrent use of resources, mutually exclusive goals achievement and, therefore, their wasteful spending, the cooperation involves harmonization of efforts [2]. The increasing role of cooperation institutions is a natural outcome of technological, cultural and institutional evolution.

In connection with the communication opportunities development with relatively low costs, which previously would require a larger effort on the part of market actors [3], the structure is being changed. The alteration of the society role, the increase in the rate of consumer expectations and preferences changes, the consumer and domestic demand formation, stimulating new developments and implementations, impact the economic environment and the proliferation of new goods and values.

The article is aimed at the investigation of government, business and science cooperation in innovation economy sectors at the regional level.

IOP Conf. Series: Materials Science and Engineering **253** (2017) 012023 doi:10.1088/1757-899X/253/1/012023

As the methodological base of economic systems and models comprehensive analysis conducting that has sufficient empirical basis to describe the dynamics of many national institutional arrangements, their relationships in the face of uncertainty with other factors, political and social development and the "traps" of the interaction, the chosen concept of the "triple helix", by H. Etzkowitz and L. Leydesdorff and its multi-helix modifications (expanded models of innovation processes), key features of which are: the ability to compromise between the differentiation and integration of components, field to the participants' interactions and relationships changing, the strengthening effect in the context of communication theory [4].

Discussion of the problem

The concept of the triple partnership between the universities (science), business and government set by the triple helix model, introduced in the middle of 1990-ies – in the form of a synthesis of institutional attitudes of sociologists and biological analogy. G. Etzkowitz and L. Leydesdorff presented the interaction of government, business and the state in the form of hybrid social structures, like DNA molecules, which is characterized by the coupling of spiral structures, and increased adaptability to changes in the external environment. In the 2000-ies this model was actively implemented in the economic practice of developed countries as a methodological basis for the formation and organization of regional innovation systems. On its basis there were being created clusters, industrial parks, centers of competence and others.

It is clear that such experience in its pure form may not be transferred to Russian ground. The nature of Russian integration processes are mixed, due to a number of reasons: institutional and regulatory instability, the contradictory goals and conflicts of economic agents, unequal access to production factors, weak management mechanisms, the threat of markets monopolization, poor property rights protection, the relatively low level of business culture.

In this regard, Russia's peculiarities require economic imbalances removal, the institutional environment improvement and regional development management system reform, namely the formation of common priorities; scientific or business environment economic agents identification and promotion in local communities; commercialization and diffusion of innovation processes maintenance; implementation of mechanisms of coordination and control over activities and change functions and areas of participants responsibility.

Considering the region as the central element in the transformation [5], the authors set a task to identify key hierarchical levels of regional "pyramid of innovations":

- innovation environment;
- innovation processes;
- mechanisms for the societal needs implementation;
- innovative activity and measures for its promotion (credits as one of the institutional determinants of consumer demand, tax and customs benefits, insurance and intellectual property protection tools);
- results and economic consequences of innovation for the subsequent filling with mathematical tools that allow to quantify and to identify "weak links", and the proposals for structural change, adapted to different formats of interaction in three main "blocks": the state, science and education, business and society - united on the basis of the strategic objectives of the industry and the traditional linkages, financing, degree of mutual influence and other factors (including life-cycle stage).

Such enterprises can take many forms and include the scientific and educational institutions, financial and government institutions, system integrators, manufacturers and/or suppliers of technologies, equipment, raw materials, manufacturers of finished products serving the production and so on. Table 1 shows examples of the most common contractual forms.

IOP Conf. Series: Materials Science and Engineering **253** (2017) 012023 doi:10.1088/1757-899X/253/1/012023

Table 1- The most common contractual forms of government, business and science cooperation in innovation sphere

No.	Form of interaction	Features of interaction
1	Concession agreement	Under the concession agreement one party (concessionaire) undertakes, at its own expense create and (or) to reconstruct defined by this agreement, the property (real estate or immovable property and movable property technologically interconnected and intended for the implementation of activities under the concession agreement) (hereinafter the object of the concession agreement), the right of ownership to which belongs or will belong to other party (Concedent), to exercise activity with using (operation) of object of concession agreement, the concessionaire shall provide the concessionaire for a period specified in this agreement, rights of possession and usings object of concession agreement for the implementation of these activities.
2	The agreement on public-private partnership (PPP)	Under the agreement the private partner undertakes to create a fully or partially at the expense of own or borrowed funds is the subject of the agreement is immovable property or immovable property and movable property technologically interconnected and intended for the implementation of activities under the agreement to provide operation and (or) maintenance of such property, and the public partner agrees to provide the private partner the right of possession and use them for implementation of the agreement activities to ensure the right of ownership of the private partner the object of the agreement, subject to the requirements of the present Federal law and the agreement. Under the agreement, the parties also undertake to perform other obligations that arise from defining the form of public-private partnerships, a form of public-private partnership elements of the agreement.
3	Special investment contracts (Spic)	Special investment contracts (SPIC) are introduced by the Federal law "On industrial policy in the Russian Federation". The industry development fund acts as the operator at the conclusion of the ACNP. A special investment contract is an agreement between the investor and the Russian Federation (or its subject), which records the obligation of the investor (to master the production of industrial products within the scheduled time) and of the Russian Federation or its subject (to guarantee the stability of tax and regulatory conditions and to provide incentives and support). The validity of Spic is equal to the term of the project to operating profit plus 5 years but not more than 10 years.
4	Grants, budget investments, budget credits, state job	Various forms of (paid and free) budget financing of scientific research, to implement medium-term projects projects of the Federal program

Forms of government, business and scientific-educational institutions cooperation can be simple or complex. Structural elements in such alliances can be a commercial affiliation program, the production of goods under license, the joint research and development, joint procurement, investment projects, enterprises, unions and associations, clusters, markets, industries, regions, etc., the composition of participants and their roles (and thus strategies) are constantly changing, which reduces the importance of rigid designation boundaries for these communities for a long period, highlighting the following problems:

IOP Conf. Series: Materials Science and Engineering **253** (2017) 012023 doi:10.1088/1757-899X/253/1/012023

- choice and analysis of the environment, which most closely is associated with the strategic objectives achievement;
- relationships definition;
- interaction processes "contact points" control.

Such forms are difficult to compare, to develop the criteria of effectiveness and efficiency.

A serious problem for the regional innovation systems development is a vertical system of cooperation between the state and scientific organizations, inherited from the Soviet Union. In the period of market transformations these problems were added by a mismatch between supply and demand on the scientific product division, base for researching and professional training, lack of science and technology policies coordination, the distribution between a number of weakly coordinating ministries and agencies, uneven financial flows and so on do not contribute to the true priorities and socio-economic sustainability. All this leads to the fact that the regional innovation system forming is extremely slow, contradictory and spontaneously.

Historically, the closest links have formed with the research structures assigned to certain departments, primarily belonging to the commodity sector of the economy, which is the one the state gives sustainable preference to. At the same time, Russian major business does not show sufficient level of activity and merging with the state leads to lobbying of local interests and reduces the overall efficiency, despite high incomes and international markets competitiveness.

Table 2 shows the structure of the agreements of the largest Russian company in the oil and gas sector, that is PJSC Gazprom.

Table 2 – The main types and the number of agreements concluded by PJSC Gazprom and regional authorities in 2017
[Official website of PJSC Gazprom, 2017]

No.	The types of agreement	The number of regions
1	Cooperation agreements (long- and short-term)	61
2	The agreement on wider use of natural gas as motor fuel	61
3	The program "Gazprom - to children"	61
4	Road map of the project "Expanding the use of high-tech products of regional science-intensive organizations, including import substitution, in the interests of "Gazprom"	43
5	Memorandum of cooperation in the sphere of the regional educational and industrial center organizing and functioning	2
6	The agreement on scientific and technical cooperation and partnership (2016-2020)	2

The data given in the table eloquently testify the weak commitment of the company to participate in the regional innovative projects. Of the 61 region with which PJSC Gazprom signed a variety of cooperation agreements, only in 2 regions (Tomsk and Omsk regions) there are signed a Memorandum of cooperation in the sphere of the regional educational and industrial center organizing and functioning and an Agreement on scientific-technical cooperation and partnership, aimed directly at the regional innovation systems development. Intermediate between cooperation agreements aimed at maintaining the investment and income in the region, and agreements aimed at regional innovative development are roadmaps, or the projects aimed at expanding the use of high-tech products of regional science-intensive organizations in the interests of "Gazprom. They contain the plan and the stages of development of high-tech industries in the regions, local producers are involved, the examination and certification of products. In some regions, the roadmaps are involved in scientific and educational organizations. However, their number is not large, only 10.

In general, assessing the degree of the basic entities (components of the triple helix) interaction development, it should be recognized that the interaction between the scientific and the business environment appears to be the most problematic. This is caused by a number of objective and subjective factors, among which the most important is the applied sciences backwardness.

To evaluate the relationship between business and other elements (state and society) as well is quite difficult, but as the overall institutional environment has not yet formed adequately, and is not favourable, we can say that the existing format of the relationships is qualitatively different from the one that adopted in foreign countries.

A significant place in the socio-economic development of constituent entities of the Russian Federation is given to the arrangement of the clusters, since they allow to move away from the previously dominant approach in the orientation exclusively on high-tech industry, to stimulate research and development to accelerate the processes of creation and diffusion of innovations through the formation of stable ties between science, business and local authorities. These relations have a decisive influence on the frequency and speed of propagation of innovation and competitiveness.

For qualitative and quantitative assessment of innovation activity the authors developed a list indicating any trends in the structural indicators, distributed in three groups: an innovative component for economic growth; the investment component for the development in the long term; component of sustainable development, ensuring the growth of competitiveness and quality of life. Through econometric modelling the peculiarities of the development of the national industry structure for different public organizations historical periods that have allowed to develop additional indicators and to offer methods of measurement in order to assess the effectiveness of ongoing reforms in the country.

Recommendations for systematic upgrading of aspects of the socio-economic system developed by the authors based on the functional economic and mathematical models and lay the foundations for a unified approach ("designer") in the formation of resilient regions, regardless of geographical, social etc. differences.

In the study of the establishment and functioning of national innovation systems and approaches are of interest: the global experience, developed on the basis of various factors and the concept of "triple helix" of national forms of interaction of elements that contribute to the development of both fundamental and applied research of an interdisciplinary nature, the emergence of start-ups, variety of entrepreneurial universities and science parks.

A deep analysis of the "Euro-Atlantic", "Asian", "alternative" and other models can be the basis for the development of Russian regional ecosystems, taking into account local specificity and "smart specialization" in order of greatest contribution to socio-economic development.

By now there have been received a number of conceptual and applied results of the "triple helix" model research that take into account both the global and Russian economic realities, and historical retrospectives, which is the base for the current state and a number of inherited elements of economic activity.

The conceptual results, clarifying a number of provisions required to adapt the model of "triple helix" to domestic conditions [6] are as follows:

- the analysis of the empirical base of the model identifying the structural units, their substantive content, the development schemes of their cooperation in the area of cross-section;
- the study the possibility of initiating the process of interaction of any party structure in the model;
- the allocation of favourable conditions for development in the triple interaction.

In general, identified through econometric modelling features of development of economy of our country in different public organizations historical periods, and fragments [7] have received a possible economic interpretation of the role and structural proportions of the industry in socio-economic development and inconsistencies of modern economic policy that creates the real basis for further use of existing indicators and suggestions and development of additional ones for methods of measurement to assess the effectiveness of ongoing reforms in the country.

Full ecosystem restructuring in the Russian context will require the overcoming of existing gaps in three dimensions: 1) level of education and science – to form innovative generation ("people with imagination"); 2) at the enterprise level – in order to build the relationship between the development of fundamental and applied research of an interdisciplinary nature and implementation of developments in production, contributing to the achievement of a balance between international innovations and home traditions; 3) at the level of society – to create consumer confidence and domestic demand, stimulating new developments and implementations.

Conclusion

The result of set in this article the problems of application of the triple helix model and its modifications in the Russian context can be formulated as following: a priority approach should be a refusal of domination by any one side in favor of transformation of the state control ways and methods, higher adaptability of the regional authorities and stimulating all types of events, aimed at a variety of economic agents.

The implementation of the principles of the "triple helix" element of the "state" can be realized as following:

- on the federal level there are the functions of determination of priorities of innovation activities and creation of institutional preconditions for its development; management and financing of innovation activities (including such forms as the state order, tender or grant); the development of innovative policies and methods, models and tools for its implementation, including the development of virtual networks; the establishment of mechanisms of commercialization of knowledge and the organization of international cooperation in this field;
- at the regional level there are the creation of an appropriate infrastructure; establishment of integrated service centers with the participation of all stakeholders, with sufficient generation capacity development on the basis of key regional competencies, providing information, consulting, investment, branding support, technology transfer, strategic and operational management of innovation chains design.

Summing up, it should be emphasized:

- founded in the model of "triple helix" network format on the basis of the principle of consent and cooperation contributes to the integral effect of continuous updates;
- for Russia, foreign experience of the concept introduction can contribute to identifying and selecting the most promising forms of multilateral regional cooperation in matters of disclosure implicit opportunities for the development of new comparative advantages of regions, use of innovation, adapted to the conditions of the regions, development of the internal market and closer interregional cooperation [8].
- organization mechanisms of the innovation management need to be based on the network interaction innovation-based factors with the goal of transformation costs in a much larger-scale socio-economic impact at the level of individual enterprises and at the level of industries, regions, and states;
- social activity is another resource of competitiveness improvement;
- in conditions of intensified competition and changing patterns of consumer behavior, the key parameter is the rate of the innovation cycle reproduction;
- in any space can be formed in a special "ecological environment", which includes mutual expectations and requirements of the elements and permits adaptation of the old forms of economic entities to new conditions at the micro-, meso-, macro levels.

Acknowledgments

The authors are grateful for financial support to Russian Foundation For Basic Research (the "Instrumental and methodical approach to the adaptation of the triple helix model for Russia given the historical retrospective" project, No. 16-06-00054).

References

[1] Smorodinskaja N V 2011 The triple helix as a new matrix for economic systems *Innovations* **4**
[2] Polterovich V M 2015 From social liberalism to the philosophy of co-operation *Social Sciences And Modernity* **4**
[3] Krupanin A A, Mjachin Ju V, Rastova Ju I 2011 Management problems of modernization (Saint-Petersburg: Russia) p 236
[4] Etzkowitz H, Leydesdorff L 1995 The triple helix of university-industry-government relations: a laboratory for knowledge-based economic development // EASST Review
[5] Pakhomova E A 2010 Methodological basis for the assessment of university influence upon regional development efficiency (Moscow: Russia) p 725
[6] Istomina S V, Lychagina T A, Pakhomova E A 2016 Russian perspectives of the triple helix development *National interests: priorities and security* **12(345)**
[7] Davtyan T B, Pakhomov A V, Pakhomova E A, Rozhkova O V 2017 An econometric analysis of some sectoral indicators of the USSR and Russia's economies in line with historical retrospect *National interests: priorities and security* **1(346)**
[8] Nikitenko S M, Goosen E V, Sablin K S Perspectives of comprehensive mineral exploitation based on the principles of public-private partnership *IOP Conference Series: Earth and Environmental Science* Volume **45**, Number **1**

IOP Conf. Series: Materials Science and Engineering **253** (2017) 012024 doi:10.1088/1757-899X/253/1/012024

Objects of Industrial Property as an Instrument for Introducing Technological Innovations in Machine Building

S M Nikitenko [1, 2, a], **M A Mesyats** [2]

[1] Federal Research Centre of Coal and Coal Chemistry of Siberian Branch of the Russian Academy of Sciences, 10 Leningradsky Ave, Kemerovo, Russia, 650065
2 Kemerovo Institute (branch) of Plekhanov Russian University of Economics, 39 Kuznetskiy Ave, Kemerovo, Russia, 650992

E-mail: [a] nsm.nis@mail.ru

Abstract. The article is devoted to the investigation of the possibility of applying new credit instruments for the implementation of technological innovations in machine building and other branches of the fuel and energy complex. The possibilities of using the new tool as an element of industrial policy in the program of import substitution of high-tech products have been considered.

Introduction

The main challenge for the world's economies today is the process of building national and regional innovation systems. The methodological basis for this is the Triple Helix Model, describing the directions and forms of cooperation between science, businesses and government in the innovation sphere and its "evolution" sequel - the Quadruple-Helix Model, which includes the civil society along with the three listed participants.

Since then the conditions have changed. New organizational and economic diffusion (transfer) tools of intellectual activity as the basis for municipalities' innovative development have appeared in the Resilient Regions. Today, technology transfer takes into account the "host regions," their resource security, economic, environmental and social component. The synergistic effect plays an important role due to innovative technologies complementarity, the latter is particularly important for the Resilient Regions.

Practice has shown that harmonization on the basis of pledge mechanisms of intellectual property management processes can significantly speed up the innovation implementation process and serve the equalization of socio-economic development levels of Resilient Region individual subjects and their economy promotion in an innovative way. This is a new phenomenon in Russia. Therefore, the experience of European countries in the intellectual property management, including intellectual property rights as a pledged asset, is so important for Russia.

In Russia the pledge as a way to enforce the obligations came much later, in comparison with countries of Romano-Germanic law system. However, in recent years an institutional leap towards the formation of institutions of investment in intellectual property has been taken in Russia. As a result, in 2016 about 20 leading Russian banks concluded 81 pledge agreements over the exclusive rights to trademarks, inventions, utility models and industrial designs. The role of regional funds to support small and medium-sized innovative companies providing on behalf of the government pledges to commercial banks has been growing.

The main purpose of commercial use of intellectual property objects (IPOs) is to generate profit. The objects of transactions are not the objects themselves, but the property rights to these objects, including patents and patent applications. They are the assets that are used as pledge for loans or other loan agreements. Formal intellectual property rights (patents and trademarks that are objects of the official registration system) and informal (copyright) are considered to be Intellectual property rights for the purposes of IP lending. Formal intellectual property rights protect the monopoly and, as a rule, they do not exist for as long as some informal intellectual property rights. In the case of registered trademarks, the rights may be used for an indefinite period, provided that the rights to IP continue to be used [3].

Such assets are especially valuable for businesses engaged in research and results commercializing. In this situation, the cost of a patent portfolio can many times exceed the cost of other assets of a company and act as the most valuable asset. To date, in developed countries investment in IPO accounts for more than 15% of GDP and almost half of all investments is made in fixed capital. The share of investment in IPO in the largest international corporations is rapidly growing [1]. So, if in the early 80's of the 20th century the share of intangible assets in the structure of the cost of American companies accounted for no more than 40%, then at the beginning of the 21st century their share exceeded 70%. At the same time, 30-40% of non-material property was not identified and reflected in corporate balances. This allows to say that the share of intangible assets can actually be even higher. A similar picture is typical for West European corporations [6]. According to Interbrand consulting company, tangible and intangible assets of the world well-known companies are correlated as follows: British Petroleum - 30:70, IBM-17: 83, Coca-Cola - 4:96. For comparison, in Russia, the volume of investment in intellectual property objects has not exceeded 1% so far [7].

It is important to note that pledge transactions involving the use of IPOs rights along with benefits carry potential high risks for the parties involved. To reduce them, countries are developing special legislation regulating transactions with IPO rights; companies and banks use such protection mechanisms as a preliminary assessment of IPOs market value, determining the best way to calculate depreciation charges and depreciation periods, revaluing the cost, using IPO as a pledge in combination with other assets of the enterprise, preliminary expert analysis of the economic efficiency of the IPO use, etc.

Discussion of the problem

In addition, many countries provide for the possibility of registering transactions with industrial property in the public registry, which guarantees the rights of parties involved in the transaction. The French Intellectual Property Code defines the rules regarding the pledge of the right to use the software. Article L 132-34 of this Code provides for mandatory registration of a pledge in the special register of the National Institute of Industrial Property (INPI) [8, 9, 10, 12]. In Austria and the United States, the exclusive rights are not the subject of the pledge, but the patent itself as movable property is. A pledge transaction must be registered with the Patent Office [5].

In the case of a pledge over formal intellectual property rights the EU countries take into account the following points:

- public registers operate a priority system granting to the first applicant a priority pledge right over Formal IP Rights;
- formalities for creating the pledge over the Formal IP Rights are regulated by the law of the country in which the pledge right needs to take effect;
- the Formal IP Rights remain vested in the company-borrower, which will continue to be responsible for the maintenance of the Formal IP Rights.
- formalities for creating a pledge over the Formal IP Rights are comparable to the ones that are traditionally provided for the immovable assets (real properties) [10].

In accordance with the pledge practice of Great Britain, the pledger will need to consider and estimate carefully the consequences of entering into the pledge agreement and scrutinise its terms. The pledgee will have the right to sell the patents in the event of a default by the pledger. If the pledgee has taken pledge over not just the patents but the other business assets, the pledgee may have other options too, such as the power to appoint an administrator with authority to run the business of the defaulting pledger. The pledgee will have the power to appoint a receiver to the patents which are subject to the pledge. Appointing a receiver is a procedure enabling the pledgee to sell the patents or collect the revenues derived from the patents in order to repay the secured debt on a default by the pledger. The pledgee will wish to ensure that the value of the pledged patents is maintained; in the agreement, it may seek to impose obligations on the pledger to comply with such a guarantee. The pledger would be well advised to seek to reserve the right to allow to lapse, or abandon, patents or patent applications it reasonably considers are no longer of value [12].

In most cases the pledger will want the freedom to continue to exploit the patent portfolio as part of its business. In this regard, it must make sure any pledge document does not its freedom to do this in any material way. It should be emphasized, that there is normally a community of interest between the pledger and pledgee in this context because the pledgee wants the business to generate revenue to repay the debt. Wherein, the pledger needs to bear in mind the fact that any future licensee will almost certainly become aware of the pledge over the patents (either from inspecting the relevant registers or as the consequence of seeking standard warranties as to the pledger's title and right to license free from encumbrance). The pledger has to negotiate the terms with the pledgee at the outset so as to make clear that any future licensee's interest is not at risk of being defeated by the pledgee exercising its right to sell the patents in the event of default on the loan. In such cases the pledger could offer as a compensation to the pledgee an access to the revenue generated by the licence either by agreeing to pay it directly to the pledgee or by giving the pledgee a pledge over the revenue stream itself [12].

The following principles of the EU regarding pledging intellectual property rights are mentioned. The Community Trade Marks (CTMs) system coexists with domestic systems. The CTM system is valid and enforceable across the European Union as a whole territory (and any new Member State).

Registration of trademarks with the Office for Harmonization in the Internal Market based in Alicante, Spain (OHIM) allows lenders/advisors to deal with: a single Administrative Centre to be checked; a single filing procedure to be carried on; a priority regime valid in all the Member States.

The effects of CTMS are governed by the provisions of EC Council Regulation No. 207/2009 (2009 EC Regulation). Article 19 of the 2009 EC Regulation expressly provides for pledge over CTMs:

1. CTMs may, independently of the undertaking, be given as a pledge or be the subject of rights in rem;

2. Upon request of one of the parties, pledged CTMs or rights in rem shall be entered in the Register of Community Trademarks (Register) and published on the Community Trade Marks Bulletin (Bulletin).

Within the framework of the European Patent System, providing for the procedure for granting the European Patent (EP), which is administered by the European Patent Office with the head office in Munich, Germany (EPO) on the basis of international rules set up in Munich in 1973 (European Patent Convention; latest update in 2010, 14th edition) (Munich EP Convention). Under the EP regime, EPs may be granted as pledge (see Article 71 of the European Patent Convention) and the relevant pledge needs to be filed with the EPO. National laws apply in relation to the effect and validity of the pledges, including formalities in order to oppose the right to third party at the local level (see Articles 74 of the European Patent Convention) [10].

In Russia, the experience of IPO pledge transactions is not big. This version of bank lending in the country began to be used only in 2009 and is at the initial stage of development (Table 1). The conclusion of any contracts (including pledge agreements) related to the disposal of an exclusive right to patented objects of intellectual property and trademarks in Russia are the subject to registration in Rospatent [4].

Table 1

Number of IPO pledge agreements registered by Rospatent in 2009-2016

	2009	**2010**	**2011**	**2012**	**2013**	**2014**	**2015**	**2016**
Pledge agreements over exclusive rights to the results of intellectual activity	-	8	16	17	20	15	13	9
Pledge agreements over exclusive rights to trademarks	16	62	52	24	60	40	58	72
Number of trademarks referring to which the pledge agreements are concluded	-	195	280	82	191	258	362	689
Total number of pledge agreements over exclusive rights to IPOs	16	70	68	41	80	55	71	81

Source: compiled by the author on the basis of [4].

The main constraint on the issue of extending the practice of bank lending against IPOs pledge is the difficulty in assessing the value of pledged assets as intangible assets [2]. From the author's point of view, the existing Russian experience with the use of IPOs as a pledge for obtaining credit and applying the experience of European countries today can serve as a solid foundation for activating lending in the banking services market; and the pledge over IP rights can be a promising tool for the development of companies actively engaged in the implementation of intellectual activity results.

Thus, the pledge over intellectual property rights is an opportunity to build a close and, most importantly, mutually beneficial interaction between science and business, which are the two participants in the Quadruple-Helix Model of building national and regional innovation systems. The third party is government. The role of government, when using the mechanism of pledging intellectual property rights as the starting point of innovative development of the economy is to ensure, including the legislative provision, the security of pledge transactions. Government that acts as a guarantor in an IPO pledge transaction significantly reduces the risks of the pledgee, thereby increasing the attractiveness of such transactions.

Government influences pledge relations through economic and administrative methods which are based on the guarantee mechanism. The basis of legal regulation of pledge relations is the government guarantees of pledgee's rights which are called upon to provide an adequate protection in case of owners' interests violation while carrying out innovative activities. Wherein, government, as a rule, guarantees first of all the stability of rights, which is very important in long-term relations.

In order to define the meaning of "government guarantee mechanism of pledging intellectual property objects" concept, it must be borne in mind that these are a form of social guarantees. These mean material and legal means that ensure the realization of the social and economic rights of society members. Consequently, legal guarantees of investment are legal means ensuring the implementation of the rights of the parties.

In the legal sense, the concept of "guarantee" is defined as a system for ensuring the reality of rights established by law. This system includes ensuring the monitoring of compliance with the legislation, the activities of relevant government authorities, as well as legislative norms that ensure the stability of public relations.

Thus, the essence of guarantees provided by national legislation is to not violate the rights of the interacting parties and ensure their implementation. The peculiarity of these guarantees lies in the fact that they come from government itself.

Government assumes the obligation to perform certain actions in relation to the pledger or refrain from actions that violate its rights and legitimate interests.

The essence of the guarantee mechanism is the specific obligations of the government to ensure the security of the pledgee's property.

If we analyze the state of Russia's notion of "government guarantees as the basis for ensuring a secure IPO pledge", the following should be highlighted:

IPO is a specific object, with a complex estimation of its value and, therefore, very risky.

For full-fledged government involvement as a guarantor, when conducting IPO pledge transactions between science and business, it is necessary:

- to expand the pledge law on the basis of the world's leading practices, international agreements on the forms and methods of protecting the rights of participants in IPO pledge transactions;
- to define clearly, preferably legally, the ultimate goal of government actions in this area, which is ensuring the security of participants' property in IPO pledge transactions and the actual realization of their rights as well.

After implementing the above measures, we will get a clear and transparent model of interaction between science, business and government (authorities) in creating innovative growth points based on the mechanism of IPO rights pledging.

At the same time, in Quadruple-Helix Model of building national and regional innovation systems, there is another participant. It is a civil society. Within the framework of the Quadruple-Helix Model concept, a society is understood as the public based on the media and culture. Culture is understood as values, traditions, etc. Media is understood as television, the Internet, newspapers, as well as news, social networks, communication, etc. The role of a society is reflected in the creation of public organizations (parties, unions, and associations) and their activities.

Within the innovation process, it is necessary that the relationships between government, business and society are balanced. The imbalance of these relationships leads to the inefficiency of the innovation process. In the conditions of totalitarianism it is difficult to carry out innovative activities because government dictates to the business what and in what volumes to produce, therefore the enterprises do not have the opportunity to realize the ideas of new products. Anarchy also does not promote innovation. At anarchy it is generally difficult to carry out production. In the situation of the "economy of individuals" it is just as difficult to innovate. This is due to the fact that the managers of enterprises are not aimed at the prosperity of the enterprises they work for. Their personal benefit is more important for them. They will be more willing to steal the enterprise's property than improving the production process, especially, creating new products.

Government, science, business and society are not separate subsystems. The same participant can be an element of two of the listed subsystems. For example, enterprises that own a significant share of government participation should be classified as both government and business; commercial structures that are owned by public organizations can be attributed as intersection of society and business. The intersection of such spheres as society and government covers the specifics of the functioning of elective authorities, parties, and the electoral system.

If we consider innovation based on the example of an enterprise, it is worthwhile to say that modernization development depends on the involvement of employees in the work process, the use of experience and abilities of employees, as well as motivation and incentives. The more employees of the company are involved in the innovation process, and each of these employees is motivated to innovate, the better the result is [14]. The same can be said about the process within the framework of the Quadruple-Helix Model concept. Only with the participation of all the participants the greatest economic effect can be achieved.

Thus, we can say that "society" is inseparable from the "three" which is science, business and government; in case of its isolation it is impossible to consider the interaction mechanisms of the "three" in any sphere, including the sphere of IPO pledges.

The role of society in developing innovative systems on the basis of IPO pledge relations is enormous, because it is connected with all the other participants in the Quadruple helix concept. IPOs are created by science for the needs of society; there is a relationship between society and science. Business in assessing the cost of IPOs will always take into account the attitude of society to the product or service created or provided using IPOs; there is a relationship between society and business. Government (authorities) in the development of guarantee mechanisms legislation within the framework of IPO pledge transactions will rely on representatives of the society (for example, deputies of various levels); there is a relationship between society and government (authorities).

Conclusion

Summarizing the above, we can conclude the following. Holding IPO pledge transactions accelerates the pace of creating innovative systems in the economy, especially in the Resilient Regions. These transactions take place within the framework of the Quadruple-Helix Model concept that involves four participants. They are science which creates, business which invests, government which guarantees the observance of the participants' rights and society which determines the type of IPOs and the rules of interaction between science, business and government.

Acknowledgments

The authors are grateful for financial support to Russian Foundation For Basic Research (the "Instrumental and methodical approach to the adaptation of the triple helix model for Russia given the historical retrospective" project, No. 16-06-00054).

References

[1] Goosen E.V. Prospects for the development of PPP projects in the Fuel and Energy Complex of Russia: assessment and current state. Fundamental research. No. 11-2, 362-366, 2016.

[2] Nikitenko S.M. The general state and development trends of business and science joint projects in the sphere of productive sectors of the resource producing region economy. Fundamental research. No. 11 (Part 1), 196-199, 2016.

[3] Nikitenko S.M., Goosen E.V, Klishin V.I. Experience of interaction of institutions of academic science with business on the principles of PPP (on the example of the Institute of Coal, SB RAS, Kemerovo). Innovations. No. 9 (179), 9-19, 2013.

[4] Report on the Rospatent activities in 2015. Available at: http://www.rupto.ru/about/reports/2015 (accessed: 08.02.2017).

[5]Ruzakova O. Exclusive rights pledge agreement. Available at: http://superpressa.ru/index.php?Itemid=82&id=105&option=com_content&task=view (accessed: 20.02.2017).

[6] Khotinskaya A.V. Intangible assets as marketing resource of the company. - Available at: http://elibrary.ru/download/elibrary_13288643_66954171.pdf

[7] Anatomy of Growth. Top Growing Brands. 2016. Available at: http://interbrand.com/best-brands/best-global-brands/2016/

[8] Code de la propriété intellectuelle (version consolidée au 23 février 2015). - Available at: http://www.wipo.int/wipolex/en/text.jsp?file_id=363403 (accessed: 29.01.2017).

[9] Dutch Civil Code, Book 3. Available at: http://www.dutchcivillaw.com (accessed: 08.02.2017).

[10] Livio Esposizione. IP rights and loan financing: European perspective. Available at: http://techlaw.org/wp-content/uploads/2011/05/110426_IP-rights-and-loan-financing_1.pdf?x13654 (accessed: 29.01.2017).

[11] Loans & Secured Financing 2016 / Contributing editor George E. Zobitz Cravath, Swaine & Moore LLP. - London: Law Business Research Ltd, 120, 2016.

[12] Marshall J., Caldwell R., Cain B. Taking security over patents. Available at: https://united-kingdom.taylorwessing.com/synapse/march14.html (accessed: 29.01.2017).

[13] Nikitenko S. M., Goosen E. V., Sablin K. S. Perspectives of the Comprehensive Mineral Exploitation Based on the Principles of the Public-Private Partnership. IOP Conference Series: Earth and Environmental Science. International Scientific and Research Conference on Knowledge-Based Technologies in Development and Utilization of Mineral Resources, 012001.45, 2016

[14] Lugovaya E.S. Innovations as the basis for modernization of modern society. Bulletin of Volgograd University, Ser. 7, Philos., № 2 (17), 103-108, 105, 2012. Available at: http://www.volsu.ru/struct/generalservices/publish/vestniki/lastmagazine/ser-7-philosophy-2-17-2012/2_%D0%9B%D1%83%D0% B3% D0% BE% D0% B2% D0% B0% D1% 8F.pdf (accessed: 14.11.14)

Impulse Control Process Arc Welding Covered Electrode

Anatoly F Knjaz'kov[1], Victor L Knjaz'kov[2], Sergei A Knjaz'kov[1], Kirill I Dementsev[3], Elena E Petrova[2, a]

[1]National Research Tomsk Polytechnic University, 634050, Tomsk, Leninast., 30
[2]T.F. Gorbachev Kuzbass State Technical University, Kemerovo, 28, Vesennyayast.
[3]AO STNG
E-mail:[a] pee5152@mail.ru

Abstract.A method and technological advantages of impulse control processes of arc welding electrodes coated further RD. Ways to overcome the shortcomings RD.

Introduction

When welding structures that are difficult to achieve due to rational automation methods, for example, when installing and repairing pipes for heating surfaces of high-pressure boilers, manual arc welding with coated electrodes (RD) is one of the leading technological processes. At the same time, it should be noted that the reserves for increasing the productivity and efficiency of the RD have been exhausted. One of the promising modifications taxiway, which makes it possible to increase the efficiency of this process, is the use of pulse modulation the welding current [1]. However, the existing methods and technical means of implementing this process realize the concept of "machine-technology" with programmatic control of the parameters welding current pulses giving the welder essentially the role of a mechanism that mechanically moves the electrode along the joint line and maintains a constant arc length. The welding rate is set by the technical means of modulation and remains unchanged during the welding process.

The concept of "machine-technology" is based on the desire to transfer the operation of dosing thermal energy to technical means of modulation, facilitating the work of the welder and reducing the requirements for his qualification [2, 3]. The technical means and methods of modulating the welding current that realize the concept of "machine-technology" can be attributed to the first generation of developments in this field.

The review of the state development methods and means of pulse modulation the first generation showed that the introduction them into the industry is hampered by a number of factors decisive importance, for example, ignoring:

- the phenomenon of mass transfer through the arc gap;
- the influence of current modulation on the physical stability of arc combustion and technological stability of the process;
- influence on the welder;
- the impossibility of active control the thermal power of the arc by the welder during the welding process.

The possibility of fulfilling all the requirements for the methods and technical means for RD with a modulated current represents the concept of "machine-man-technology" implemented in a second-generation pulse modulator created at the OTSP NI TPU department [4]. At the same time, automatic control of the thermal power, and, consequently, the rate of welding, while simultaneously improving the quality of the welded joint during welding in all spatial positions and reducing the load on the eyesight is carried out.

Results and Discussion

In Fig. 1 shows the developed algorithms for automatic control of the thermal power arc with the adaptation of the welding current pulse parameters.

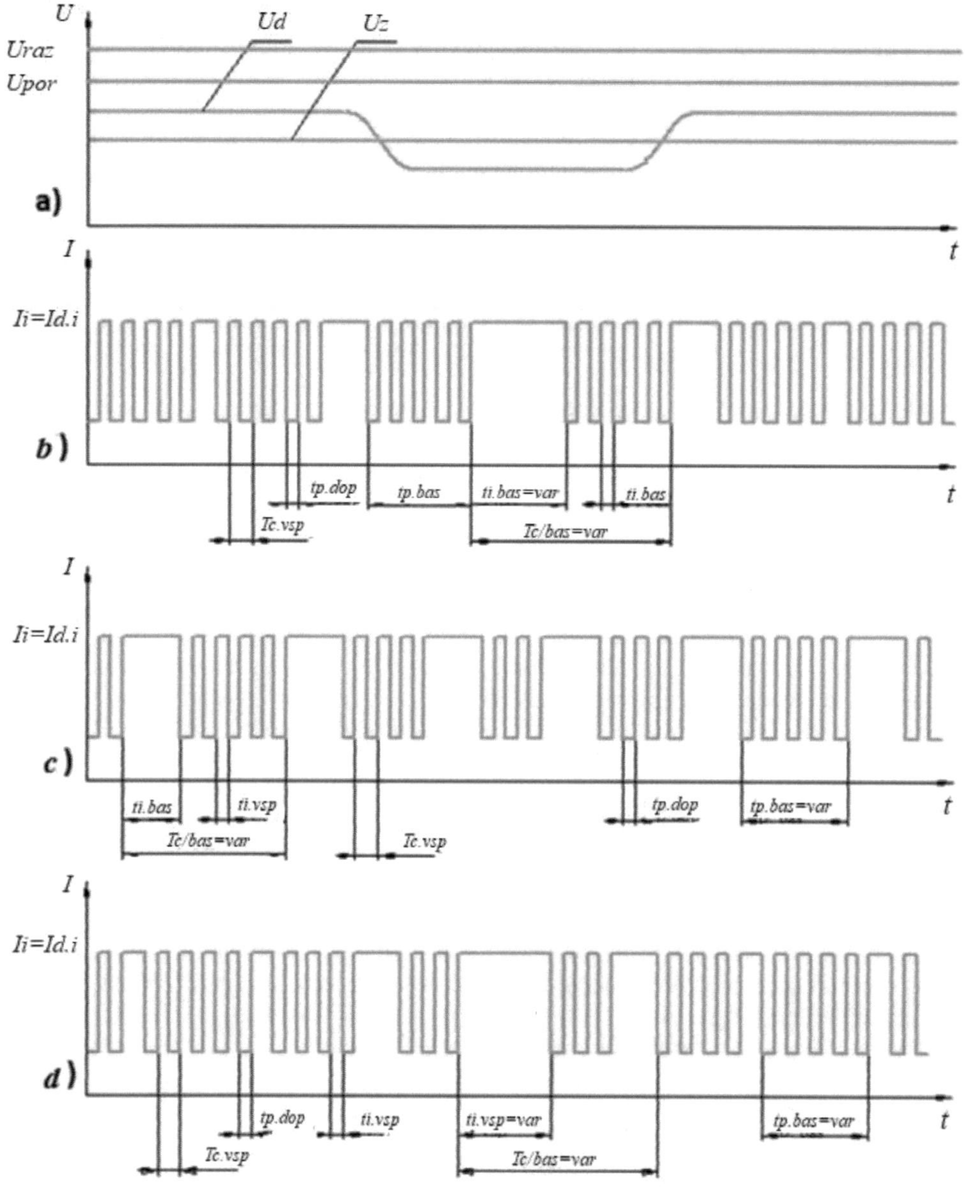

Fig.1. Cyclogramm of the modulated current with additional pulses: I_i - impulse current; I_b - base current (pause); $I_{d.i.}$ - auxiliary pulse current; $t_{i.bas.}$- the duration of the main pulse; $t_{p.bas.}$- duration of the main pause; $t_{i.dop}$ - the duration of the additional pulse; $t_{p.dop}$- duration of additional pause; $T_{c.bas}$- duration of the main cycle of modulation of the welding current; $T_{c.dop.}$- the duration of an additional cycle of modulation of the welding current

IOP Conf. Series: Materials Science and Engineering **253** (2017) 012025 doi:10.1088/1757-899X/253/1/012025

In Fig. 1a shows the arc voltage plot;

In Fig. 1b shows the current cyclogramm during modulation of the duration main pause;

In Fig. 1c is a current cyclogramm with simultaneous modulation of the durations main pulses and pauses;

In Fig. 1d shows the current cyclogram with modulation according to the level of the deviation average voltage arc gap from the set at the relay mode of regulation.

In the process of welding with pulse modulation welding current automatically as a function of the deviation average voltage arc gap from the set value, the duration of the main pulses and pauses is regulated by varying the arc length, and additional pulses are applied to the current during the main pause with a repetition frequency of at least 50 Hz and a duration from 0.5 to 2 ms.

Control of the thermal power of the welding arc by changing the length of the arc (1.5 ÷ 2) V, which does not violate the gas-slag protection of the welding zone - by extending or shortening the arc length within these limits, the welder, taking into account the situation in the weld zone and his capabilities, adapts the parameters of the regime to these conditions for Pulse modulation of pulse parameters.

Current modulation makes it possible to use the amplitude value of the current of the main and additional pulses equal to the nominal value of the current each particular electrode, in which there is no short circuit in the continuous mode and the drop of electrode metal is transferred to the weld pool during the pulse also without short circuits with minimal spattering. Fig. 2 shows the oscillograms of the welding current and the voltage of the RD using SSSI-13/55 electrodes with a diameter of 3 mm. According to the oscillograms, the region of the rated current at which the best welding-technological properties of these electrodes manifest themselves is in the range from 170 to 180 A [5].

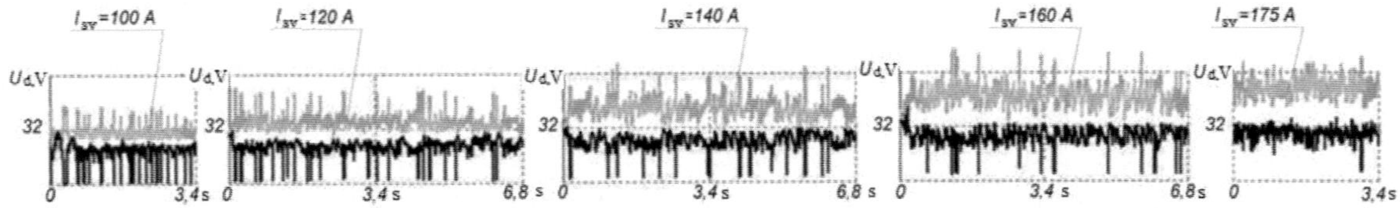

Fig. 2. Oscillograms of currents and voltages recorded at different values of the current of the RD by SSSI-13/55 electrodes with a diameter of 3 mm [4]

The use of a nominal current in the pulse (primary and secondary) allows to significantly increase the modulation efficiency by increasing the depth of modulation and reducing the average value of the welding current to (30 ÷ 40) A while maintaining the best welding and technological properties of the coated electrodes. The transition of drops during a short circuit in the main pause is guaranteed by one additional pulse with minimal spattering. In Fig. 3 shows the forced transfer of a drop by an additional pulse during a pause period [2, 3, 5].

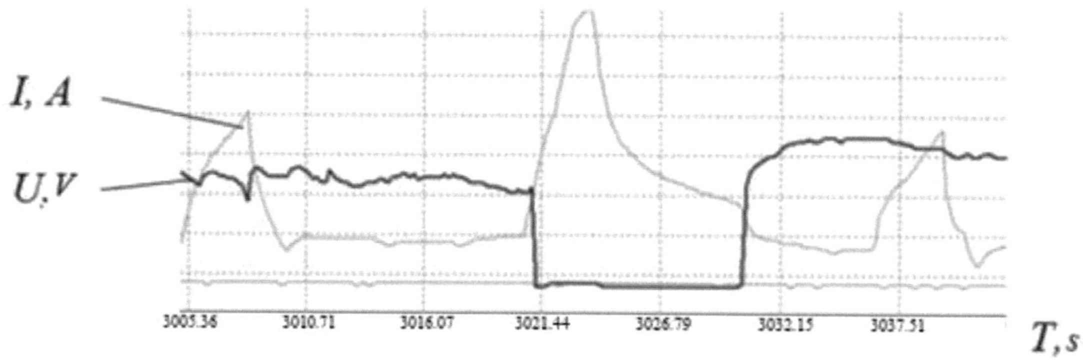

Fig. 3. Current and voltage oscillograms for modulated current welding, showing forced transfer of an electrode metal droplet during a pause by the energy of one additional pulse

Figure 4 shows a part of the welded joint, as a practical result of using modulated current welding in the framework of the concept of "machine-man-technology" [5, 6, 7 ,8].

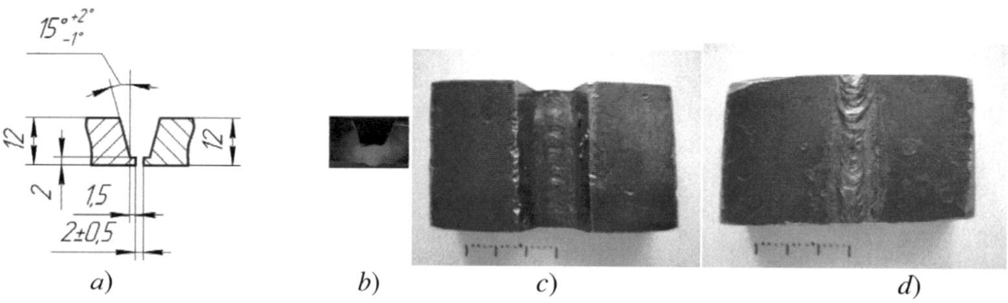

Fig. 4. Welded seam made by modulated current welding within the framework of the concept "machine-man-technology"; Electrodes SSSI-13/55 with a diameter of 3 mm; Plate 220x70x12mm (09GS); The spatial position when making the seam - ceiling; Modulation parameters: Ion and i = 1 and ≈ 170 A; Ip ≈ 30 A; T.n. and 55 ms; T.ns.p ≈ 200 ms; T.i≈ 6 ms; Thickness and ≈ 380 ms; Td.i≈ 20 ms (50Hz).

a) configuration of cutting edges;b) macrosection of the root suture;c) the front side of the seam;d) reverse side of the seam.

Conclusions

1. The application of the pulse-controlled welding process according to the concept of "machine-man-technology" allows the welder to regulate the volume of the weld pool, and therefore perform welding in all spatial positions of the weld, including electrodes, welding in spatial positions other than the lower one, such As corrosion-resistant alloys on the basis of nickel, wear-resistant and other materials with special properties.

2. The regulation of the volume of the weld pool is performed by the welder by changing the arc length with respect to the specified voltage. Accordingly, the welder is able to control the welding speed by changing the pulse time of the main pulse, or the main pause, or by a joint change in the durations of the main pulse and pause.

3. Melting of coated electrodes at rated current makes it possible to provide their best welding and technological properties, namely, to influence the oxidation-reduction processes during electrode melting, during the transfer of electrode metal droplets and at the stage of the weld pool.

4. Ensure the stability of arc burning during the period of the pause current (Ip = 15 ÷ 30 A), by superimposing additional pulses with a frequency of at least 50 Hz and amplitude equal to the amplitude of the main pulses.

IOP Conf. Series: Materials Science and Engineering **253** (2017) 012025 doi:10.1088/1757-899X/253/1/012025

5. A concomitant effect of the application of additional pulses is the elimination of the effect of the variable light flux of the impulse process on the visual function of the welder.

References

[1] Zaruba I. I., Lebedev V. K., Shejko P. P., BasanskijV.V. Svarkamodulirovannymtokom // Avtomaticheskayasvarka. 1968. 11. p. 35-40.

[2] Knyazkov V.L. Povysheniееffektivnostiruchnojdugovojsvarkitruboprovodov / V.L. Knyazkov, A.F. Knyazkov, GU KuzGTU, Kemerovo 2008. 104 p.

[3] Kuznetsov V.D., Stepanov D.V. Strukturaisvojstvametallasvarnogoshva, modifitsirovannogonanooksidami. // Avtomaticheskajasvarka. 2015. 11. p.19 - 24.

[4] Pat. rf 2268809, MPK V 23 K 9/095. Sposobelektrodugovojsvarki s impulsnojmodulyaciejtoka / A. F. Knyazkov, V. L. Knyazkov, S.A. Knyazkov - Zayavl. 12.02.2004.

[5] Dementsev K. I. Povysheniееffektivnostirabotysvarochnyxpreobrazovatelejinvertornogotipazaschetmodulyaciisvarochnogotoka: diss. …kand. texn. nauk. Tomsk, 2010. 172 p.

[6] Sidoruk V.S., Dudko D.A., Zatserkovnyj S.A., Taraborkin L.A. Vlijanieparametrovrezhimaruchnojdugovojsvarkimodulirovannymtokomnapervichnujustrukturumetallashva. // Svarochnoeproizvodstvo. 5. 1988. p.4 - 6.

[7] Dudko D.A., Zatserkovnyj S.A., Sidoruk V.S., Taraborkin L.A., Mahlin N.M. Zavisimost' himicheskogosostavametallashvaotparametrovrezhimadugovojsvarkimodulirovannymtokom. // Avtomaticheskajasvarka. 1989. 2. p.27 - 29.

[8]Larionov V.P., Sleptsov O.I., Mihajlov V.E. Osobennostitehnologiisvarkinizkolegirovannyhstalejpovyshennojivysokojprochnostiprinizkihtemperaturah // Dokl. na XIX komissiistran-chlenov S`EV poprobleme "Svarki". PNR, 1986, 44 p.

Prospects for the Development of Innovative Technology of Supersonic Gas-Powder Surfacing Coatings from Alloys of the System Ni-Cr-B-Si

Mikhail V Radchenko [a], Vadim S Kiselev [b], Yuri O Shevtsov [c], Tatyana B Radchenko [d]

Altai State Technical University after I.I. Polzunov, Lenin av., 46, Barnaul, 656038, Russia

E-mail: [a] mirad_x@mail.ru, [b] smallal@inbox.ru, [c] yuoshevtsov@mail.ru, [d] radchnko_tb@mail.ru

Abstract. The article is devoted to the research and development of technological recommendations of supersonic gas-powder surfacing, an example of the practical use of the developed technology is given. Market research indicates that demand for these products is stable due to the growing quality requirements of components and parts produced by the largest machine-building enterprises of Russia.

Introduction

The worldwide practice of the application of protective coatings is widely used in industry due to technical and economic considerations. In Russia a number of companies and organizations deal with the implementation of different methods of applying protective coatings. So the Altai Federal research and production center "Altai" (Biysk city) and the Association "Interregional Centre for Nanoindustry" engaged in research and development in the field of technologies of galvanic processes of application of anti-corrosion and hard coatings [1]. In addition, galvanic coating methods are widely implemented at JSC "Altai precision products plant ", LLC "Altai fuel equipment plant", JSC AAP "Rotor" (Barnaul city).

However, galvanic methods of deposition of protective coatings also have features, limiting their broader use in industry.

Quite effective alternative of galvanic coating are gas-thermal methods of applying protective coatings: technology of plasma, gas-flame methods of spraying and surfacing, detonation spraying, spraying with subsequent fusing tamifluswine materials [2]. However, the relatively high cost of the proposed equipment (mostly manufactured abroad), not always acceptable characteristics of the coating and considerable energy consumption in its production prevent the widespread implementation of the proposed technology.

To obtain new technological characteristics of the coating in the LLC "Research Institute of High technologies" (Barnaul city) in the framework of the Federal program "START" equipment and technology of supersonic gas-powder surfacing (SGP-surfacing) are developed and patented [3-6]. To implement this process a complex of studies was performed, including calculations of Laval nozzles with different Mach numbers and the production of a series of nozzles and their experimental testing in the process of welding durable powder alloys of the system Ni-Cr-B-Si on heavily wear in the operation process of the workpiece surface.

Results and Discussion

At a later stage, a comprehensive analysis of the derived process of supersonic gas-powder surfacing was made [7], the algorithm is presented in figure 1.

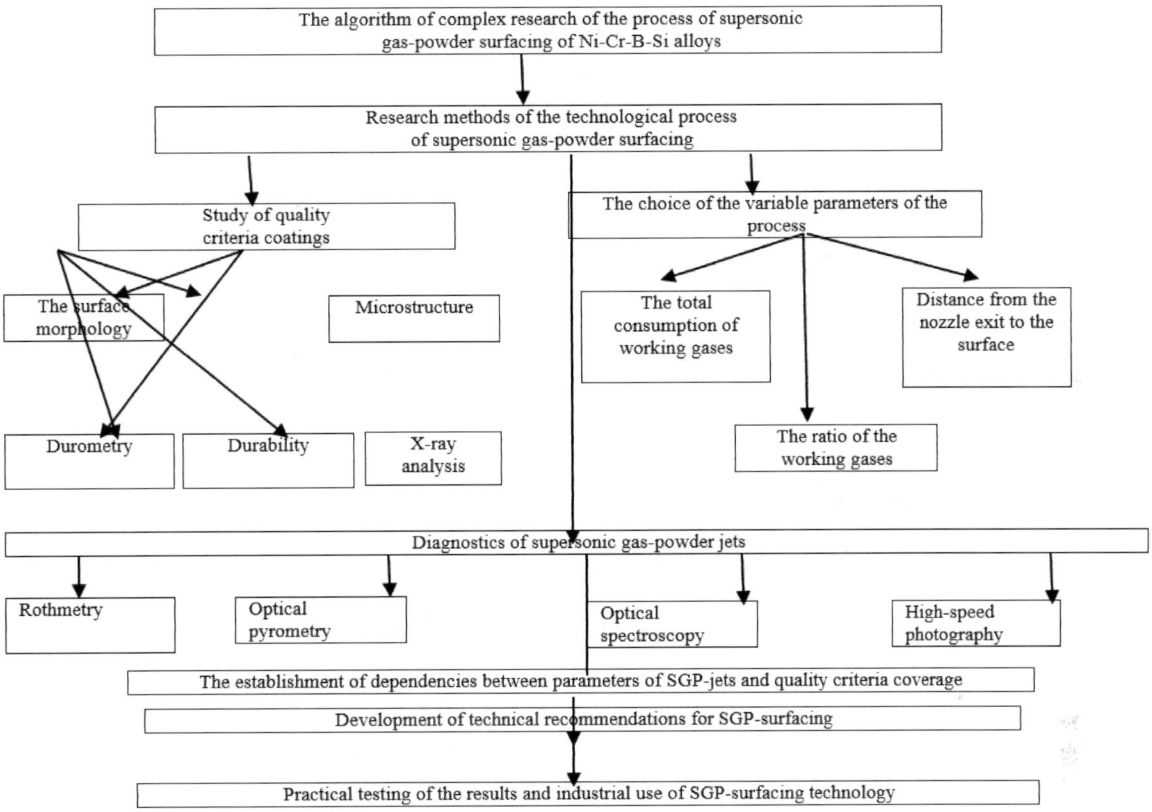

Figure 1 – Algorithm of complex research of the process of supersonic gas-powder surfacing

The conducted complex research allowed us to formulate technological recommendations for SGP-powder surfacing of alloys of the system Ni-Cr-B-Si, which are shown in table 1.

Table 1 - Technological recommendations for SGP-surfacing

1. The range of pressures of the working gases:
Oxygen: 4.5 to 5 ATM; Propane: 1.0...3.0 ATM.
2. Consumption of working gases: Oxygen: 1.9...6.5 l/min Propane: 1.5...5 l/min.
3. The coefficient β of the ratio of the working gases: 1.1...1.3.
4. The working distance L from the nozzle exit to filler items: 20...30 mm.
5. Powder alloy: the system Ni-Cr-B-Si brand PG-SR3.
6. The fraction of alloy powder: 40...100 microns.
7. The performance of the powder feeder: 50 g/min.
8. The moving speed of the burner: 3...5 mm/s.

LLC "Energotech" (Barnaul city) implemented the technology of supersonic gas-powder surfacing wear-resistant coatings on most wear surfaces of the cavitation pump grinders produced by this company [8, 9]. The cavitation mechanism of grinding leads to extremely intensive wear of the blades of the primary node installation - the impeller (figure 2).

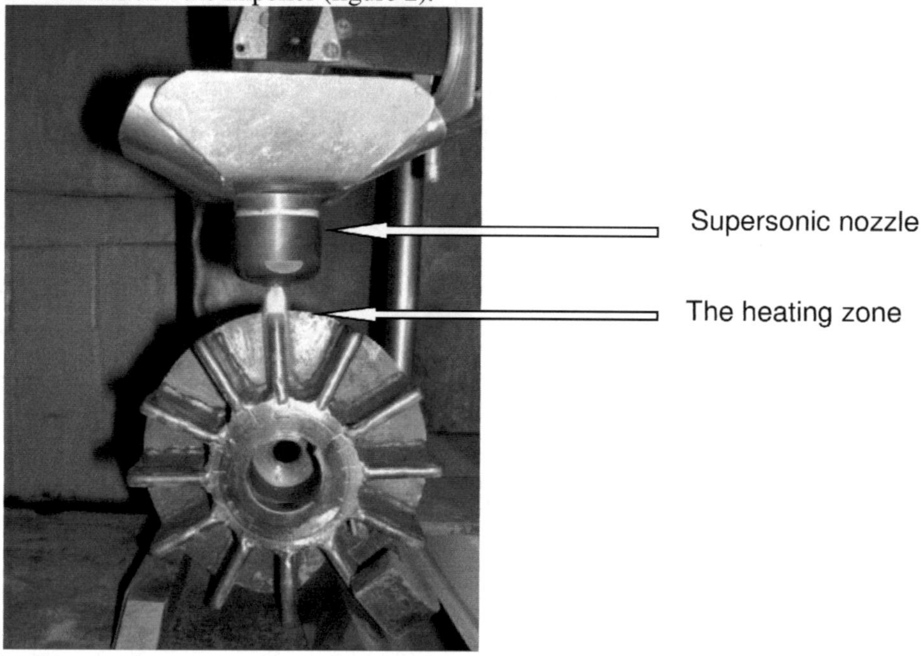

Figure 2 – Supersonic gas-powder surfacing of the mechanical and radial surfaces of the impeller blades

Under the contract between LLC "Research Institute of High technologies" and the factory LLC "Energotech" surfacing of wear-resistant alloy powder PG-SR3 on the end and radial surfaces of the impeller blades most exposed to water-jet cavitation wear was produced on the technical task. Surfacing of all the planes of the impeller was performed in the lower position (figure 2).

The wear resistance of the surfaces of the pump grinder with the deposited coating as compared with 45 steel was increased by 6...8 times. The estimated economic effect of the introduction of the developed technologies on a single impeller is 16.000 rubles for one pump per year.

As a further development of the technology of SGP-surfacing employees of the LLC "Research Institute of High technologies" considered the possibility of creation of technology of supersonic gas-powder surfacing of protective coatings using the combined nanostructured materials based on tungsten carbide and with the use of industrial diamonds based on Nickel and cobalt binder. These technologies will allow us to apply the method of SGP-surfacing not only to protect wearing parts of various engineering equipment, but also to create sample instruments for metal processing with the desired properties of the processing surface.

As a result of the research prototypes of deposited coating alloy system Ni-Cr-B-Si by SGP-surfacing with the subsequent introduction of a cluster of diamonds of varying fractions (40/28, 50/40, 63/50) into the surface coating layer were produced. The saturation of the Nickel matrix with diamond grains allowed us to obtain a new type of combined coating that combines the advantages of flame and diamond-galvanic coverage.

The base material is fused with the coating, thereby creating a monolithic connection, and the introduction of the strengthening phase in the form of diamonds has increased the hardness and wear resistance by one order. Figure 3 shows photographs of the surfaces of the Nickel-diamond coatings obtained in various ways.

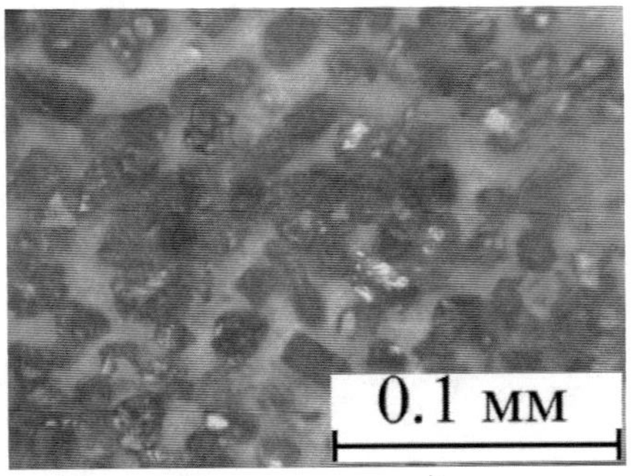

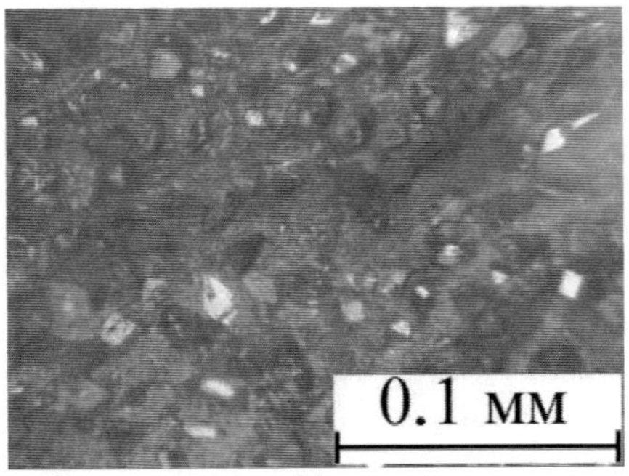

Figure 3 - the surface of the Nickel-diamond coatings obtained by means of:
a) electroplating b) SGP-surfacing

Conclusions

The analysis showed that for similar properties of a surface, the coating obtained by means of SGP-surfacing has no tendency to peel due to the formation of chemical bonds with the substrate, as well as there is no tendency for chipping of diamonds, since the introduction of diamonds into the coating at high temperature leads to the transition of diamond to graphite and its dissolution in the Nickel matrix with the formation of the thinnest boundary layer between the diamond grain and the coating material. The temperature characteristics of the flame and the temperature of the substrate in the rework process indicate the qualitative penetration of the substrate and coating with embedded diamond grains [10].

Market research indicates that demand for these products is stable due to the growing quality requirements of components and parts produced by the largest machine-building enterprises of Russia. Prospects of development of the market allow us to predict the increase in sales of products and services coating.

References

[1] Information on http://altainano.ru.
[2] Information on http://www.tspc.ru.
[3] Borisov Yu. S., Petrov S. V. the Use of supersonic jets in the technology of thermal spraying // Automatic welding. 1 (1995) 41-44.
[4] Khromov V. N., Vertov V. G., Korovin, A. Y., A. Y. Alasheev From subsonic to supersonic gas-flame spraying of coatings with the restoration and hardening of machine parts (review). Welding production. 2 (2001) 39-48.
[5] Baldaev L. H., Shesterkin N. G., Lupanov V. A., Shatov A. P. The features of high-speed flame spraying. Welding production. 5 (2003) 43-46.
[6] Frolov V. A., Poklad V. A., Ryabenko B. V., Viktorenko D. V. Technological features methods of thermal spraying by supersonic (overview). Welding production. 11 (2006) 38-47.
[7] M.V. Radchenko, Yu.O. Shevtsov, T.B. Radchenko, V.V. Ignatyev Patent 2346077 Russia, Method of supersonic gas-powder surfacing of protective coatings (2007).
[8] M.V. Radchenko, Yu.O. Shevtsov, A.J. Suranov, V.S. Kiselev, V.S. Smolin, S.G. Uvarova, Diagnostics of supersonic gas jets in the processes of gas-powder surfacing of protective coatings, Welding and diagnostics. 2 (2009) 31-36.
[9] Information on http://www.enerteh.ru/.
[10] M.V. Radchenko, D.A. Nagorny, Yu.O. Shevtsov, T.B. Radchenko, Development and manufacture of the device for a supersonic gas-powder surfacing and analysis of its technological characteristics. Polzunovskii Vestnik. 4/2 (2013) 55-59.

IOP Conf. Series: Materials Science and Engineering **253** (2017) 012027 doi:10.1088/1757-899X/253/1/012027

Research Nature of Damage Parts Machines for the Purpose of Optimization Technology Recovery

Victor L Knjaz'kov[1], Elena E Petrova[1, a], Maksim V Pimonov[1], Nicholas V Melkozernyh[2]

[1] T.F. Gorbachev Kuzbass State Technical University,
28, Vesennyayast., Kemerovo, 650000, Russia
[2] AO Holding Company "SDS-Ugol"

E-mail: [a] pee5152@yandex.ru

Abstract: The results of metallographic studies and microhardness damaged sections detail "axis" the excavator P & H 2800 XPC, subjected to accelerated deterioration the local. Lots of studies to be removed during the repairs identified by magnetic noise control and acoustic spectral analysis. A repair technology.

Introduction

High productivity of mining equipment operating under extreme conditions is provided by a complex of mechanical and special properties machine parts such as resistance to impact loads at low temperatures, abrasion resistance of the friction surface [1, 2, 3, 4, 5].

The investigated "axle" part of the P & H 2800 XPC excavator is the most loaded element in the hub-axis system, which receives unevenly distributed loads during operation, which leads to their accelerated wear (Fig. 1). The technology of manufacturing the "axis" part is much more complicated and more expensive than the details of the bush, so it is more economical to restore the "axis" part.

a) b) c)

Fig. 1. Excavator P & H 2800 XPC - a;the nature of the damage is divided by the axis-b, the nature of the damage of the detail of the bush - c

The purpose of the work is the timely and accurate determination part boundary at the stage previous destruction, the development of the technology applying the wear-resistant layer that is optimal in size and properties, reducing the cost of repair, and improving its quality.

IOP Conf. Series: Materials Science and Engineering **253** (2017) 012027 doi:10.1088/1757-899X/253/1/012027

In accordance with the technical documentation, the warranty period for the "axis" part provided by a foreign manufacturer is 1 year.

The high cost of the "axis" component (more than 1.000.000 rubles) is the rationale for developing an effective technology for repairing this part.

Materials and techniques of the experiment

To determine the chemical composition of the main and welded metal, the "axis" part (before repair, during the selection of the wear-resistant layer material and after repair) was performed by spectral analysis using the Q4 TASMAN optical emission spectrometer.

Spectral acoustic analysis, with the automated acoustic system ASTRON, was used to identify the zones with a changed structure, as well as the degree of its change, measurements were made on a transverse sample at 11 points at a distance of 9 mm.

The magnetic noise control method was performed with the INTROSCAN instrument, with a voltage analyzer and a metal structure. The studies were carried out on a transverse sample at 9 points at equal distances from each other.

For the metallographic study, the plane of the sample was aligned by successive grinding, polishing and etching. Micro-alysis were performed on an Axio Observer microscope with an increase of × 130, × 500, × 800. The etching of the samples was carried out with a 4% solution of HNO_3 and 2% HNO_3 in ethyl alcohol.

The microhardness was measured on a PMT-3M device [6] with an increase of 100 times by indenting the tip (quadrilateral pyramid with a square base), under a load of 1.96 N applied for 20 s, the analysis of the data was carried out according to [7].

Results and discussion

The chemical composition of the factory surfacing is determined, %: C – 0.34; Si – 0.37; Mn – 1.2; Cr – 5.0; Mo – 0.9; Ni – 1.0; V – 0.15; W- 0.9, the thickness of which is about 10 mm. The chemical composition of the base metal %: C – 0.43; Si – 0.22; Mn – 0.5; Cr – 1.2; Mo – 0.3; Ni – 4.1.

Below are the results of investigations the surface layer a part by a magneto-noise method (Fig. 2) and acoustic emission analysis (Fig. 3).

At point 8 of the measurement delay time acoustic waves (Fig. 3), changes in their lengths can be observed. At point 6 (Fig. 2) there is a jump in the indices of the magnetic noise. Indications may indicate the presence of structural changes and possible microcracks.

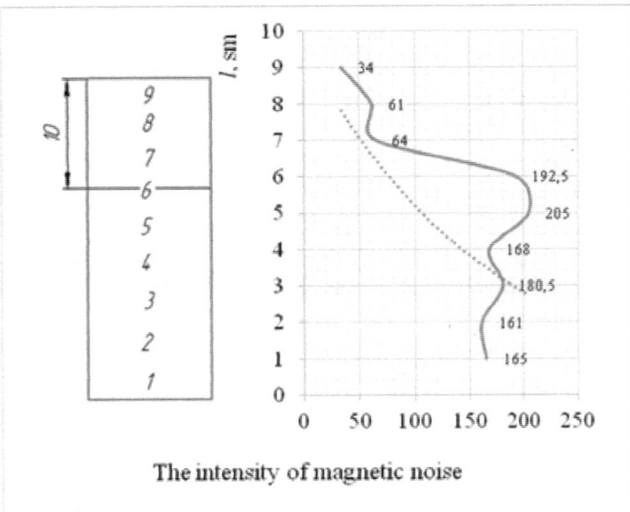

Fig. 2. The results of measurements of the magnetic noise level of the sample surface

Zones with a change in the amplitude of acoustic waves and the intensity of magnetic noise on a sample with an imported surfacing were found, samples for microstructure research and microhardness measurements were cut from this zone, the following results were obtained.

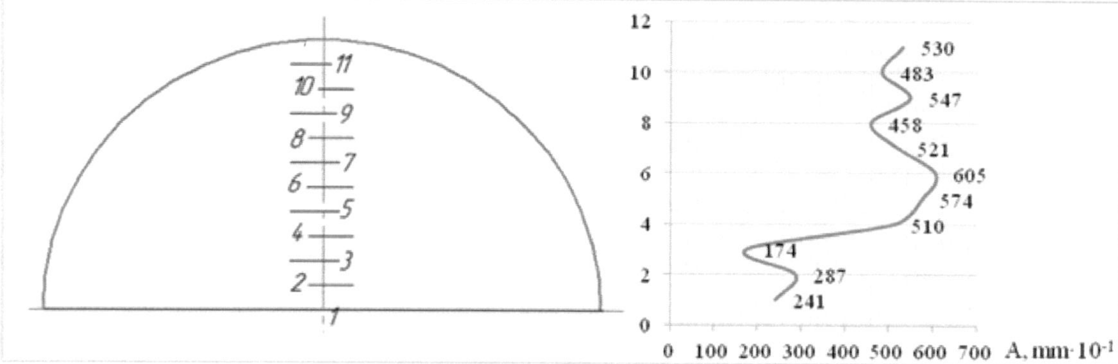

Fig. 3. Graph of values of delay time of amplitudes of acoustic waves

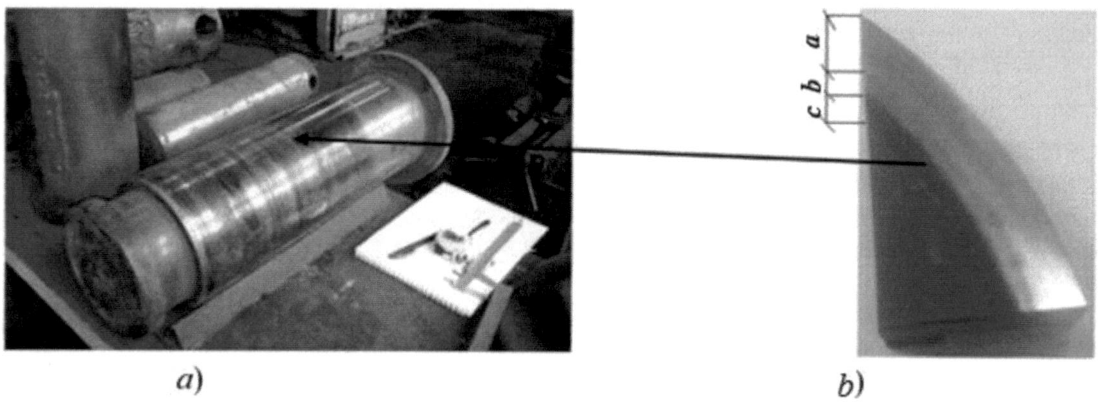

Fig. 4. Worn detail of the "axis" of the pair of bushing-axis of the P & H 2800 XPC excavator excavator – a, a sample with an imported surfacing – b

In the metallographic study, the interface between the layers was revealed, as well as the microstructure of the deposited and base metals (Fig. 4, b). In the cross-section, the initial sample of the "axis" part consists several layers (Fig. 4, b) a = 1.12 mm, b = 2.61 mm, c = 7.92 mm, which differ in their degree of etching, microstructure and size Grains. The structure of surfacing at the boundary of the HAZ and the base metal is a ferritic cement mixture with dendrites that transform into a weld metal, the microhardness is from 305 to 501 HV. The boundary of the HAZ and the welded metal layer, is characterized by high microhardness heterogeneity, which is from 438 to 590 HV, the structure is martensite. Increased, relative to other areas of surfacing, hardness is probably due to an increased cooling rate of the lower part of the liquid phase of the bath.

Microcracks (Fig. 5, a) with a size of 0.14 were found in the section "c" (Fig. 4, b); 0.05; 0.06 and 0.09 mm, the layer is characterized by a sharp transition of microhardness from 305 to 501 HV (HAZ). Microcracks are oriented in one direction in the direction of tensile stresses, but at different angles. In the zone "b" (Fig. 4, b), microcracks were also detected, but in a smaller quantity than in the section "c" (Fig. 4, b), the size reached 0.2 mm, while the microhardness values were From 540 to 438 HV (Figure 5, d, e).

The etching of layer "a" (Fig. 4, b), subjected to shock loads, was performed in a 2% solution of HNO_3 ethyl alcohol and investigated at an increase of 130 and 800 times, a deformed metal structure was revealed.

Based on the results of metallographic studies, micro cracks were detected in areas with acoustic waves and the level of magnetic noise, and the boundaries site to be sampled were determined.

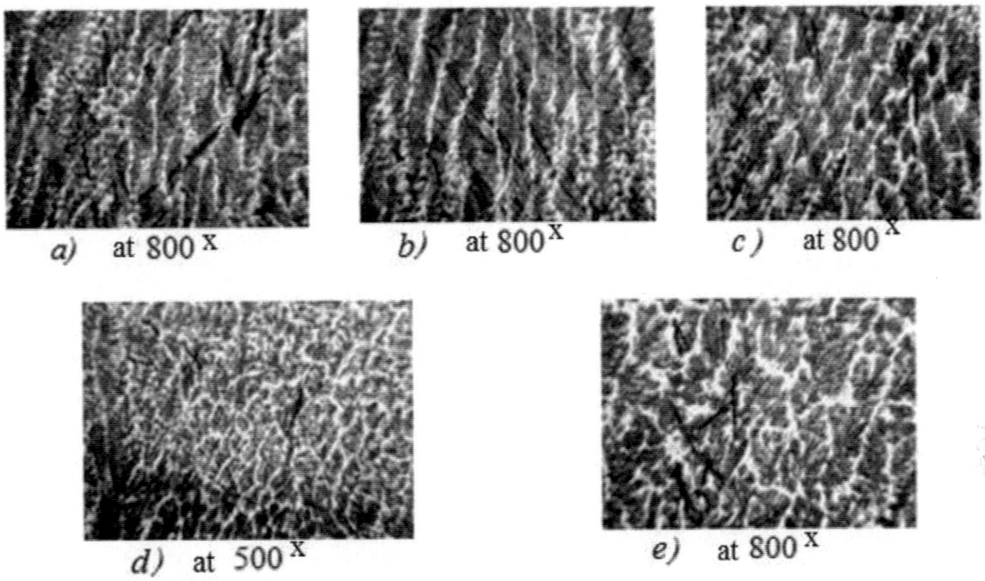

Fig. 5. Microstructure of HAZ and weld metal with microcracks:
a, b - microhardness from 305 to 501 HV; c - microhardness from 499 to 416 HV;
d - microhardness from 416 to 590 HV; e - microhardness from 540 to 438 HV

To restore the details, the flux-cored wire wasused by NIIMONTAZH, Krasnodar city. Surfacing was carried out under a layer of AN-60 flux, 35X5GFNVM flux-cored wire 2.0 mm in diameter, current - 250 A, voltage 26-28 V. Before surfacing, the original layer was not removed.

The results of the studies by spectral-acoustic analysis and the magneto- noise control method are presented below (Fig. 6, 7).

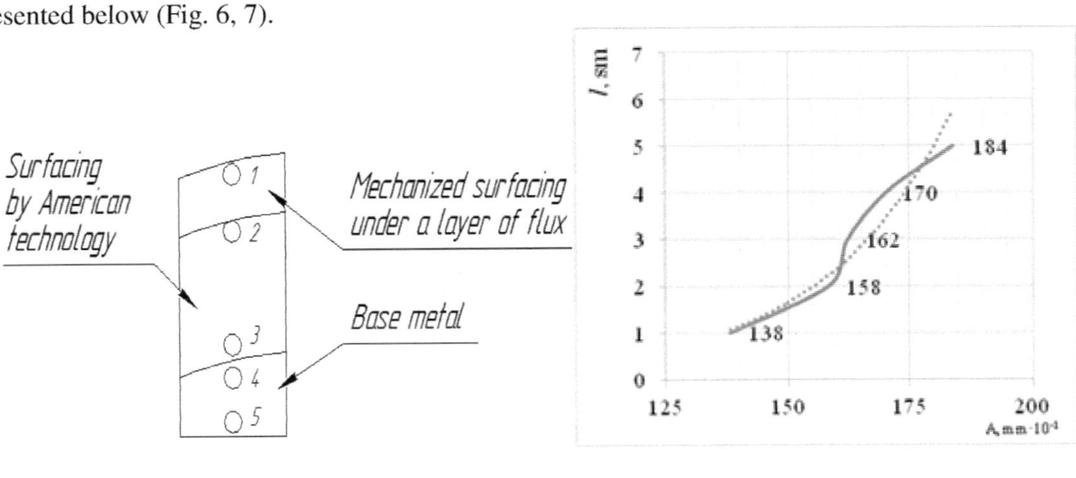

Fig. 6. Graph of the delay time values of the amplitude of acoustic waves, the reconstructed part

The graphs (Figures 6, 7) show that there are no zones with a sharp change in the lengths of acoustic waves and the level of magnetic noise in the section with mechanized surfacing. This indicates that there is no change in structure and mechanical properties.

The metallography of the deposited layer was performed, the structure is fine-grained, the grain number is 7 [8]. Defects in the fusion zone of the initial (import) and reductive surfacing were found. The values of the microhardness deposited layer reach, on average, 450 HV.

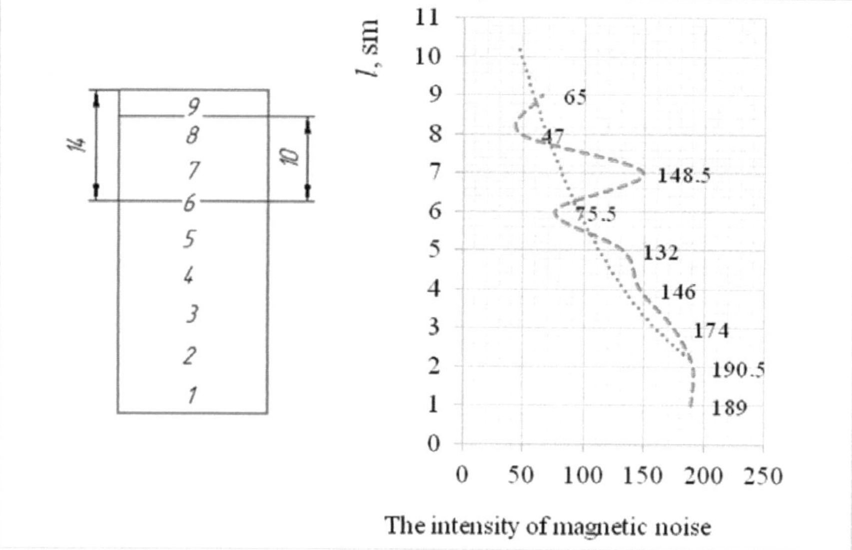

Fig.7. Results of measurements of the level of magnetic noise, the surface of the restored part

Conclusions

1. The untimely repair of the details of the LOW machines leads to their accelerated wear.
2. The boundaries of separation of layers, identified by acoustic method and metallographic studies, coincide.
3. The use of a magnetic-noise control method and spectral-acoustic analysis helps to quickly determine the boundaries of the damaged area, reduce the amount of weld metal, reliably determine the quality repair.

This work was supported by a grant from the RNF, agreement № 14-09-00724 and a grant from the President of the Russian Federation to support young candidates of sciences, MK-1341.2017.8.

References

[1] Slepcov O.I. perspektivy razvitiya metallurgii v yakutii v svyazi s sozdaniem novyx pokolenij stalej severnogo ispolneniya/ B.S. Ermakov, O.I. Slepcov, P.P. Petrov// NAUKA I OBRAZOVANIE. 2015. 3. p. 59-63.

[2] Naplavochnyj material s `effektom vtorichnogo tverdenija dlja povyshenija stojkosti metalloobrabatyvajuschego instrumenta / E.N. Eremin, A.S. Losev, A.E. Eremin, G.P. Rumjantsev // Porodorazrushajuschij i metalloobrabatyvajuschij instrument - tehnika i tehnologija ego izgotovlenija i primenenija: Sb. nauch. tr. K.: INM im. V.M. Bakulja NAN Ukraïni, 2011. Vip. 14. p. 453 - 455.

[3] Larionov V.P., Sleptsov O.I., Mihajlov V.E. Osobennosti tehnologii svarki nizkolegirovannyh stalej povyshennoj i vysokoj prochnosti pri nizkih temperaturah // Dokl. na XIX komissii stran-chlenov S`EV po probleme "Svarki". PNR, 1986, 44 p.

[4] N.V. Kobernik, R.S. Miheev, A.S. Pankratov, A.A. Linnik. Sovremennye predstavlenija o modifikatsii naplavlennogo metalla i metalla shva nanorazmernymi chastitsami. // Svarka i diagnostika. 2015. 5 p. 13 – 18.
[5] S.O. Gordin, A.N. Smirnov, V.L. Knjaz'kov. Sostav `elektrodnogo pokrytija dlja iznosostojkoj naplavki. // Vestnik Kuzbasskogo gosudarstvennogo tehnicheskogo universiteta. 2015. 2. p. 106 108.
[6] GOST R ISO 6507-1-2007. Metallyisplavy. izmerenie tverdosti po Vikkersu. Chast 1. Metodizmereniya. Moskva: Izd-vo Ctandartinform, 2008. 2 p.
[7] GOST R ISO 6507-4-2009. Metallyisplavy. Izmerenie tverdosti po Vikkersu. Chast 4. Tablicy opredeleniya tverdosti. Moskva: Izd-vo Ctandartinform, 2010. 25 p.
[8] GOST 5639-82 Staliisplavy. Metody vyyavleniya i opredeleniya velichiny zerna. Moskva: Izd-vo IPK Izdatelstvo Standartov, 2003. p. 21.

The Management of Short Circuits' Mechanism of CO_2 Shielded Welding

Anatoliy F Knyaz'kov, Sergey A Knyaz'kov, Artur M Tabanov

National Research Tomsk Polytechnic University, 30, Lenin Street, Tomsk, 634050

E-mail: kaf@tpu.ru

Abstract: It is observed the mechanism of short circuits' management in the process of which drops of electrode material move to welding pool and all the phases of drops' formation process are under control. It is shown the role of drop of electrode material and welding pool in the decreasing arc space, providing forced short circuit of arc space at the time of sharp decrease of welding current. Obtained results evidence that it is necessary to consider the movement of welding pool at the time of creation of model of electrode materials movement and creation of process management's algorithms, especially considering changeable position of welding pool, as the role of drop and welding pool in the decreasing arc space is almost the same.

Introduction

The process of CO_2 shielded consumable electrode welding [1] got a wide use in the industry due to its advantages. But existing disadvantages in some cases limit the extension of its application area.

The analysis of the process of CO_2 welding shows that large exposure of short circuits' mechanism to external disturbances is the main reason of these disadvantages. The process of metal transfer during short-arc welding exists during short circuits (further named SC). The study of SC's mechanism and the possibility of its management is an actual aim because the process of CO_2 welding with SC is one of the mass welding processes.

According to modern concepts, SC's mechanism can be imagined in the following way. The main reason of SC is that welding is led along the length of arc just the same as the size of drop. Decreasing length of arc, leading to short-circuiting of arc space, is determined by the following reasons:

- Increase of drop's size;
- Drop's intention to take coaxial position with an electrode;
- Electrode feed;
- Counter movement of welding pool before short circuit in the process of its oscillation under forces' impact, appearing during SC and rupture of short-circuiting jumper.

The study of current's oscillograms and voltage of arc space shows that SC's duration and current at the moment of its termination are different during every certain SC, so extension of arc space during short-circuiting jumper's rupture is unstable for different periods of drops' formation that it generates non-periodicity of drops' formation and leads to different primary conditions of drops' formation process. It leads to impairment of welded seam's formation especially in the positions different from low one and increase splashing. It should be mentioned the significant role of welding pool's oscillationsin the short circuits' mechanism. Welding pool has ultimate amplitude and frequency of its oscillations,

IOP Conf. Series: Materials Science and Engineering **253** (2017) 012028 doi:10.1088/1757-899X/253/1/012028

which cannot be exceeded as it can lead to interruption of process. It is especially dangerous in the positions of welding pool different from low one.

It is difficult to describe the process mathematically and to calculate the moment of short circuit's start, so while studying CO_2 welding process with SC people use the methods of mathematical statistics.

The primary conditions, appropriate to the moment of SC's start, are unstable (the sizes of drops, their coaxiallity with an electrode, the phases of welding pool's oscillations, current level, arc space's length, etc.). It leads in its turn to instability of regime's parameters at the time of SC as well as at the time of drop melting, producing in its turn instability of the primary conditions. Due to this welding process is characterized by:

- large exposure to external disturbances and connected with this one instability of process (period of drops' formation);
- low regulating abilities, as process proceed normally in the narrow range of arc space's voltages;
- increased splashing;
- non-optimal form of seam.

Materials and methodsof research

Many works are known in the literature that directed to removal of these disadvantages. they can be divided in some groups according to principle of impact on welding process:

- stabilization of average parameters of regime ($U_d I_d f_k$);
- impact on regime's parameters during particular phases of period of drops' formation [6,7,9,11];
- stabilization of period of drops' formation at whole and at the same time of particular stages based on impulse supply.

In our opinion the last variant is more perspective. the meaning of process is forced providing of short circuits by impulse management of energetic parameters of welding arc.

Nowadays priority directions of removal of the mentioned disadvantages and of improvement of CO_2 shielded metal welding process are creation of methods and means, providing management of melt kineticsand electrode metal's transfer during period of drops' formation by instantaneous values of regime's parameters [10,11].

So in the work [2] it was proposed to enter periodically current decrease for forced formation of SC with the aim of drop's transfer to welding pool and to providing their almost periodic transfer. Further the amount of algorithms of drops' transfer management increased notably. It was developed so many methods of management of short circuits' mechanism by particular phases of drops' formation as well as by period of drops' formation at whole in all the phases.

For creation of models of short circuits' mechanism, of algorithms of their management and systems of welding arc's supply it is necessary to know the role of drop and welding pool in the decreasing arc space during sharp current decrease, providing forced SC and drop's transfer to welding pool. But in scientific and technical literature except [3], there is not detailed analysis of drop's movement and welding pool and their shared parts in arc space's decrease and SC's creation during sharp current decrease. In some works it is shown the results of welding pool's movements during ordinary CO_2 welding process with SC, but there are not such data during controlled SC's mechanism [12]. While reliable information about this process is very important for studying kinetics of melting and transfer of electrode metal, especially in the positions different from low one.

The aim of this work is determination of the role of drop and welding pool in the creation of SC of arc space during CO_2 welding.

For this purpose we should observe the process [4], for which in the picture 1 synchronous kinograms with the oscillograms of current and voltage are represented. At the present time this process abbreviated as STT is widely used in the welding practice.

In this process for drop's transfer in the welding pool forced SC are created by current decrease until the meaning of a few dozens of ampere. At the same time the forces that influence on the drop and welding pool acutely fall. Metal of welding pool along the surface of crystallization's front, partly crystallized, under its own weight start to return under an electrode. The drop, pressed by forces' action, aims to take coaxial position with an electrode. In the results of counter movements of drop and welding pool, their contact appears in the first touching, as welding current is minor. The creation of stable jumper appears between continuous electrode and welding pool.

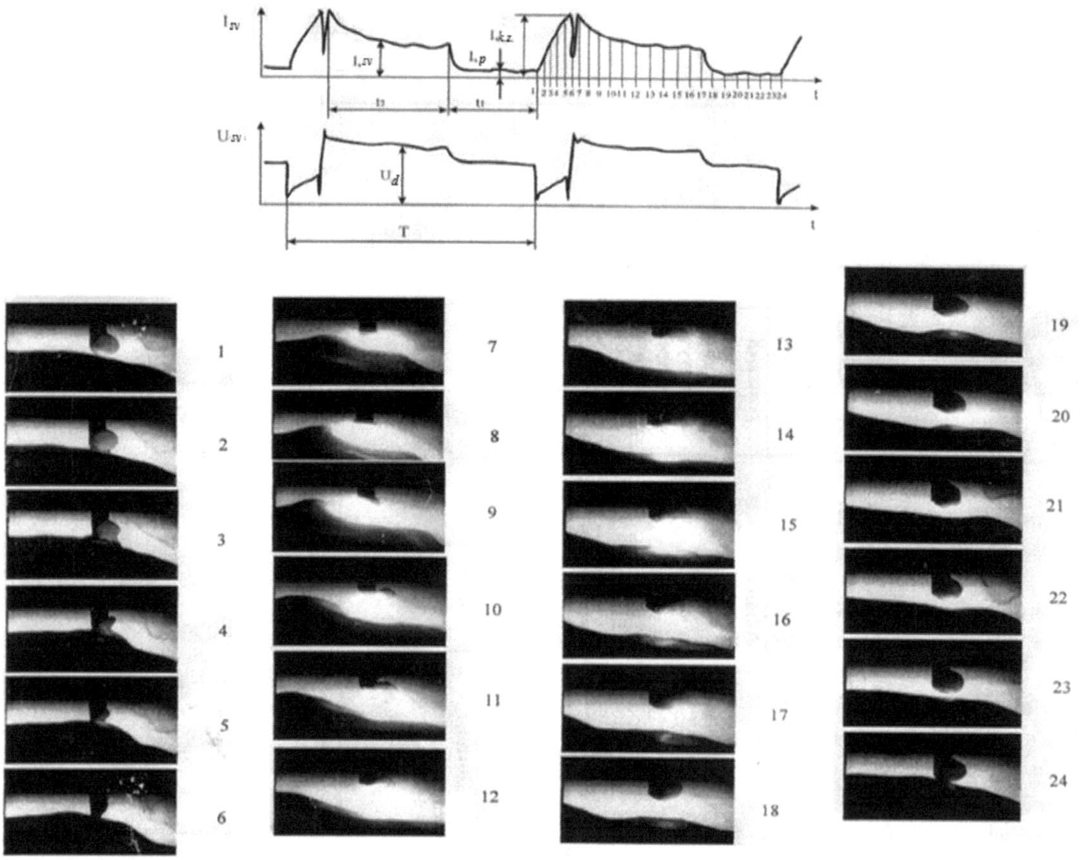

Fig.1 Oscillograms of current and voltage of welding cycle with controlled SC's mechanism synchronously with kinograms (U_i=37 V, U_a=26-26.5 V, I_w=155-165 A, V_w=15 m/h, L_e=11 mm.)

The drop of electrode metal transfers into welding pool under influence of the complex of forces. The jumper appears at the time of rupture of which current decrease until 15-30 ampere.

From the moment of jumper's rupture and arc starting at the moment of time t_7 the energy is dosed, that is used for drop's melting (the variant of dosing by melting's time). After melting drop of certain size at the moment of time t_{17} welding current decrease until the meaning of a few dozens of ampere, providing forced short circuit. On the interval from t_7 to t_{17} it is decreasing the length of arc space till its shorting by supply of wire, counter movement of welding pool and movement melting drop to the coaxial position with an electrode.

As on this interval arc pressure is minor due to small amount of current, so the role of welding pool's movement (as well as drop) greatly increases compared ordinary process.

At the moment of time t_{24} (as well as at the moment of time t_1) next SC happens, and as to this moment primary conditions almost similar for every process of drops' formation, so far as on the interval from t_7 to t_{17} energetic parameters are unchangeable, so current curve of SC will be almost the same for every SC, as well as SC's duration.

So kinetics of processes of drops' formation is drawn into synchronism with the change of energetic parameters. The complex of actions, describing above, provides similar primary conditions at the moment of SC's startand during the whole periodic process.

The results of the study and their discussion

This process as well as ordinary one is characterized by average current and arc space's voltage and is controlled by indicated instruments. It is necessary to install the following parametres:

U_i- voltage of idling (V); V_f- speed of electrode feed (m/h); V_w- speed of welding (m/h); L_e- electrode extension (mm); t_a- time of arcing for drop's melting (msec); $I_{p.a.}$ - pilot arc's current (A); α - angle of electrode's inclination (°).

The following parameters will be derivative:

I_a- average current, $U_{a.a.}$ - average voltage of arc, I_{max} - maximal current of short circuit, I_{min} - minimal arcing, $U_{a.max}$ -maximal voltage of arcing, $U_{a.min}$ - minimal voltage of arcing, $I_{a.a}$ -average current during welding cycle (period of drop's transfer), $U_{a.a}$ - average voltage of arcing during welding cycle (period of drop's transfer), t_{sc} - duration of short circuit, t_p - duration of pause before SC, b - seam's width, g - height of seam's strengthening , h - depth of penetration.

Regime's installation on equal conditions should be made by two parameters of regime U_a and t_a. The peculiarity of welding process with management of short circuits' mechanism is overvoltage of idling compared with ordinary process, that allows to regulate the seam's width in a wide range. The pause of current before SC determines current increase on the interval of arcing, that increases intensity of self-regularity and impact on the main metal, strengthens welding pool's oscillations and contributes some settingof welding pool from an electrode. Due to current is lower at the moment of SC's start than in the ordinary process, it can be admitted high speeds of current increase of SC.

Scheme of measurement, represented in the picture 2, is used for identification of drop and welding pool's participation in the creation short circuit. For measuring the point of crossing of electrode's axis and top edge of frame is considered as fixed point of countdown, as far as top edge of frame is fixed concerning image, as well as side edges of frame concerning electrode's axis.

But it is not convenient to measure this point, as charts, taken by traditional position of axis (when zero is point of crossing of axis), are inverted, and it is difficult to interpret experimental data. So imaginary low point of countdown, which is lower than sample, is on the crossing of imaginary line parallel to top edge of frame and electrode's axis. All the measurements according to the following scheme (fig.2) was made from this point of crossing [5].

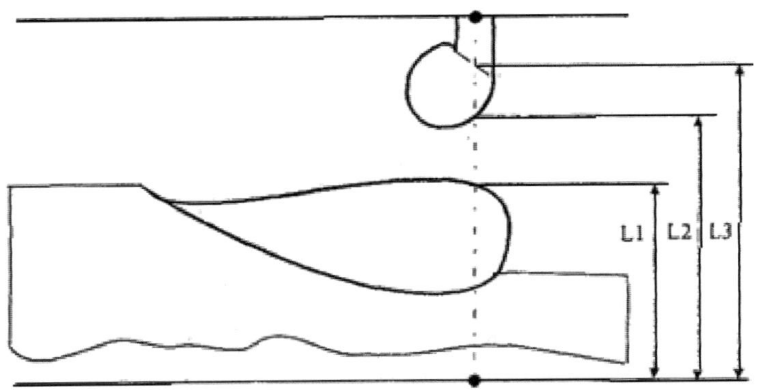

Fig 2. The scheme of measurements: L1 - space from base to crossing of welding pool's surface with electrode's axis; L2 -space from base to crossing of low surface of electrode metal's drop with electrode's axis; L3 - space from base to crossing of fusion's front with electrode's axis

Because of this fact that space L3 isn't seen clearly in the picture, it was found in the following way: it was determined positions on the left and the right sides of electrode, where it started to deviate from diameter and to transfer into drop, and they were joined by direct line, which was considered the line of electrode and drop's fusion.

As large-scale coefficient we took diameter of welding wire, which was known initially. So increase of diameter of welding wire at ten times in the picture leads to increase the whole image at ten times.

The measurements were made by the following way. Photographed kinogram was regarded, and only such welding cycles were taken that contained no failures and approximate by amount of cadres. Despite amount of cadres in the cycle we made measurements over three or five cadres. The charts were built according to the results of measurements.

The discussion of obtained results

The analysis of obtained charts and comparison of them with the cadres and oscillograms, represented in the figure 1, shows precise correlation between the change of welding regimes during the cycle of drop's transfer and oscillation of welding pool's surface, and also electrode metal's drop on the face of wire electrode.

During welding and fusion in low position one of the main physics factors that influence on periodic oscillation of welding pool and electrode metal's drop is arc pressure $R_a=cI^2$, where c - coefficient, depended on type of welding. For consumable electrode welding $c=(3.12 \div 5.62)10^{-5}$ g/a^2.

So, by changing, power of welding arc, according to chart represented in the picture 1, affects on the surface of welding pool's metal and volume of electrode metal's drop by pressure of arc.

The measurement was made from the moment of jumper's rupture between drop and electrode and to next jumper's rupture. In the charts of picture 3 the moment of jumper's rupture is clearly observed, when welding arc burns up like explosion, the pressure of welding arc increases instantly (gas-dynamic impact), welding pool is thrown away from electrode for maximal distance, maximal length of arc. Welding pool due to its inertness with little lateness starts to influence on increasing arc's length. Also in the charts it is noticed the influence of molten metal's wave, rising on the side surface of welding pool on the primary stage of arcing. During the process of arcing , arc pressure decreases, electrode metal's drop grows, welding pool flows under arc, moving on the surface of crystallization under impact of hydrostatic pressure, this all decreases arc length.

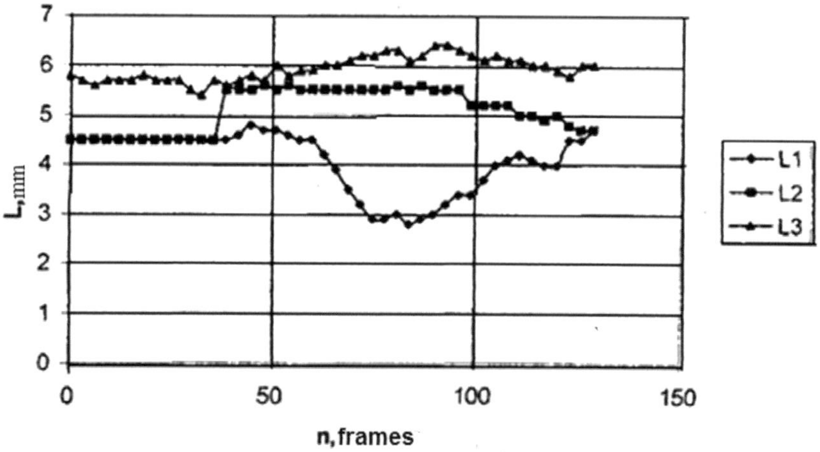

Fig 3. Experimental charts of movement of front of drop's fusion, electrode and welding pool.

After sharp decrease of welding current to ordinary one electrode metal's drop tries to take coaxial position with electrode, moves to welding pool, which in its turn starts to rise to electrode. In the charts of figure 3 it is clearly observed that this movement is more slow due to welding pool has large weight and partly so larger lag compared with metal's drop, and welding pool needs some time to flow from the surface of crystallization into crater and fill it, and then on the anterior front by taken potential energy it starts to rise to electrode. Short circuit appears.

According obtained data, during management of process of CO_2 welding by changing energetic parameters of arc welding pool takes direct part in the creation of short circuit on a level with electrode metal's drop.

Conclusion

1. The obtained results evidence that creating models of electrode metal's transfer during short circuits and creating algorithms of process management, especially changing space position of welding pool, it is necessary to take account the movement of welding pool, as role of drop and welding pool in decrease of arc space is almost the same.
2. Decrease of welding current to normal one leads to sharp acceleration of counter movement of welding pool and drop.
3. The role of drop and welding pool in short circuits' mechanism is almost the same.
4. By changing regime's parameters on the stage of electrode's fusion, it is possible to force or to weak welding pool movement.
5. Line of front of electrode fusion behaves conservatively in relation to changing thermal flow.

References

[1] Novozhilov N.M., Suslov V.N. *Svarka plavyashimsya elektrodom v uglekislom gaze* [Consumable electrode CO_2 welding] Moscow, Mashgiz, 1958. 194 p.

[2] Sh'ekin V.A., D'urgerov N.G., Sagirov H.N., Nebilitsin L.E. *Prinuditel'niy perenos metalla pri svarke modulirovannim tokom v uglekislom gaze* [Forced metal's transfer during module current CO_2 welding]. Svarochnoe proizvodstvo, 1973, no.3, pp.23

[3] Chernishov G.G., Spitsin V.V. Dinamicheskoe vozdeistvie svarochnoy dugi, goryash'ey v CO_2 na svarochnuyu vannu [Dynamic impact of CO_2 welding arc on welding pool]. Svarochnoe proizvodstvo, 1971, no.4, pp. 8-9.

[4] No.1371821. Timoshenko A.K., Saraev U.N., Knyaz'kov A.F. Sposob elektrodugovoy svarki s korotkimi zamikaniyami dugovogo promezhutka i ustroistvo dlya ego osushestvleniya [The way of arc welding with short circuits of arc space and device for its providing]. Pub. 07.02.88, bul. no.5

[5] Knyaz'kov A.F., Petrikov A.V. Metodika provedeniya issledovaniya dvizheniya svarochnoi vanni [The methodic of exploration of welding pool's movements] Trudy of XIV nauchnoy konferencyi, posvyashennoy 300-letiu inzhenernogo obrazovaniya v Rossii. Urga, 2001. 207 p.

[6] Zaruba I.I., Andreev V.V. osobennosti primenyaemyh sposobov ogranicheniya toka korotkogo zamykaniya pri svarke v uglekislom gaze [The peculiarities of used ways of current limitation of short circuit during CO_2 welding]. Avtomaticheskaya svarka, 1978, no.1, pp. 16-19.

[7] Zaruba I.I., Bargamen V.P., Andreev V.V., Sidorenko M.N. Vliyanie metoda ogranicheniya toka korotkogo zamykaniya na formirovanie vertikal'nyh i potolochnyh shvov pri svarke v uglekislom gaze [The influence of current limitation of short circuit's method on formation of vertical and overheld welds during CO_2 welding]. Avtomaticheskaya svarka, 19723, no.4, pp. 64-67.

[8] Zaruba I.I., Andreev V.V., Stepahno V.I., Koritskiy V.A. Puti povysheniya tehnologicheskoy effektivnosti vypryamiteley dlya mekhanizirovannoy svarki i naplavki [The ways of improvement of technological efficiency of rectifiers for mechanized welding and fusion]. Avtomaticheskaya svarka, 2011, no.11, pp. 45-49.

[9] Zaruba I.I. Usloviye ustoichivosti protsessa svarki s korotkimi zamykaniyami [The condition of welding process stability with short circuits]. Avtomaticheskaya svarka, 1971, no.2, pp. 23-25

[10] 1-4 A.s. 1310140 SSSr, MPK V 23 K 9/00. Zaruba I.I., Saraev U.N., Knyaz'kov A.F., Timoshenko A.K. Sposob dugovoy svarki s korotkimi zamykaniyami dugovogo promezhutka i ustroistvo dlya efo osushestvleniya [The way of arc welding with short circuits of arc space and device for its providing].Pub. 15/05/87; bul.no.18

[11] Ioffe U.E., Kvasov F.V. Novye vysokotekhnologichnye sistemy poluavtomaticheskoy svarki kompanii "Linkol'n Elektrik" [The new high technologic sustems of semi-automatic welding by company "Lincoln Electric"]. Svarochnoe proizvodstvo, 1997, no.4, pp. 40-43.

[12] Getskin O.B., Poloskov S.I., Erofeev V.A., Vit'ko O.P. Phiziko-matematicheskaya model' sistemy "istochnik pitaniya-duga" dlya svarki plavyash'imsya elektrodom v zash'itnyh gazah [Physical and mathematical model of system "power source - arc" for consumable electrode shielded welding]. Tyazheloemashinostroenie, 2008, no.6, pp. 18-20.

IOP Conf. Series: Materials Science and Engineering **253** (2017) 012029 doi:10.1088/1757-899X/253/1/012029

Forming a Single Plastic Imprint of the Indenter under Ultrasonic Surface Hardening

Kharis M Rakhimyanov [a], Yulia S Semenova [b], Anna S Eremina [c]

Novosibirsk State Technical University, 20, Prospekt K. Marksa, Novosibirsk, 630073, Russia

E-mail: [a] x.raximyanov@corp.nstu.ru, [b] yu.semenova@corp.nstu.ru, [c] a.eremina@corp.nstu.ru

Abstract. The results of the experimental investigation of forming the geometry of a single imprint appearing on the machined surface during a single tool stroke in ultrasonic surface hardening are given. The equations of regression for calculating the geometrical parameters of the imprint depending on the machining modes such as the static force and the diameter of the deforming element are obtained. Unlike the existing models of forming the imprint, the appearance of the material pressed out around the tool during its penetration in the surface layer of the machined detail is taken into account. The pressed-out metal looks like a hill. The equations obtained allow us to determine the height of the hill as well as the depth and diameter of the imprint taking into consideration the hill dimension. The calculation of these parameters is necessary to find out the boundary conditions of starting the appearance of the plastic flow wave under ultrasonic plastic deforming of metals and alloys. These conditions determine the criterion of forming waviness and present the restrictions of the limiting values of feeding and speed in machining the cylindrical surfaces.

Introduction

Nowadays more and more interest in ultrasonic plastic deforming as hardening and finishing operations of details is taken [1-9]. Due to this fact, the questions of the technological quality assurance of the surface under the conditions of discrete deformation impact are considered to be vital. It is known that waviness appears on the machined surface under certain conditions in ultrasonic plastic deforming of natural steels. Paper [10] presents the theoretical investigations on finding out the conditions of appearing waves of plastic flow of metal under the discrete impact by the ultrasonic tool. This condition is a system of inequalities as restrictions determining the limiting values of the speed and feeding under which the waves appear.

$$\begin{cases} V \le \dfrac{d_{pl}}{2 \cdot 60 \cdot f}, \\ S \le \dfrac{d_{pl}}{10} \end{cases} \qquad (1)$$

where V– the speed of the detail rotation, m per minute; d_{pl} - the diameter of the plastic imprint with the pressed-out metal round the tool during its penetration in the surface, mm, f– the frequency of ultrasonic oscillations of the tool, Hz; S – the feeding, mm per rotation. (Fig. 1).

Conversely, the dependency of the plastic imprint diameter on the mode parameters of machining is not presented in the paper. Traditionally the oscillation amplitude (A), static force (F_{st}) for pressing the ultrasonic tool to the detail and the diameter of the deforming element (D_C) are used as the machining modes influencing the imprint parameters in ultrasonic plastic deforming. The purpose of the paper is estimating the influence of the technological machining modes mentioned above on the geometrical parameters of the imprint.

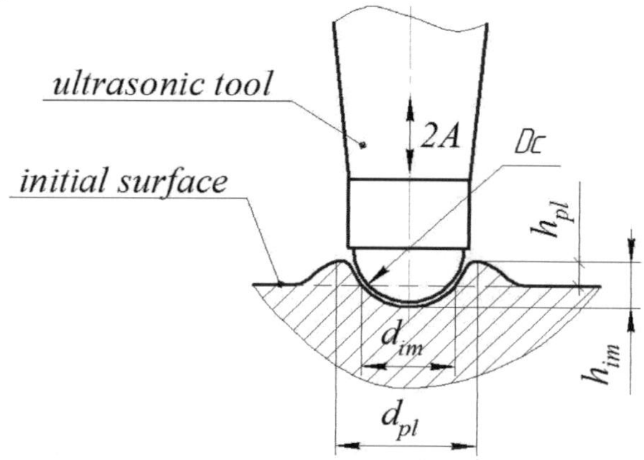

Fig. 1. Geometrical parameters of a separate imprint formed as a result of plastic flow of material

Experimental Investigation of Forming the Imprint
Goals of the experiment

To exclude the overlapping of the imprints one on another, it is necessary to determine the corresponding modes in machining. Kinematic parameters of machining such as the speed and the feeding affect the location of the imprint centers, and deformation parameters (the amplitude and the frequency of ultrasonic oscillations, the indenter diameter, the static force of pressing the tool to the detail) influence the imprint dimension. According to the deformation model described in Paper [11] the imprint diameter can be determined by the following formula:

$$d = \int_{t_0}^{t_1} 2\sqrt{D_c \cdot h_1(t)}dt - \left(h_{\max} - \int_{t_1}^{t_2} 2\sqrt{D_c \cdot h_2(t)}dt \right), \quad (2)$$

where $h_1(t)$ and $h_2(t)$ – the functions which describe the depth change of the tool penetration in the respective phase of its direct and reverse movement in the material, D_c- the diameter of the spherical part of the indenter, h_{max} – the maximum depth of the indenter penetration, t_0 – the time during which the contact of the deforming element with the detail surface takes place, t_1 – the time of the maximum penetration, t_2 – the time of completing the contact between the deforming tool and the machined surface.

Calculating the imprint dimensions, it is necessary to take into account the tool sliding during its contact with the machined surface, as a result of which the imprint takes the form of ellipse. Paper [12] traces the correlation between the imprint parameters and geometrical, kinematic and force parameters of ultrasonic plastic deforming. According to the paper mentioned, in machining the cylindrical surface the ellipse dimensions depend on the indenter geometry (the diameter of the spherical part, r), the time of contact between the detail and the tool (Δt), the speed of the detail rotation (V_d), the depth of the deformer penetration (h) as well as elastic and plastic properties of the machined surface (Fig. 2). In this case the dimensions of the ellipse semi-axes are determined by the following equations:

$$a \approx \sqrt{D_c(2h+h_y)-h^2} \tag{3}$$

$$b \approx \sqrt{D_c(2h+h_y)-h^2} + 0.5 \cdot \Delta t \cdot V_d \tag{4}$$

where the parameters r, h, h_y, h are calculated by the deformation model [11]. The calculations showed that the value of the large ellipse axis ($2b$) did not exceed two diameters of the imprint (d_{im}) at a maximum possible speed of the detail rotation, taking into consideration the real conditions of machining.

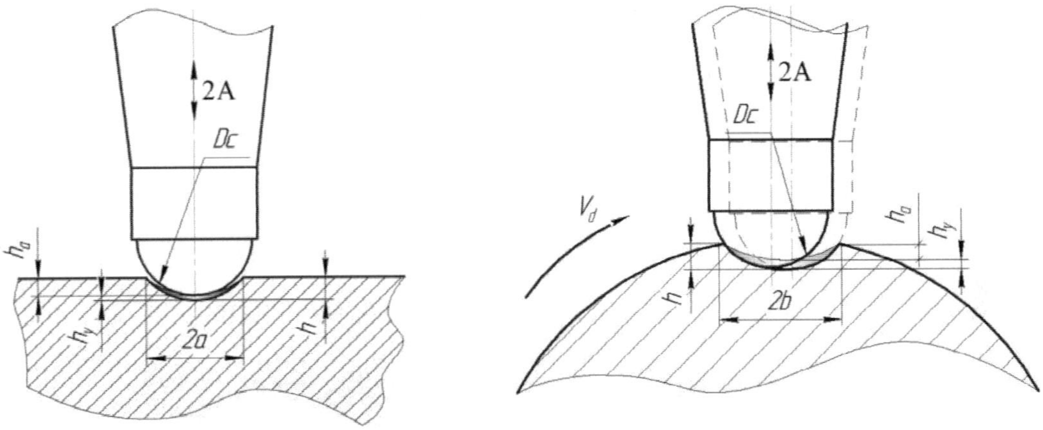

a b

Fig. 2. The scheme of contacting between the ultrasonic tool and the cylindrical surface (a – in the longitudinal section, b – in the cross section)

Because the model used does not take into account the appearance of the pressed-out material around the imprint, we can consider that it is necessary to provide the distance of more than $2d_{im}$between the imprints in order to ensure the absence of overlapping the imprints. As the dimension of the pressed-out material is small in comparison with the imprint diameter, the distance in $3d_{im}$ will be sufficient. The distance between the imprints in the feeding direction must be not less than $2d_{im}$. Thus, to obtain separate imprints on the machined surface in ultrasonic plastic deforming, the condition mentioned above can be presented in the following way:

$$\begin{cases} l_V \geq 3d_{im}, \\ l_S \geq 2d_{im} \end{cases} \tag{5}$$

where l_v – the distance between imprints in the direction of the speed of the detail rotation, l_S– the distance between the imprints in the direction of feeding.

The distance between the imprints in the direction of the speed is presented in Paper [12] as follows:

$$l_V = \frac{V}{60 \cdot f}, \tag{6}$$

where f – the frequency of ultrasonic oscillations, Hz.

The value of feeding for one revolution is the distance l_S. Thus, the restriction for ultrasonic plastic deforming modes, which provide the formation of separate imprints on the machined surface, can be presented by the following system of inequalities through simple transformations:

$$\begin{cases} V \geq 3d_{im} \cdot 60f \\ S \geq 2d_{im} \end{cases} \tag{7}$$

Methods of the Experiment

Cylindrical samples with a diameter of 50 mm made from steel were chosen for carrying out the experiment. The samples were preliminary turned to obtain the initial roughness of the surface Ra 2.5.

The sample machining with the successive change in each parameter (the static force – from 50 to 250 N and the deformer diameter – from 5 to 20 mm) was provided on the modes chosen according to the restriction (Eq.6). Unchanged parameters in machining were the frequency of ultrasonic oscillations (22 kHz), the amplitude (20 micrometers), the speed of the detail rotation (280 m per min) and the feeding (1.3 mm per rotation).

The typical picture of the machined surface obtained by microscope Nikon MM-400 is presented in Fig. 3. Separate imprints in the form of ellipse, evenly distributed on the surface at a distance l_v and l_S in the directions *V* and *S* respectively, can be seen in the picture.

The profile diagrams (5 for each mode) were taken from each surface after machining, after that the geometrical parameters of the imprint were determined. The typical profile diagram with the geometrical parameters is presented in Fig. 4.

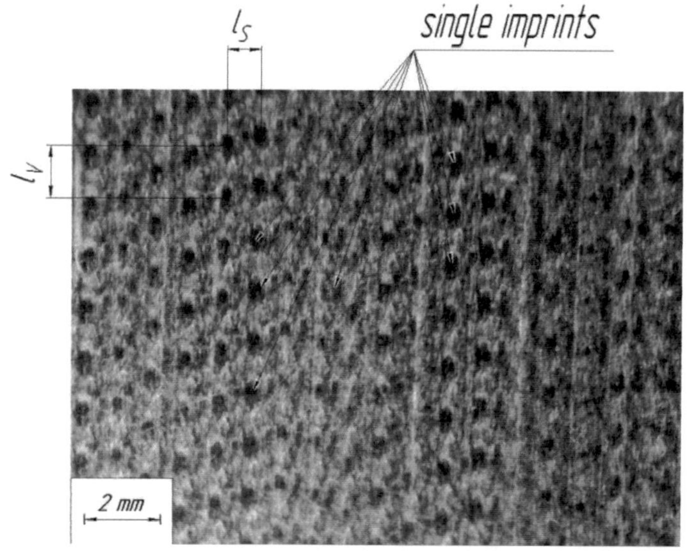

Fig. 3.Typical view of the surface after ultrasonic plastic deforming at modes providing the formation of separate imprints without overlapping

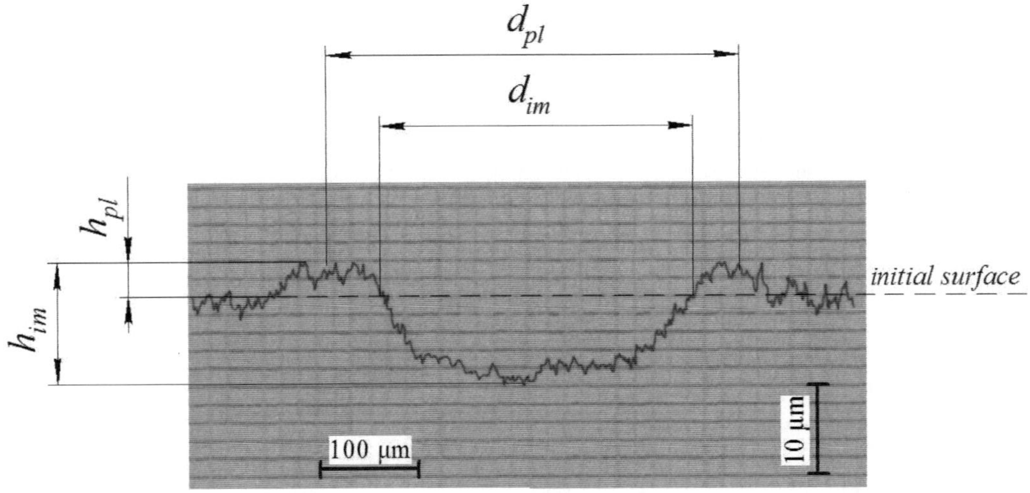

Fig. 4.Typical profile diagram of a single imprint with the pressed-out metal

Results and Discussion

According to the experimental results the equations of regression for calculating d_{pl}, h_{pl}, h_{im} were obtained under different mode parameters.

The diameter of the plastic imprint (d_{pl}) depending on the static force (F_{st}) and the deformer diameter is determined according to one of the Eq.8-10.

$$d_{pl} = -0.0051F_{st}^2 + 3.155\ F_{st} + 504.7(Dc=20\ mm) \quad R^2 = 0.9955,$$
$$d_{pl} = -0.0037F_{st}^2 + 2.6491\ F_{st} + 474.7 \quad (Dc=10\ mm) \quad R^2 = 0.9972, \quad (8)$$
$$d_{pl}= -\ 0.0039F_{st}^2 + 2.6214\ F_{st}+417.4(Dc=5\ mm) \quad R^2 = 0.9944.$$

The dependence of the imprint depth (h_{im}) on the static force and the deformer diameter is presented as:

$$h_{im}=-8\cdot10^{-5}\ F_{st}^2 + 0.0654\ F_{st} + 2.0348 \quad (Dc=20\ mm) \quad R^2=0.9938,$$
$$h_{im}=-4\cdot10^{-5}\ F_{st}^2 + 0.0463\ F_{st} + 2.0148 \quad (Dc=10\ mm) \quad R^2=0.9968, \quad (9)$$
$$h_{im}=-4\cdot10^{-5}\ F_{st}^2 + 0.0411\ F_{st} + 1.3569 \quad (Dc=5\ mm) \quad R^2=0.9933.$$

The dependence of the pressed-out material height (h_{pl}) on the static force and the deformer diameter is described in the following way:

$$h_{pl}=-4\cdot10^{-5}\ F_{st}^2 + 0.0273\ F_{st}+1.1044\ (Dc=20\ mm) \quad R^2=0.9940,$$
$$h_{pl}=-2\cdot10^{-5}\ F_{st}^2 + 0.0199\ F_{st} + 1.0583 \quad (Dc=10\ mm) \quad R^2=0.9969, \quad (10)$$
$$h_{pl}=-2\cdot10^{-5}\ F_{st}^2 + 0.0179\ F_{st} + 0.755\ (Dc=5\ mm) \quad R^2=0.9935.$$

Conclusions

The obtained equations (8-10) describe the dependence of the geometrical parameters of the imprint on the static force and the indenter diameter during ultrasonic plastic deforming taking into account the appearance of the pressed-out material. A high validity of approximation allows us to use these dependences in engineering calculations. The established interconnections between the geometrical parameters of the imprint and machining modes give the possibility to determine the boundary condition of forming the waviness during ultrasonic plastic deforming.

References

[1] B.A. Kudryashov, Improvement of service characteristics of brake disks by methods of ultrasonic surface plastic deformation, Science intensive technologies in mechanical engineering. 41 (2014) 14-18. (in Russian)

[2] V.P. Gileta, Formation of microgeometry of a surface at ultrasonic processing, Proceedings - KORUS 2000: 4th Korea-Russia International Symposium on Science and Technology (27 June - 1 July 2000), Vol. 3, № 866075 (2000) 150-154.

[3] V.P. Gileta, G.A. Iskhakova, Investigation into regularities of the surface microgeometry formation during diamond ultrasonic strengthening finishing, Sverkhtverdye Materialy, Iss. 1, (1992) 45-50.

[4] V.P. Gileta, G.A. Iskhakova, Simulation of regular relief formation in ultrasonic diamond burnishing, Soviet journal of superhard materials. Vol. 10, Iss. 6 (1988) 64-69.

[5] A.I. Beznedelnyy, V.B. Asanov, V.P. Gileta, Technological inheritance influence on surface layer forming in condition of ultrasonic plastic deformation of hardened steels, Obrabotka metallov (tekhnologiya, oborudovanie, instrumenty) 4 (2012) 19-22 (in Russian).

[6] V.P. Gileta, A.I. Beznedelnyy, V.B. Asanov, Centerless finishing and hardening processing with ultrasonic instrument, Systems. Methods. Technologies. 28 (2015) 49-54 (in Russian).

[7] Sand, A., Garca Navas, V., Gonzalo. O. Surface state of Inconel 718 ultrasonic shot peened: Effect of processing time, material and quantity of shot balls and distance from radiating surface to sample. Materials and Design, Vol. 32 (2011) 2213-2220.

[8] Bagherifard, S., Guagliano, M. Fatigue behaviour of a low-allow steel with nanostructured surface obtained by severe shot peening. Engineering Fracture Mechanics, 81 (2012) 56–68.

[9] Kh. M. Rakhimyanov, Yu. S. Semenova, Technological support of geometrical parameters of surface quality with ultrasonic plastic deformation, Obrabotka metallov (tehnologiya, oborudovanie, instrumenty) = Metal Working and Material Science, 3 (56) (2012) 33–36. (in Russian)

[10] Kh. M. Rakhimyanov, Yu. V. Nikitin, Yu. S. Semenova, The condition of waviness forming at ultrasonic impact treatment of metals, Obrabotka metallov (tehnologiya, oborudovanie, instrumenty) = Metal Working and Material Science, 54 (2012) 4–9 (in Russian).

[11] Kh.M. Rackhimyanov, Yu.V. Nikitin, J.S. Semyonova, Surface preparation of machine parts and instruments by ultrasonic impact treatment before coating, The Third International Forum on Strategic Technologies, Novosibirsk-Tomsk, Russia, june 23-29 (2008)108-114.

[12] K. M. Rakhimyanov, I. S. Semenova, Surface state control by ultrasonic plastic deformation at the final machining stage, Materials and Manufacturing Processes. Vol. 31, iss. 6 (2016) 764-769.

IOP Conf. Series: Materials Science and Engineering **253** (2017) 012030 doi:10.1088/1757-899X/253/1/012030

Technological Restrictions on Mode Parameters of Ultrasonic Plastic Deforming

Kharis M Rakhimyanov [a], **Yulia S Semenova** [b], **Anna S Eremina** [c]

Novosibirsk State Technical University, 20, Prospekt K. Marksa, Novosibirsk, 630073, Russia

E-mail: [a] x.raximyanov@corp.nstu.ru, [b] yu.semenova@corp.nstu.ru, [c] a.eremina@corp.nstu.ru

Abstract. The paper considers technological restrictions that appear under real conditions of machining during ultrasonic plastic deforming. Possible defects of machining which are formed at the regimes, not suitable for the conditions described, are presented. The reasons for their appearance because of the possibilities of the technological equipment are found out. The ways of preventing the appearance of defects in developing the finishing and hardening operations as well as in preliminary machining of surfaces are suggested.

Introduction

Nowadays the interest in the technology of ultrasonic plastic deforming as an effective way of modifying the surface layer of machine parts is growing. The advantages of ultrasonic plastic deforming in comparison with the competing methods of finish machining are shown in papers [1-7]. The authors present different schemes of using ultra sound in the hardening and finishing machining. The main characteristics of quality provided at the finish stage of parts machining are the depth and degree of hardening the surface layer material, the value, sign and character of distributing residual stress as well as surface micro-geometry. Besides, ultrasonic plastic deforming of metals and alloys of low hardness under certain conditions can be accompanied by forming roughness on the surface that can be classified as waviness according to the ISO 4287:1997 standard.

The main mode parameters of ultrasonic plastic deforming are the amplitude and frequency of oscillations of the ultrasonic tool, geometrical dimensions of its working part, i.e. the deforming element, the static pressure force of the tool applied to the machined surface. Such parameters of ultrasonic plastic deforming define the deformation conditions in the machined zone but the rotating speed and feeding - kinematic parameters of the process. A certain combination of deformation and kinematic parameters of ultrasonic plastic deforming provides the hardening and finishing effects during machining.

The regularities that allow us to predict the state of the surface layer after ultrasonic plastic deforming and suggest the optimal modes to provide the required quality of machining are described in papers [8-12]. However, these regularities do not allow us to determine the restrictions existing in practice for setting the mode parameters of ultrasonic deforming which can appear in reality. So, the paper is devoted to theoretical reasons and experimental confirmation of actions to determine the technological restrictions in setting the mode parameters of ultrasonic plastic deforming.

IOP Conf. Series: Materials Science and Engineering **253** (2017) 012030 doi:10.1088/1757-899X/253/1/012030

Results and Discussion

Frequency of ultrasonic oscillations. In conducting the experiments it was found out that even a slight change (up to 5%) in the frequency of ultrasonic oscillations induced by the generator leads to the change in forming the waviness. Fig. 1 shows changes in the geometrical parameters of waves under ultrasonic plastic deforming caused by changing the frequency of ultrasonic oscillations as a result of self-adjustment of the generator for the resonance frequency at the initial stage of its operation. Paper [9] presents the mathematical model of forming the waves of the plastic deformation after ultrasonic plastic deforming. According to the model the inclination angle of waves depends on the combination of the ultrasonic oscillation frequency, the number of spindle revolutions and the number of strokes which form a single wave crest. Instability of the frequency characteristic of ultrasonic oscillations under other consistent modes defines the topography of the formed waviness.

a b

Fig. 1. Formation of waviness during the generator operation with constant (a) and changing (b) frequencies of the tool ultrasonic oscillations

Stabilizing the frequency characteristic of ultrasonic oscillations is accompanied by forming the surface geometry with a constant inclination angle of waves of the plastic deformation. Thus, to provide the stability of the wave geometrical parameters, it is necessary to follow the conditions of stable frequency characteristics of the ultrasonic generator. Paper [8] shows that the stability of the generator operation at a constant value of the frequency characteristic happens after the first ten minutes of the operation after switching on.

Amplitude of ultrasonic oscillations. The parameter under consideration is one of the components in controlling the deformation process during machining. The depth and the degree of the material deformation in the surface layer as well as the change in the dimension of a single imprint of the deforming element on the machined surface can be regulated by changing the amplitude. However, the excessive intensification of the deformation process can cause the destruction of the deforming element. Paper [9] presents the results of the investigations dealing with the influence of the ultrasonic tool shape on the coefficient of intensifying the amplitude. Thus, the use of the cone tool made from the titanium alloy resulted in increasing the amplitude in the deformer zone up to 100 micrometers. At such a value of the amplitude and even at a minimal value of the static force ultrasonic machining caused the destruction of the deforming element made of synthetic diamond. Such a fact confirms the existence of restrictions on the maximum value of the oscillation amplitude of the ultrasonic tool. So, in reality the values of the amplitude used during ultrasonic plastic deformation are limited by the range of 5- 30 micrometers.

Static force. The static force of the deforming element applied to the machined surface is also a mode parameter that defines the deformation processes during machining. The minimum value of static force is determined by the design features of the technological equipment that provides the stability of the force closure of the ultrasonic tool and the machined part under the elastic contact. The presence of certain friction losses in the connection joints of the equipment at an insufficient value of the static force can lead to the irregularity of the force closure. As a result, there appear zones on the machined surface that are not subject to the deformation impact (Fig. 2).

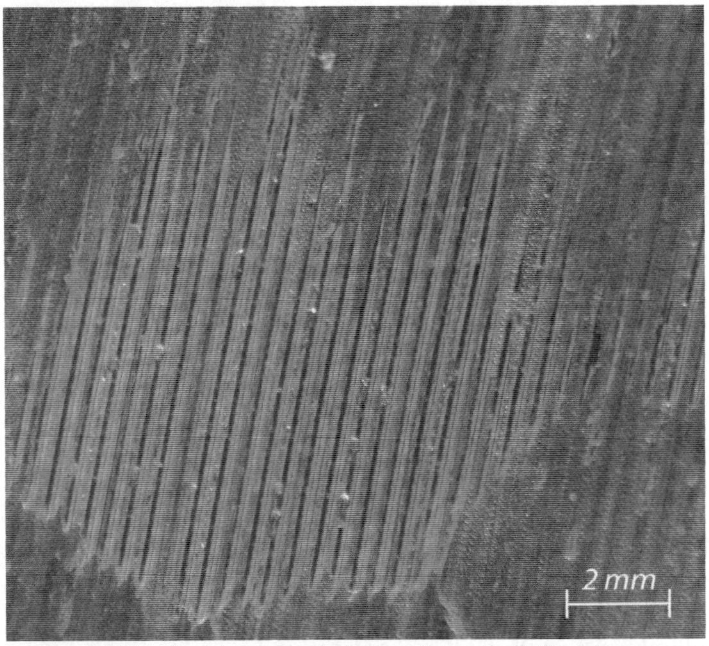

Fig. 2. Zone of the non-machined surface after ultrasonic static deforming as a result of insufficient static force

The maximum value of static force is limited by the necessity of obeying certain conditions to accomplish a discrete impact of the ultrasonic tool on the machined surface. The quantitative estimation of such a restriction is described in detail in investigating the deformation model of ultrasonic plastic deforming in paper [8]. The discrete impact with changing periods of intensive plastic deformation and pauses for the relaxation of stresses in the material creates favorable conditions to implement the hardening effect in machining. At the same time due to the fractional impact of the deforming element on the machined surface, a regular micro-relief is formed on it as single plastic imprints located on the surface with a coefficient of overlapping $0<k<1$. Under real conditions the values of static force used varies from 50 N to 300 N during ultrasonic plastic deforming of metals and alloys. The specified value of the parameter is chosen depending on the properties of the machined material and the required characteristics of the hardening and finishing effects during machining.

Initial roughness of the machined surface. The role of ultrasonic plastic deforming in creating the finishing effect is in providing a new, fully regular micro-relief on the machined surface without traces of the previous machining. Because of this fact, there are certain restrictions in the value of the initial roughness before ultrasonic plastic deforming. Otherwise, a part of traces of the preceding machining will remain on the formed micro-geometry. Fig. 3 shows a scanned three-dimensional image of the surface after ultrasonic plastic deforming. The traces of the preceding lathe machining are seen on the surface as grooves from the cutting tool located at a distance of feeding (S) used during lathe turning. To form a fully new and regular micro-relief without traces of the preceding machining, the height of the initial roughness must not exceed a certain value. Experiments showed that during machining of unhardened steel by ultrasonic plastic deforming, it was possible to form a new micro-geometry when $R_{max}<2$ micrometers. This statement corresponds to the conditions when the oscillation frequency is 22 kHz, the amplitude is 25 micrometers, the static force is 250 N, the speed of the part rotation is 110 m/min., the feeding is 0.1 mm per rotation, the diameter of the deforming element is 8 mm.

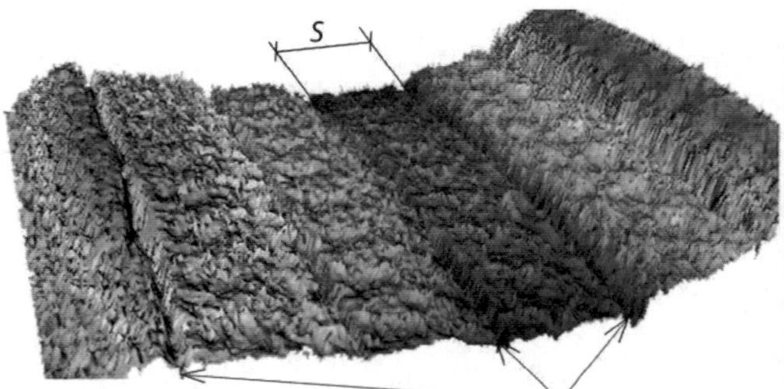

a

b

Fig. 3 Formation of the surface micro-geometry after ultrasonic plastic deforming with traces of the previous machining

Conclusion.

As a result of the investigations conducted, the recommendations concerning the mode parameters of ultrasonic plastic deforming were formulated by taking into consideration the restrictions described.

The restriction on the amplitude of ultrasonic oscillations by a value of 30 micrometers was suggested to exclude the destruction of the deforming element made of synthetic diamond.

When setting a minimum value of the static force, it is necessary to take into account the friction losses in the machining attachment that provides the force closure of the tool and the machined surface according to the elastic contact. The maximum value of the static force is determined by the necessity of preserving the discrete loading because of the pauses of intensive deforming and unloading.

The stability of the frequency characteristic of the ultrasonic process is provided by enduring the ultrasonic generator after its turning on during ten minutes, which allows us to exclude the instability of the geometrical parameters of plastic flow waves.

The necessity of forming a new, fully regular micro-geometry of the surface after ultrasonic plastic deforming limits the maximum value of the initial roughness of the surface.

References

[1] K. M. Rakhimyanov, I. S. Semenova, Surface state control by ultrasonic plastic deformation at the final machining stage, Materials and Manufacturing Processes Vol. 31 iss. 6 (2016) 764-769.

[2] Kh.M.Rakhimyanov, G.A.Iskhakova, A.Kh. Rakhimyanov, The role of ultrasonic plastic deformation in spark alloying, Fizika i Khimiya Obrabotki Materialov. 1 (1996) 68-72.

[3] V.P. Gileta, G.A. Iskhakova, Formation of the component surface layer in ultrasonic diamond burnishing, Soviet journal of superhard materials, Vol.9, Iss. 5 (1987) 66-71.

[4] Sand, A., Garca Navas, V., Gonzalo. O. Surface state of Inconel 718 ultrasonic shot peened: Effect of processing time, material and quantity of shot balls and distance from radiating surface to sample. Materials and Design, Vol. 32 (2011) 2213-2220.

[5] A.I. Beznedelnyy, V.B. Asanov, V.P. Gileta, Technological inheritance influence on surface layer forming in condition of ultrasonic plastic deformation of hardened steels , Obrabotka metallov (tehnologiya, oborudovanie, instrumenty) = Metal Working and Material Science, 4 (57) (2012) 19–22 (in Russian).

[6] A.I. Beznedelnyy, V.B. Asanov, V.P. Gileta, Increasing the durability of the exhaust punches (Povyshenie stojkosti vytjazhnyh puansonov), Obrabotka metallov (tehnologiya, oborudovanie, instrumenty) = Metal Working and Material Science, 4 (37) (2007) 11–12 . (in Russian).
[7] V.P. Gileta, A.I. Beznedelnyy, V.B. Asanov, Centerless finishing and hardening processing with ultrasonic instrument, Systems. Methods. Technologies. 28 (2015) 49-54 (in Russian).
[8] Kh. M. Rakhimyanov, Yu. S. Semenova, Technological support of geometrical parameters of surface quality with ultrasonic plastic deformation, Obrabotka metallov (tehnologiya, oborudovanie, instrumenty) = Metal Working and Material Science, 3 (56) (2012) 33–36 (in Russian).
[9] Kh.M. Rakhimyanov, Iu.S. Semenova, Modelling of a regular microrelief formation by ultrasonic plastic deformation, Uprochnyayuschie Tehnologii I pokrytiya (Hardening Technologies and Coatings). 2(62) (2010) 3-9 (in Russian).
[10] V.P. Gileta, Formation of microgeometry of a surface at ultrasonic processing, Proceedings - KORUS 2000: 4th Korea-Russia International Symposium on Science and Technology (27 June - 1 July 2000), Vol. 3, № 866075 (2000) 150-154.
[11] V.P. Gileta, G.A. Iskhakova, Investigation into regularities of the surface microgeometry formation during diamond ultrasonic strengthening finishing, Sverkhtverdye Materialy, Iss. 1, (1992) 45-50.
[12] V.P. Gileta, G.A. Iskhakova, Simulation of regular relief formation in ultrasonic diamond burnishing, Soviet journal of superhard materials. Vol. 10, Iss. 6 (1988) 64-69.

IOP Conf. Series: Materials Science and Engineering **253** (2017) 012031 doi:10.1088/1757-899X/253/1/012031

Application of Electric-Arc Pulsed Mode for Obtaining Surfacings of Fe–Mo–Cr–Ti–Ni–Cu–Al-System

Anatoly F Knyaz'kov [1, a], Ivan A Ditenberg [2,3], Aleksey N Gavrilin [1], Konstantin V Grinyaev [2,3], Sergey A Knyaz'kov [1], Ivan V Smirnov [2,3]

[1] National Research Tomsk Polytechnic University, Lenin Av. 30, Tomsk, Russia, 634034
[2] National Research Tomsk State University, Lenin Av. 36, Tomsk, Russia, 634050
[3] Institute of Strength Physics and Materials Science SB RAS, Akademicheskii av. 2/4, Tomsk, Russia, 634055

E-mail: [a] kaf@tpu.ru

Abstract: Method of electric-arc melting with nonconsumable electrode in a protective argon atmosphere with using a pulsed mode was used to show possibility in principle of obtaining multi-component metallic surfacing of Fe–Mo–Cr–Ti–Ni–Cu–Al system on uncooled steel substrate. Features of structure and elemental composition were studied using scanning electron microscopy methods. Component intermixing of the surfacing is observed. Surfacing microhardness is 5 times greater than that of the substrate. «Surfacing-substrate» boundary has high adhesion strength.

Introduction

One of the directions for obtaining materials with high strength properties is the development and creation of multicomponent systems based on a combination of different metals [1-7]. To obtain such systems, various variations of multiple vacuum remelting are often used, followed by ultrafast quenching. Essential disadvantages of such methods are the technological complexity of the processing process, which requires material heating, cleanliness of samples surface and vacuum, low productivity and small dimensions of the obtained samples [8-15].

In the present work, the structure and strength properties of multicomponent metal surfacing of Fe-Cr-Mo-Ti-Ni-Cu-Al system obtained by the electric arc melting method with a non-consumable electrode in a protective argon medium using a pulsed mode have been validated [16-21].

Methods of experimental research

Samples of steel 35 were used as a substrate. Surfacing was obtained by melting a composite billet from metal wires (Fe, Cr, Mo, Ti, Ni, Cu, Al) by an electric arc with a non-consumable electrode. A pulsed regime providing the dynamic character of arc burning was realized: the amplitude of the pulses was 350 A, the duration was 70 microseconds, and the frequency was 2000 Hz. The formation of a protective atmosphere is ensured by the continuous supply of argon to the region of arc formation.

Structural studies were performed using a Tescan Vega 3 SBH (30 kV) scanning electron microscope and an FEI Quanta 200 3D (30 kV) electron-ion microscope with an attachment for determining the elemental composition by the energy-dispersive analysis.

Microhardness (Hμ) was determined in the end section by the Vickers method on a Neophot 21 device with a load of 1 N during 15 seconds.

Results and discussion:

Figure 1 shows an overview scanning electron microscopic image of end section of Fe-Cr-Mo-Ti-Ni-Cu-Al system surfacing on a steel 35 substrate. The asterisks denote the microhardness measuring regions (1 – 3) and region for element composition determination by energy-dispersive analysis (1 – 4) at different distances from the substrate. As can be seen from Table 1, the microhardness values of surfacing in region 3 exceed the Hμ of the substrate by 4.7 times, and near the surface (region 1) its value is greater by a factor of ~ 5.2.

Table 1 – Microhardness values (Hμ, GPa) at various distances from the interface "surfacing-substrate".

Region	1 (3.5 mm)	2 (2.5 mm)	3 (1.5 mm)	Substrate
Hμ, GPa	7.23	6.81	6.64	1.41

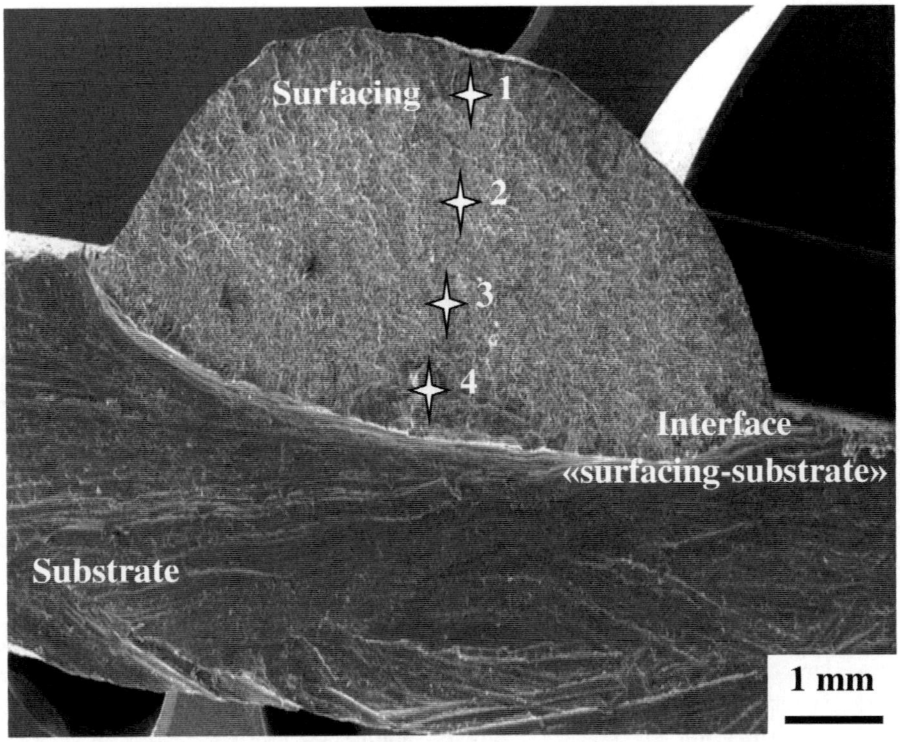

Fig. 1. Electron-microscopic image of end section of Fe-Cr-Mo-Ti-Ni-Cu-Al system surfacing on a steel 35 substrate.

Table 2 shows the results of qualitative-quantitative energy-dispersive analysis. The distribution and the corresponding concentrations of elements, as well as the absence of liquation inhomogeneities, attest to the intensive mixing of the material of the formed surfacing during melting. The increased Fe content in regions 2, 3, 4 is a consequence of mixing with the substrate material.

The lenticular shape of the interface "surfacing-substrate" (Fig. 1), determined by the geometric dimensions of the weld pool, also indicates intensive mixing of the formed surfacing with the substrate. As a result of the bending test at room temperature, there was no detachment or cracking of the formed surfacing from the substrate, which indicates a high adhesive strength.

Table 2 – Elemental composition of surfacing at different distances from the interface "surfacing-substrate".

Region	Elements concentration, wt. %						
	Fe	Cr	Mo	Ti	Ni	Cu	Al
1 (3.5 mm)	36.85	7.26	1.85	4.73	24.24	19.13	5.95
2 (2.5 mm)	50.36	5.02	1.42	3.74	18.02	16.85	4.59
3 (1.5 mm)	53.44	4.52	1.74	2.98	16.39	16.63	4.31
4 (0.5 mm)	48.72	4.72	2.30	5.25	18.55	15.95	4.50

The fractograph of the surfacing break is shown in Figure 2. Near the interface "surfacing-substrate", a viscous character of fracture is observed. At a distance of about 0.3 mm from this boundary, the fracture mechanism changes from viscous to brittle with elements of intercrystalline fracture, which is typical for the surfacing bulk. At the same time, in near-surface layers (0.3 mm) intercrystalline fracture changes to intracrystalline fracture, as evidenced by the river relief characteristic of fracture by cleavage.

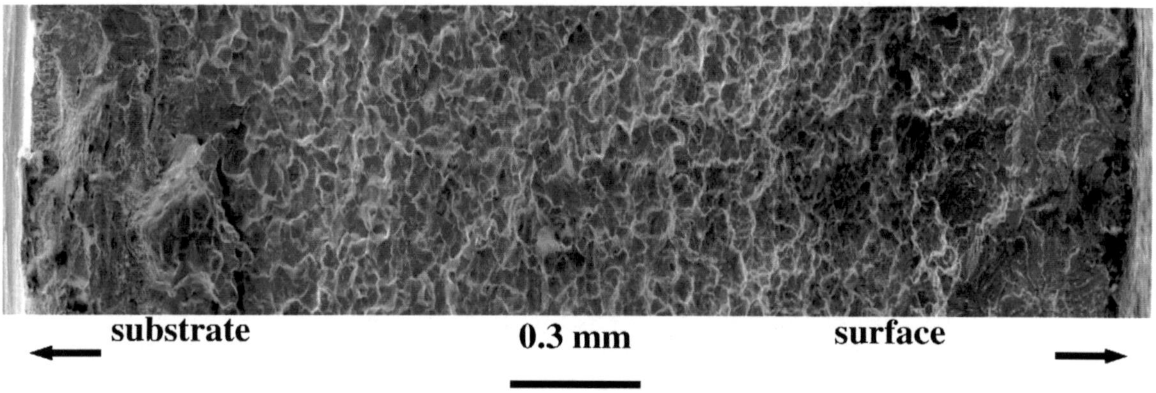

Fig. 2. Fractograph of surfacing break after the bending test at room temperature. Panoramic electron microscopic image

In the process of detailed analysis in the region of intercrystalline fracture, it was revealed that the characteristic sizes of the crystallites reach 30 microns (Fig. 3a), and on their surface (boundaries) a slightly-marked river relief is observed. As a result of a complex analysis of electron microscopic images and maps of the distribution of elements, it was established that the composition of crystallites is characterized by a solid solution of Fe-Cr-Mo-Ni system (regions 2 and 3 in Fig. 1, Table 2), which is close to elemental composition of heat-resistant high-alloy steels. Within the crystallites, particles of 1 to 3 microns in size are found, which are Ti-Al system based compounds (indicated by white arrows in Fig. 3a), and the boundaries between the crystallites are enriched in copper (Figure 3b).

Inside the crystallites of near-surface layer (region 1 of Figure 1, Table 2), a significant increase in the concentration of Cr, Mo, and Ni occurs with a significant decrease in the Fe concentration. In turn, the decrease in the size of crystallites, which leads to an increase in the specific area of the intercrystallite boundaries, is the reason for an increase in the concentration of copper along the boundaries. In addition, an increase in the particle density on the basis of the Ti-Al system was observed inside the crystallites.

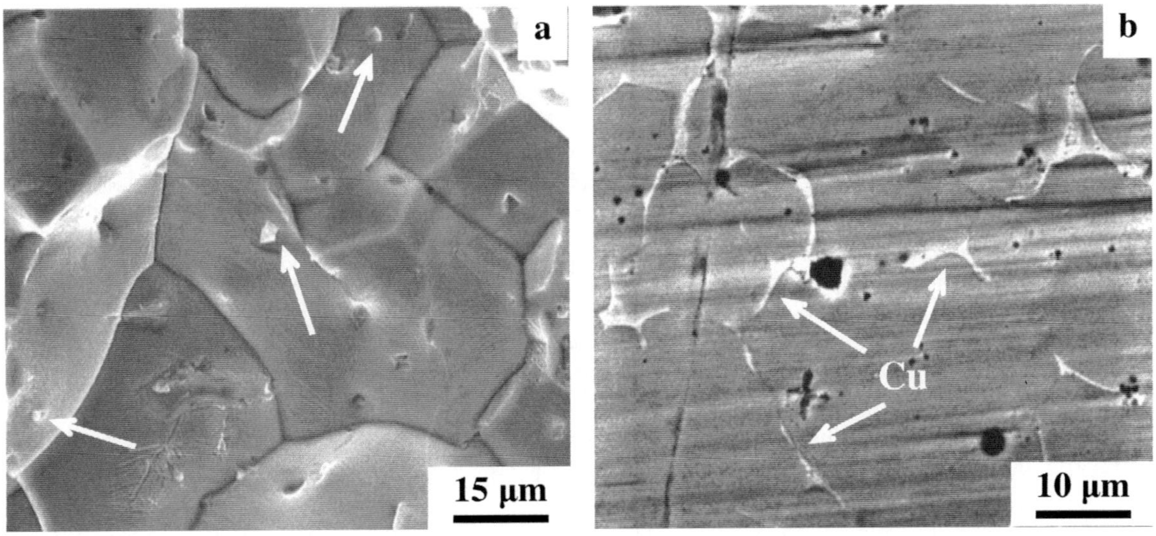

Fig. 3. Electron microscopic images of crystallites with Ti-Al-system based particles (a) and copper-enriched regions (b)

Thus, the surfacing formed is characterized by gradients of the grain structure, element composition and strength properties. The viscous nature of fracture, which is characteristic for plastic state, determines the high adhesion strength of the "surfacing-substrate" interface near the substrate. A gradual change in the elemental composition in the bulk of the material determines a smooth increase in microhardness from the substrate to the surface of surfacing, characterized by maximum values.

It is important to note that heat-resistant high-alloy steels (type EI680), which are close in elemental composition, are characterized by microhardness of 1.8 GPa [22], which is almost 4 times lower than the microhardness of the resulting surfacing.

Conclusions

By the example of Fe-Cr-Mo-Ti-Ni-Cu-Al system, the fundamental possibility of obtaining multicomponent metal surfacing on uncooled steel substrate using a pulsed regime of arc burning is demonstrated.

The resulting gradient surfacing, on the one hand, provides a highly reliable connection with a less strong substrate, on the other, allows the formation of high-strength states in the near-surface layers subjected to the strongest external effects in practical applications.

The work was carried out using technological equipment of NR TPU and research equipment of the Tomsk Materials Science Collective Use Center of NR TSU.

References

[1] Ivchenko M.V., Pushin V.G., Uksusnikov A.N., et al. Specific features of cast high-entropy AlCrFeCoNiCu alloys produced by ultrarapid quenching from the melt // The physics of metals and metallography. 2013. V. 114. No 6. P. 503–513.

[2] Kourov N.I., Pushin V.G., Korolev A.V., et al. Structure and physical properties of the high-entropy AlCrFeCoNiCu alloy rapidly quenched from the melt // Physics of the solid state. 2015. V. 57. No 8. P. 1616-1626.

[3] Firstov S.A., Gorban V.F., Krapivka N.A. Mechanical properties of cast multicomponent alloys at high temperatures (in Russian) // Current problems of physical materials science. 2009. No. 18. P. 140-147.

[4] Firstov S.A., Gorban V.F., Krapivka N.A. New class of materials – high-entropy alloys and coatings (in Russian) // Tambov University Reports. Series Natural and Technical Sciences. 2013. V. 18. No. 4. P. 1938-1940.
[5] Naplavochnyj material s `effektom vtorichnogo tverdenija dlja povyshenija stojkosti metalloobrabatyvajuschego instrumenta / E.N. Eremin, A.S. Losev, A.E. Eremin, G.P. Rumjantsev // Porodorazrushajuschij i metalloobrabatyvajuschij instrument - tehnika i tehnologija ego izgotovlenija i primenenija: Sb. nauch. tr. K.: INM im. V.M. Bakulja NAN Ukraïni, 2011. Vip. 14. P. 453 - 455.
[6] Larionov V.P., Sleptsov O.I., Mihajlov V.E. Osobennosti tehnologii svarki nizkolegirovannyh stalej povyshennoj i vysokoj prochnosti pri nizkih temperaturah // Dokl. na XIX komissii stran-chlenov S`EV po probleme "Svarki". PNR, 1986,- 44p.
[7] 4. Sidoruk V.S., Dudko D.A. Kriterii, harakterizujuschie teplofizicheskuju obstanovku pri svarke modulirovannym tokom. // Svarochnoe proizvodstvo. 1984. No 6. P.42 - 43.
[8] Dudko D.A., Sidoruk V.S., Ivaschenko G.A., But V.S., Zatserkovnyj S.A., Parhomenko I.Ju. Struktura i tverdost' metalla ZTV stali 45 pri dugovoj svarke modulirovannym tokom. // Avtomaticheskaja svarka. 1990. No 3 P. 10 - 12.
[9] Sidoruk V.S., Dudko D.A., Zatserkovnyj S.A., Taraborkin L.A. Vlijanie parametrov rezhima ruchnoj dugovoj svarki modulirovannym tokom na pervichnuju strukturu metalla shva. // Svarochnoe proizvodstvo. No 5. 1988. P.4 - 6.
[10] Dudko D.A., Zatserkovnyj S.A., Sidoruk V.S., Taraborkin L.A., Mahlin N.M. Zavisimost' himicheskogo sostava metalla shva ot parametrov rezhima dugovoj svarki modulirovannym tokom. // Avtomaticheskaja svarka. 1989. No 2. P. 27 - 29.
[11] S.O. Gordin, A.N. Smirnov, V.L. Knjaz'kov. Sostav `elektrodnogo pokrytija dlja iznosostojkoj naplavki. // Vestnik Kuzbasskogo gosudarstvennogo tehnicheskogo universiteta. 2015. 2. P. 106–108.
[12] Dementsev K.I. Povyshenie `effektivnosti raboty svarochnyh preobrazovatelej invertornogo tipa za schet moduljatsii svarochnogo toka: diss. …kand. tehn. nauk. Tomsk, 2010. 172 p.
[13] Zaruba I.I., Lebedev V.K., Shejko P.P., V.V. Basanskij. Svarka modulirovannym tokom // Avtomaticheskaja svarka. 1968. No 11. P. 35-40.
[14] Novozhilov N.M., Suslov V.N. *Svarka plavyashimsya elektrodom v uglekislom gaze* [Consumable electrode CO_2 welding] Moscow, Mashgiz,1958. 194 p
[15] Chernishov G.G., Spitsin V.V. Dinamicheskoe vozdeistvie svarochnoy dugi, goryash'ey v CO_2 na svarochnuyu vannu [Dynamic impact of CO_2 welding arc on welding pool]. Svarochnoe proizvodstvo, 1971, No.4, pp. 8-9.
[16] Knyaz'kov A.F., Petrikov A.V. Metodika provedeniya issledovaniya dvizheniya svarochnoi vanni [The methodic of exploration of welding pool's movements] Trudy of XIV nauchnoy konferencyi, posvyashennoy 300-letiu inzhenernogo obrazovaniya v Rossii. Urga, 2001. 207 p.
[17] Ioffe U.E., Kvasov F.V. Novye vysokotekhnologichnye sistemy poluavtomaticheskoy svarki kompanii "Linkol'n Elektrik" [The new high technologic sustems of semi-automatic welding by company "Lincoln Electric"]. Svarochnoe proizvodstvo, 1997, No.4, pp. 40-43.
[18] Getskin O.B., Poloskov S.I., Erofeev V.A., Vit'ko O.P. Phiziko-matematicheskaya model' sistemy "istochnik pitaniya-duga" dlya svarki plavyash'imsya elektrodom v zash'itnyh gazah [Physical and mathematical model of system "power source - arc" for consumable electrode shielded welding]. Tyazheloemashinostroenie, 2008, No.6, pp. 18-20.
[19] Sh'ekin V.A., D'urgerov N.G., Sagirov H.N., Nebilitsin L.E. *Prinuditel'niy perenos metalla pri svarke modulirovannim tokom v uglekislom gaze* [Forced metal's transfer during module current CO_2 welding]. Svarochnoe proizvodstvo, 1973, No.3, pp.23
[20] Zaruba I.I., Bargamen V.P., Andreev V.V., Sidorenko M.N. Vliyanie metoda ogranicheniya toka korotkogo zamykaniya na formirovanie vertikal'nyh i potolochnyh shvov pri svarke v uglekislom gaze [The influence of current limitation of short circuit's method on formation of vertical and overheld welds during CO_2 welding]. Avtomaticheskaya svarka, 19723, No.4, pp. 64-67.

[21] Ioffe U.E., Kvasov F.V. Novye vysokotekhnologichnye sistemy poluavtomaticheskoy svarki kompanii "Linkol'n Elektrik" [The new high technologic sustems of semi-automatic welding by company "Lincoln Electric"]. Svarochnoe proizvodstvo, 1997, No.4, pp. 40-43.
[22] GOST 5949-75 (Russian). Profiled steel and calibrated corrosion-resistant, heat-resistant. Technical conditions. Enter. 1975-09-16. – High-quality steel: Coll. GOST. – Moscow: IPK Publishing House of Standards, 2002.

IOP Conf. Series: Materials Science and Engineering **253** (2017) 012032 doi:10.1088/1757-899X/253/1/012032

Estimating the Parameters of Deformation Action by Ultrasonic Surface Hardening of Metals

Kharis M Rakhimyanov [a], Konstantin Kh Rakhimyanov [b], Andrey Kh Rakhimyanov [c]

NovosibirskStateTechnicalUniversity, 20 Prospekt K. Marksa, Novosibirsk,630073, Russia

E-mail: [a] x.raximyanov@corp.nstu.ru, [b] raximyanov@corp.nstu.ru, [c] a.raximyanov@corp.nstu.ru

Abstract. Developing the effective technologies of detail machining greatly depends on understanding the processes laid down in their basis. The technological methods based on electro-physical processes are considered to be attractive. These are the methods of surface plastic deforming which use the energy of ultrasonic oscillations. The peculiarities of these methods are characterized by high intensity and impulse character of the ultrasonic action. The paper presents the results of mathematical modeling of deformation processes under the impact of the ultrasonic tool on the surface layer of metals and alloys. The theoretical approach to studying the process of ultrasonic deforming allowed us to determine the mode parameters of impact and their quantitative correlations with the main characteristics of the deformation process.

Introduction

Ultrasonic plastic deforming as a separate technological method of the hardening and finishing operations is a modification of surface hardening, i.e. diamond smoothening, which is widely used in practice. The presence of ultrasonic oscillations during diamond smoothening leads to intensifying the plastic deformation of the surface layer of the material. It gives the possibility to provide machining at smaller values of static load, which considerably increases the technological perspectives of using such a method [1-7]. To extend further the sphere of using ultrasonic plastic deforming of different materials, from non-ferrous metals and alloys to hardened steels and details with coatings, it is required to determine the range of the mode parameters of the process mentioned. It will allow us to control both hardening and finishing effects in machining.

Theory

The process of ultrasonic plastic deforming should be considered in the following way: the movement of the deformer as a component of the oscillation system after its contact with the machined surface is the excitation of free oscillations [8]. Oscillations are of a harmonic sinusoidal character, which is confirmed by the results of high-speed photography of the process [9]. The movement of the deformer after its contact with the machined surface can be described as follows:

$$\frac{\partial A(t)}{\partial t} = V_{K_0}(x,t) + \frac{\partial A_0(x,t)}{\partial t} + \frac{F_D(t)}{\rho c S}, \qquad (1)$$

where $A(t) = A(0,t)$ – the deformer displacement;

$V_{K_0} = \frac{\partial A(x,t_0)}{\partial t}$ – the distribution of the oscillation speed at time t_0 (the deformer contact with the machined surface);

$A_0(x,t) = A(x,t_0)$ – the function of displacement along the wave guide length at time t_0;

$F_D(t)$ – the dynamic force, the function of which is determined by deforming conditions, namely, by the character of the depth change of the deformer penetration $(h(t))$ in the machined material. The penetration depth is the difference between the instant deformer displacement $A(t)$ and the initial value $A_0= A(0,t_0)$;

ρ – the density of the wave guide material;

c – the distribution speed of longitudinal sound waves in the wave guide material;

S – the cross section area of the wave guide.

The correlation between the dynamic force ($F_D(t)$) and the deformer movement ($h(t)$) can be presented as:

$$F_D(t) = \pi \cdot k \cdot HB \cdot d_c \cdot h(t), \tag{2}$$

where HB – the hardness of the machined material;

$k = 1.5 – 2.25$ – the acceleration capacity coefficient;

d_c – the diameter of the deformer sphere.

The ultrasonic oscillation system during deforming is tightened to the machined surface by the static force (F_{ST}). The correlation between static and dynamic force can be presented in the following way:

$$F_{ST} \cdot T = \int_{t_0}^{t_2} F_D(t) \cdot dt, \tag{3}$$

where T – the period of ultrasonic oscillations;

t_2 – the time corresponding to the completion of the deformer contact with the machined surface.

By solving the wave equation (1) of the deformer movement, we obtained the main dependencies that characterize the deformation process:

$$A(t) = A_{max}[sin(\omega t + \varphi_1) - e^{-q\omega t} sin\varphi_1 + sin\varphi_0]; \tag{4}$$

$$h(t) = A_{max}[sin(\omega t + \varphi_1) - e^{-q\omega t} sin\varphi_1]; \tag{5}$$

$$F_D(t) = \pi \cdot k \cdot HB \cdot d_c \cdot A_{max}[sin(\omega t + \varphi_1) - e^{-q\omega t} sin\varphi_1], \tag{6}$$

where A_{max} – the amplitude of ultrasonic oscillations;

ω – the circular frequency of ultrasonic oscillations;

$q = \pi \cdot HB \cdot d_c / \omega \cdot \rho \cdot c \cdot S;$

$\varphi_1 = \varphi_0 + arctg\ q;$

φ_0 – the trajectory angle corresponding to time t_0.

The maximum depth of the deformer penetration (h_{max}) in the machined material and the peak value of the dynamic force correspond to time t_1, which value is determined by solving the equation:

$$\cos(\omega t_1 + \varphi_1) + q \cdot e^{-q\omega t} sin\varphi_1 = 0. \tag{7}$$

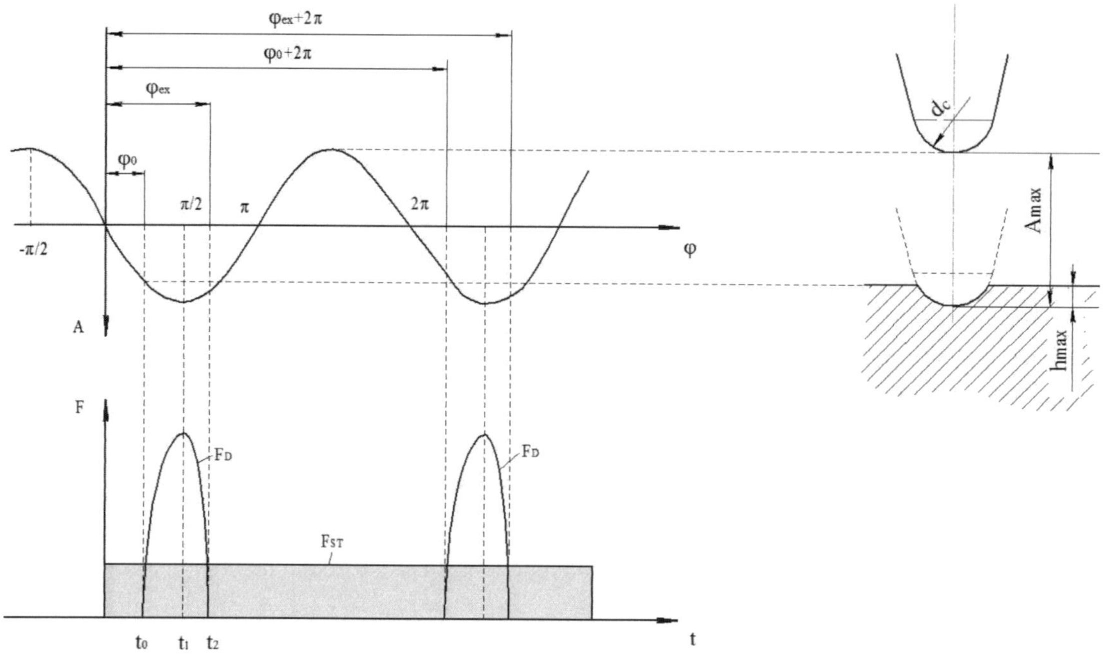

Fig.1. Scheme of locating the stroke impulses during ultrasonic oscillations

Results and Discussion

The theoretical solutions can be presented graphically by the following scheme.

The shape of the stroke impulse (Fig. 1) shows the change in the dynamic force during contacting (t_0-t_2) the deformer with the machined surface. The change in the dynamic force (F_D) during the penetration phase (t_0-t_1) corresponding to the material plastic deforming is described by the dependency (6). The regularities of changes in the dynamic force during the exit phase of the deformer from the contact with the material (t_1-t_2) which corresponds to the elastic relaxation of the elastic deformation of the imprint are not considered in the present investigations but are described in detail in paper [8]. The time of starting the contact (t_0) corresponds to the angle (φ_0), the time of the maximum penetration of the deformer (t_1) – to the angle (π/2), the time of the deformer exit from the contact (t_2) – to the angle (φ_{ex}) in the period of ultrasonic oscillations. The magnitude of the angle (φ_0) is determined by the value of the static force (F_{ST}) and can be changed from (-π/2) to (+π/2). In this case the displacement of the angle (φ_0) to minimum values takes place in increasing the static force (F_{ST}). When the angle (φ_0) reaches the minimum value (-π/2), the further increase in the static force is not reasonable because the character of deforming the surface ceases to be discrete, which is typical of and necessary for ultrasonic machining. The start of contacting at the trajectory angle of oscillations equal to (-π/2) excludes the deformer exit over the surface at the succeeding stroke. The calculated value of the maximum static force depends on conditions and modes of machining (the material hardness, frequency and amplitude of oscillations, deformer diameter). The displacement of the angle (φ_0) to the maximum values takes place in decreasing the static force and reaches the maximum value (+π/2) at $F_{ST} = 0$. It means that the physical contact between the deformer oscillating with the ultrasonic frequency and the machined surface disappears.

Figures 2 and 3 show the dependency of angle changes φ_0 and φ_{ex} on the oscillation amplitude A of the deformer at different values of the static force. According to the results of the mathematical model, we can claim that in certain combinations of the deformer oscillation amplitude and the static force it is possible to provide ultrasonic machining at the minimum value of the angle φ_0=(-π/2). The maximum value of the angle φ_{ex} decreases with increasing the amplitude and reducing the static force.

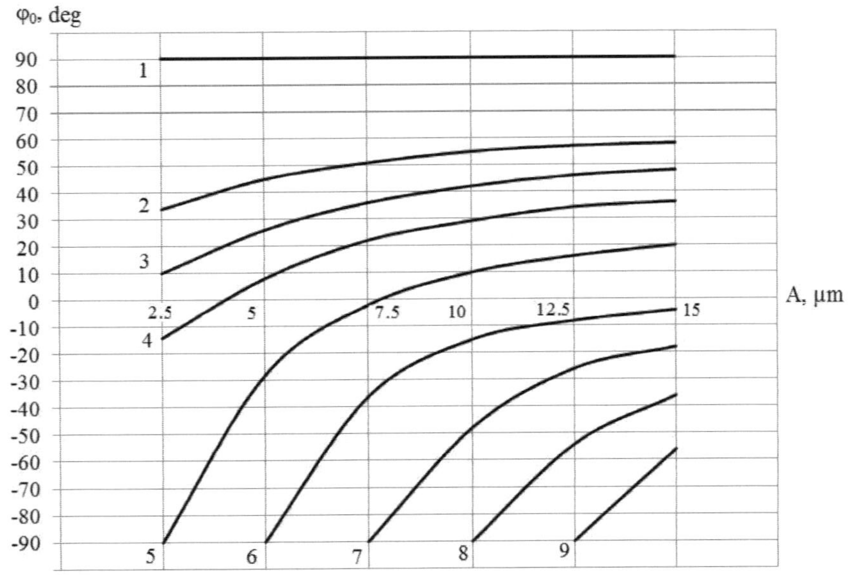

Fig. 2. Influence of the oscillation amplitude and the static force on changing the angle of starting the deformer contact (φ_0) with the machined material(HB=2070 MPa, f=20 kHz, d_c=8 mm): 1 – F_{ST}=0 N; 2 – F_{ST}=10 N; 3 – F_{ST}=25 N; 4 – F_{ST}=50 N; 5 – F_{ST}=105 N; 6 – F_{ST}=210 N; 7 – F_{ST}=310 N; 8 – F_{ST}=415 N; 9 – F_{ST}=510 N.

The phase of contacting between the deformer and the machined surface during oscillations of the ultrasonic tool will be determined by the difference in angles (φ_{ex}-φ_0).

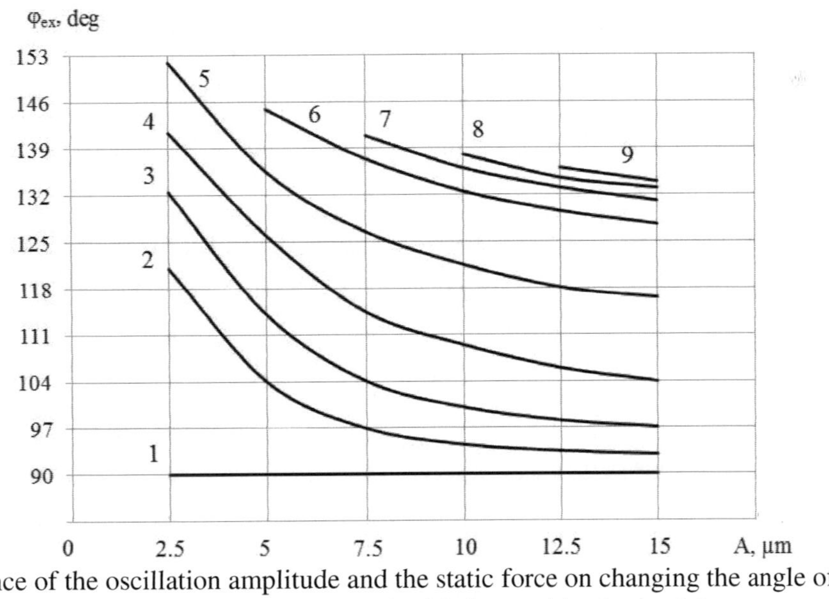

Fig. 3. Influence of the oscillation amplitude and the static force on changing the angle of completing the deformer contact (φ_{ex}) with the machined material (HB=2070 MPa, f=20 kHz, d_c=8 mm): 1 – F_{ST}=0 N; 2 – F_{ST}=10 N; 3 – F_{ST}=25 N; 4 – F_{ST}=50 N; 5 – F_{ST}=105 N; 6 – F_{ST}=210 N; 7 – F_{ST}=310 N; 8 – F_{ST}=415 N; 9 – F_{ST}=510 N.

IOP Conf. Series: Materials Science and Engineering **253** (2017) 012032 doi:10.1088/1757-899X/253/1/012032

The dependence of the contacting time (τ_c) on the oscillation amplitude at different values of the static force is shown in Figure 4.

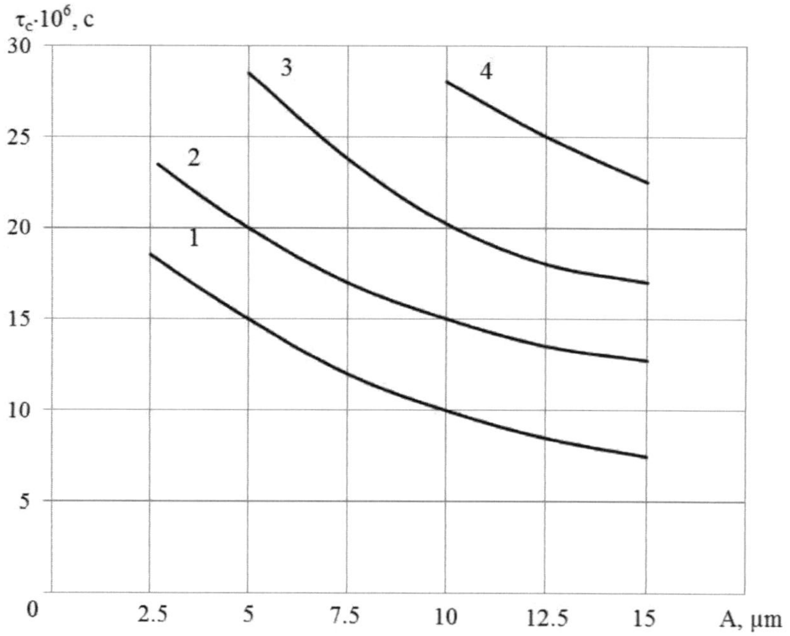

Fig. 4. Dependence of the time of the deformer contact with the machined surface on the ultrasonic oscillation amplitude at different values of the static force (*HB*=2070 MPa, *f*=20 kHz, d_c=8 mm): 1 – F_{ST}=50 N; 2 – F_{ST}=100 N; 3 – F_{ST}=210 N; 4 – F_{ST}=410 N.

The transformation of the static force (F_{ST}) constantly acting during machining into force impulses (F_D) of a certain duration and the amplitude value (Fig. 1) provides intensive deforming at moderate values of the static force (F_{ST}<600 N) through achieving considerable values (F_D>2000 N) by the dynamic force. The correlation between F_{ST} and F_D in the investigated amplitude range is shown in Figure 5. Considering the correlation of the limited values F_{ST} and F_D, we can note their insignificant changes (from 3.49 for A = 5 μm to 3.58 for A = 15μm).

Figure 5 presents the maximum values of the dynamic force corresponding to completing the penetration phase (t_1) of the deformer. Solving the equation of the deformer movement allows us to calculate the value of the dynamic force at any stage both during the penetration phase ($t_0 - t_1$) and the deformer exit ($t_1 - t_2$).

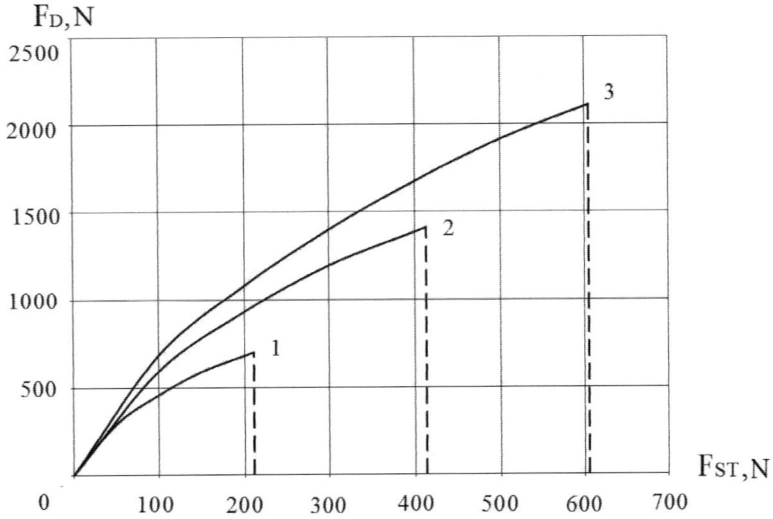

Fig. 5. Interrelation between the dynamic and static force at a different value of the oscillation amplitude (*HB*=2070 MPa, *f*=20 kHz, d_c=8 mm):
1 – *A* = 5 μm; 2 – *A* = 10 μm; 3 – *A* = 15 μm

Conclusions

By solving the equation of the deformer movement, oscillating with ultrasonic frequency during its contacting with the hardened surface under certain conditions of machining (the frequency of ultrasonic oscillations f = 20 kHz, the hardness of the machined material HB = 2070 MPa, the diameter of the deformer sphere d_c = 8 mm), we could obtain the quantitative correlation between the static force and the oscillation amplitude. It allowed us to determine the area of the mode parameters for implementing the process as discrete deforming. The quantitative estimation of such process parameters as angles of starting (φ_0) and completion (φ_{ex}) of contacting allows us to find the location of the stroke impulses during ultrasonic oscillations, taking into consideration the time of contact (τ_c), the penetration and exit of the deformer.

The correlation between the maximum value of the dynamic and static force in the investigated mode parameters is found. The considerable excess (by 3.5 times) of the dynamic force over the static force confirms the efficiency of using ultrasonic deforming of thin-walled details, constructions of small hardness and details with coatings.

References

[1] K.M. Rakhimyanov, I.S. Semenova, Surface state control by ultrasonic plastic deformation at the final machining stage, Materials and Manufacturing Processes. 31 (2016) 764-769.
[2] V.P. Gileta, Formation of microgeometry of a surface at ultrasonic processing, Proceedings - KORUS 2000: 4th Korea-Russia International Symposium on Science and Technology (27 June - 1 July 2000), Vol. 3, № 866075 (2000) 150-154.
[3] Kh.M. Rakhimyanov, G.A. Iskhakova, A.Kh. Rakhimyanov, The role of ultrasonic plastic deformation in spark alloying, Fizika I Khimia Obrabotki Materialov. 1 (1996) 68-72.
[4] V.P. Gileta, G.A. Iskhakova, Investigation into regularities of the surface microgeometry formation during diamond ultrasonic strengthening finishing, Sverkhtverdye Materialy. 1 (1992) 45-50.
[5] V.P. Gileta, G.A. Iskhakova, Simulation of regular relief formation in ultrasonic diamond burnishing, Soviet journal of superhard materials. 10 (1988) 64-69.
[6] G.A. Iskhakova, V.P. Gileta, Formatiom of the component surface layer in ultrasonic diamond burnishing, Soviet journal of superhard materials. 9 (1987) 66-71.

[7] A.I. Beznedelnyy, V.B. Asanov, V.P. Gileta, Technological inheritance influence on surface layer forming in condition of ultrasonic plastic deformation of hardened steels, Obrabotka metallov (tekhnologiya, oborudovanie, instrumenty) (in Russian). 4 (2012) 19-22.
[8] O.V. Abramov, V.I. Dobatkin, V.F. Kazantsev, Impact of powerful ultrasonic on the interphase surface of metals, Moscow, 1986.
[9] O.P. Solonenko, A.P. Alkhimov, V.V. Marusin, A.M. Orishich, Kh.M. Rakhimyanov, R.A. Salimov, V.G. Shchukin, V.F. Kosarev, High-Energy Processes of Materials Treatment, Nauka, Novosibirsk, 2000.

Functional Coating of Biocompatible Powder Composites Obtained by the Method of Detonation-Gas Spraying

Anastasiya A Popova [a], **Vladimir I Yakovlev** [b]

Altai State Technical University after I.I. Polzunov, Lenin av., 46, Barnaul, 656038, Russia

E-mail: [a] anast_82@mail.ru, [b] anicpt@rambler.ru

Abstract. In this paper we investigate the structure of composite biocompatible coatings with the composition: calcium hydroxyapatite - titanium nickelide, obtained by the method of detonation-gas spraying.

Introduction

Skeletal system of a living organism is formed and maintained as a result of complex biochemical reactions. The loss or destruction of bone leads to the use of implants for replacement. Implants consist generally of metallic materials based on alloys, stainless steel or titanium.

The use of implants, when physico-chemical and mechanical properties of metallic substrates and bone differ greatly can lead to an active rejection of the implant. To reduce the negative influence of these factors it is necessary to create a transition zone between the bone and the implant, which, along with the strong bond with the implant surface and the bone tissue should have acceptable for organism macro-and microstructure [1-2]. Such a zone can be obtained by coating with a specific porosity and it may have a composite structure. It is assumed that the elemental and phase composition of the coating should match the composition of natural bone and/or promote formation of bone tissue on its surface. Hydroxyapatite (HA) - $Ca_{10}(PO_4)_6(OH)_2$ is usually applied as a starting material for bioactive coatings. When you create a biocompatible coating, special attention is paid to the formation of a topography (roughness) on the implant surface.

There are various methods of forming the calcium-phosphate coating on metal surfaces: plasma spraying, slip method, the method of micro-arc oxidation, magnetron sputtering, the method of detonation-gas spraying (DGS), etc. [3-4]. The development of new coatings requires their full certification (surface morphology, roughness, phase and elemental composition, etc.). The method of detonation-gas spraying has good prospects for the production of biocompatible coatings due to the identity of the phase composition of sprayed material and the coating [5-7].

The task of creating a solid biocompatible coatings can be solved by the introduction of the powder mixture of hydroxyapatite calcium hyperelastic material with high biochemical and biomechanical compatibility [8]. In this work, as such material titanium nickelide was used, which positively recommended itself in medical implantology. High porosity coatings of titanium nickelide (80-90%) promotes good ingrowth of hard and soft tissues of the body. In these composites, one component (titanium nickelide) has superelasticity, and the other HA — retains the properties of bioceramics. This will allow us to obtain a class of materials possessing high mechanical properties [9].

IOP Conf. Series: Materials Science and Engineering **253** (2017) 012033 doi:10.1088/1757-899X/253/1/012033

In a research experiment composite coatings were obtained with the composition: hydroxyapatite-calcium – titanium nickelide by the method of detonation-gas spraying.

The experimental procedure

In experimental studies powdered mixture of composition HA +TiNi was used as a material. The average particle size of the biological powder of calcium hydroxyapatite was 150-300 μm, the average particle size of powder of titanium nickelide brand PNi55Ti45 – 50-100 μm.

To obtain a laminated composite consisting of HA+ TiNi the method of mechanical alloying treatment (MA) was used as an effective method for producing composite materials. Preliminary mechanical activation (MA) of the initial mixtures of HA + TiNi in the ratio: 50HA +50 wt., % (25 vol.,%) TiNi and 70HA +30 wt., % (15 vol., %) TiNi was carried out in planetary ball mill AGO–2 with water cooling. Time of mechanical activation is selected on the basis of analysis of literary sources and amounted to 3, 7, 15 and 30 min [8].

As a substrate for deposition a titanium plate (W-1.0) of size 20×20×3 mm was used. The surface of titanium plates was pre-treated with air-abrasive mixture at the blasting installation. Processing quality was determined visually according to the degree of dullness of the surface [9].

For application of composite coatings on a titanium base the installation of detonation-gas spraying of powder materials "Katun M" was used [10-11]. The thickness of the sprayed layer was measured after 50 spraying cycles for each sample. The deposition ended when reaching the coating of 150 microns.

The study of surface morphology was carried out on an optical microscope Carl Zeiss AxioObserver Z1m. Spectral analysis was performed on scanning electron microscope Carl Zeiss EVO50 equipped with microprobe EDS X-Act (Oxford Instruments) with a Si-drift detector. The study of the phase composition and structural parameters of mechanically activated composites was carried out using x-ray diffraction x-ray on General purpose DRONE-6 diffractometer.

Experimental results and discussion

At the stage of activation processing of the powder mixture to 3 min MA the mixing and dispersion of components take place. After 3 minutes of mechanical activation there is a significant broadening and decrease of intensity of diffraction peaks of reflections of hydroxyapatite and intermetallic compound, indicating the increase of non-equilibrium defects in the product of grinding and reducing the size of the crystallites, with the transition to the nanocrystalline state. With increasing MA time up to 15 min the diffraction reflections of the intermetallic compound virtually disappear, a transition of TiNi into x-ray amorphous state can be observed. Additional compounds in the process of mechanical activation are not formed (Fig.1 b, d).

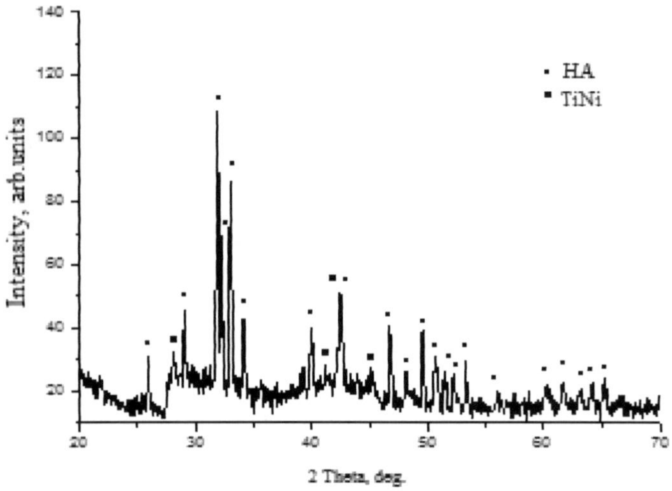

a)

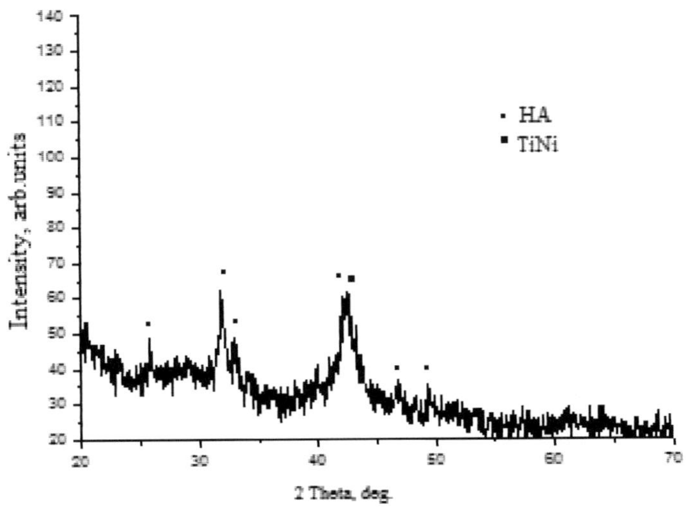
Intensity, arb.units
2 Theta, deg.
HA
TiNi
140
130
120
110
100
90
80
70
60
50
40
30
20
20
30
40
50
60
70

b)

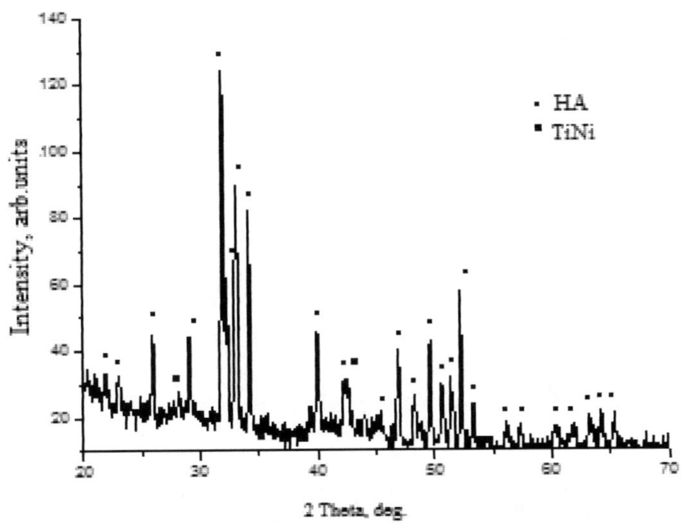
Intensity, arb.units
2 Theta, deg.
HA
TiNi
140
120
100
80
60
40
20
20
30
40
50
60
70

c)

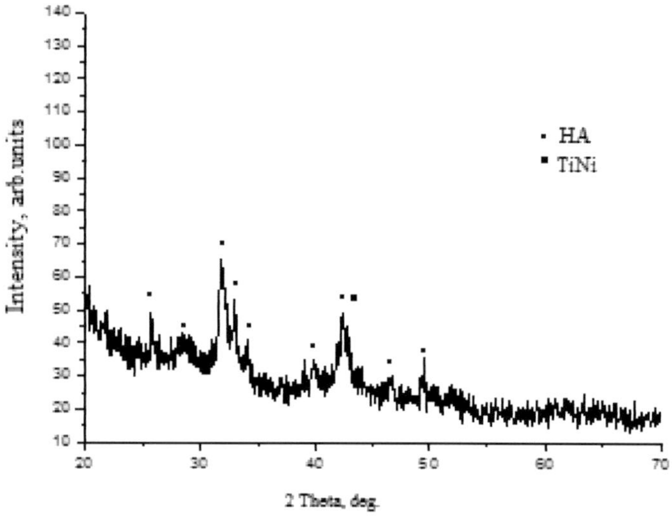

d)

Figure 1. Diffractograms of the powder mixtures: a - initial powder mixture 50HA +50TiNi (mass, %); b - 50HA +50TiNi (wt, %) min at 15 min. MA; c - the original powder mixture 70HA +30 TiNi(mass, %); d - 70HA +30 TiNi(wt, %) at 15 min MA.

Fig. 2 shows the SEM images of composites 50HA +50TiNi (wt.,%) at 15 and 30 min of MA. Fig. 2 shows the SEM images of composites 50HA +50TiNi (wt.,%) at 15 and 30 min of MA. At 15 min MA, particles of composite HA and TiNi conglomerate in integrated composition. The bulk of the particles fall within the range 50-100 μm (Fig. 3a), the average particle size of composite is 112.5 μm. With longer MA time (30 min.) there is a significant consolidation of conglomerates, the average particle size is 196.4 μm (Fig. 3b). Composites of such large size groups (over 200 μm) is not recommended for use in the process of detonation-gas spraying according to the technological requirements. For the mixture with the composition 70HA +30TiNi (wt.,%) the agglomeration of particles in conglomerates with prolonged (more than 15 min) time MA take place as well. Therefore, the most efficient activation time of the mixtures 50HA +50TiNi (wt.,%) and 70HA +30TiNi (wt.,%) before detonation-gas spraying should be considered as 15 minutes.

a) b)

Figure 2. SEM images of mechanically activated mixture of HA/TiNi:
a – 50HA +50TiNi (wt.,%), time of activation – 15 min;
b – 50HA +50TiNi (wt.,%), time of activation – 30 min

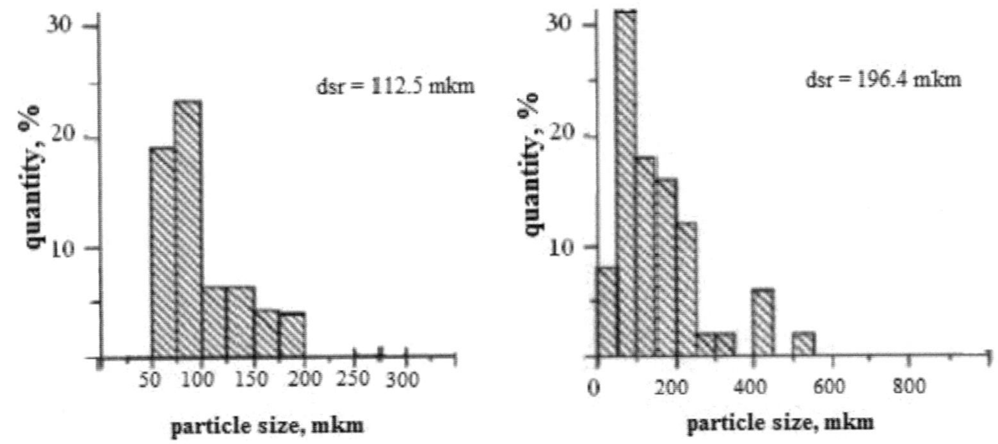

a) b)

Figure 3. Histograms of size distribution of particles of composite:
a – 50HA +50TiNi (wt.,%), time of activation – 15 min;
b – 50HA +50TiNi (wt.,%), time of activation – 30 min.

Fig. 4 shows the SEM image of the coating based on the composite caused by the detonation-gas method. It is seen that the coatings have a porous structure and a very pronounced relief, the nature of which does not change at different ratio of components in composite. The coatings consist of particles of composite HA+TiNi (wt.,%), in some cases these particles are fused under the influence of the stream flow detonation. The size of pores is from 2 to 16 microns.

Figure 4. SEM image of the coating based on the composite HA+TiNi caused by the detonation-gas method: a, c – 50HA +50TiNi (mass., %);
b, d – 70HA +30TiNi(mass., %)

IOP Conf. Series: Materials Science and Engineering **253** (2017) 012033 doi:10.1088/1757-899X/253/1/012033

Elemental composition analysis showed that the coatings consist of calcium, phosphorus, Nickel, titanium, and oxygen with a small admixture of carbon, i.e. the composition of the coating corresponds to the powder composite.

The growth of bone tissue at the implant osseointegration is significantly affected by phase composition of the coatings. X-ray analysis of coating of calcium hydroxyapatite showed that the coatings revealed no other lines besides the main phase of HA, which indicates its homogeneity. The study of phase composition of coatings of composites showed that composite coatings applied by the detonation-gas spraying method are in the x-ray amorphous state.

Conclusions

1. The amorphous composite calcium-phosphate coatings were obtained by the detonation-gas spraying method.
2. The coating contains only chemical elements of the initial HA powder and TiNi.
3. As a result of experimental studies of the morphology of the coating, the size of pores from 2 to 16 microns was determined.

References

[1] B. I. Beletskiy, V. I. Shumskiy, A. A. Nikitin, E. B. Vlasova, Biocomposite calcium-phosphate materials in bone-plastic surgery, Steklo i keramika, [Glass and ceramics], 9, (2000) 35-37. (in Russian)
[2] V. N. Lyasnikov Properties of plasma coated powder coatings, Perspektivniye materialy [Advanced materials], 4(1995) 61-67. (in Russian)
[3] J. Delecrin, N. Passuti, J. Poyer, G. Daculsi, Y. Maugars, Biphasic calcium phosphate as a bone graft substitute for spine fusion: stiffness evaluation, 4th World Biomaterials Congress, Berlin Abstract,(1992), 644-645.
[4] V. F. Pichugin, Poverkhnost. Rentgen., Sinkhrotron., Neitron. Issled. [Surface. X-ray, synchrotron and neutron studies], 11, (2007), 67–72 (in Russian)
[5] A. A. Popova, V. I. Yakovlev, E. V. Legostaeva, A. A. Sitnikov, Yu. P. Sharkeev Influence of the granulometric composition of the powder of calcium hydroxyapatite on the structure and phase composition of coatings deposited by the method of detonation-gas spraying, Izvestiya vuzov. Physics, 55, 11 (2012) 41-45. (in Russian)
[6] M. Kh. Shorshorov and Yu. A. Kharlamov, Physical-Chemical Fundamentals of Detonation Gas Spraying of Coatings, M. Kh. Shorshorov, ed., Nauka [Science], Moscow (1978). (in Russian)
[7] A. A. Sitnikov, V. I. Yakovlev, A. A. Popova Preparing an initial powder of calcium hydroxyapatite for detonation-gas spraying on a titanium base, Polzunovskii vestnik, [Polzunovskii Bulletin], 1/1(2012) 269-272. (in Russian)
[8] A. A. Popova, V. I. Yakovlev, A. A. Sitnikov, M. V. Loginova, A.V. Sobachkin Structural-stress state of mechanocomposite "hydroxyapatite-titanium nickelide" designed to create biocompatible coatings on medical implants, Fundamentalniye problem sovremennogo materialovedeniya, [Fundamental problems of modern materials science], 12,2 (2015) 179-183. (in Russian)
[9] V. I. Itin, O. G. Terekhova, T.E. Ulyanova, V. A. Kostikova, N.A. Shevchenko, D. V. Berdnikova, The influence of mechanoactivation on regularities of sintering of Nickel-titanium and composites "bioceramics-titanium nickelide", Pisma v JTP, [Technical physics Letters], 26,10 (2000) 73-79. (in Russian)
[10] L. I. Tushinskii, Methods of Investigation of Materials: Inorganic Coating Structure, Properties, and Deposition Processes, Mir [Peace], Moscow (2004). (in Russian)
[11] V. I. Yakovlev, Experimental-Diagnostic Complex for Physical Investigations on SHS Powder Materials in Detonation Spraying, Author's Abstract of Cand. Engineering Scie. Dissertation, Altai State Technical University, Barnaul, Russia (2003), 19. (in Russian)

IOP Conf. Series: Materials Science and Engineering **253** (2017) 012034 doi:10.1088/1757-899X/253/1/012034

Dependence of Manganese Content in the Weld Metal on the Velocity of Active Shielding Gas Flow

D A Chinakhov, E G Grigorieva, E I Mayorova, S A Solodsky, V F Torosjan

Yurga Institute of Technology, National Research Tomsk Polytechnic University, 652055, Leningradskaya st., 26, Yurga, Kemerovo region, Russia

E-mail: chinakhov@tpu.ru

Abstract. The paper reports findings of the study on influence of changing active shielding gas flows on consumable electrode welding and manganese content in the weld metal. A procedure to predict manganese content in the weld metal according to the velocity of a shielding gas flow, manganese content in electrode wire and base metal is proposed. A regression dependence of manganese content in the weld metal on the velocity of active shielding gas flow from the nozzle is determined.

Introduction

Gas-shielded arc welding follows the principle of pushing air aside from the weld zone by a shielding gas flow. To date, a number of gas-shielded welding techniques are available and used in manufacture [1]. Noble (argon, helium), active (nitrogen and CO_2) gases and their mixtures are applied as shielding gases. Properties of shielding gases are crucial for technological characteristics of arc and weld forms.

Jet shielding is the most wide-spread procedure of local protection in welding. Crystallization velocity of the weld metal depends on shieling gas consumption and a gap from the nozzle exit to the surface of material to be welded. When jet gas-shielded welding it is the fusion zone, which is protected. In local jet shielding the flow of gas is laminar but it turns into a turbulent one (with heavy whirls) as the velocity of shielding gas flow increases critically and a laminar flow fails. Here, up to 50% air is admixed into a shielding gas jet at a distance of several millimeters (6-10mm) [1].

The authors [2-8] studied the effect of changing shielding gas flows on consumable electrode welding and emphasized the improvement of weld zone protection, weld shaping, and welding stability owing to increased gas flow velocity.

In works [5, 9] it is described how the velocity of shielding gas supply influences transfer of electrode metal drops, and its value, furthering separation of a drop, is determined (it depends on dimensions of a drop and electrode metal). The researchers [3, 10] point at the improvement of welding stability, shaping and density of a weld when supplying of shielding gases (Ar, He) is alternated in consumable electrode welding.

In comparison with conventional shielding, two-jet gas shielding [11] provides a reliable protection of a weld pool, improves stability and frequency of drop transfer from the electrode into a weld pool, furthers refinement of weld metal structure and smooth transition from the weld metal to the base one, enhances mechanical characteristics of weld joints, and decreases chemical inhomogeneity of the weld metal due to a positive gas-dynamical effect on processes in the weld zone and better remixing of fused metal in the weld pool.

Methodology

As the velocity of active shielding gas flow from the nozzle increases, dynamic impact on a drop of electrode metal goes up in two-jet gas shielding, transfer of drops from the electrode into the weld pool gets stable and more frequent, and metallurgical processes on the surface of a drop become more intense. When welding constructional steels it is required to use a welding wire with manganese content exceeding that in the base metal. The reason is that manganese evaporates and gets oxidized intensively in welding [1, 12]. Therefore, welding wire Sv-08G2S (Tabl. 1) with a diameter of 1.2 mm is applied widely in Russian industry.

Table 1

Chemical composition of welding wire Sv-08G2S (GOST 2246-70)

Mass share of elements, %						
C	Mn	Si	S	P	Cr	Ni
0.05-0.11	1.8-2.1	0.7-0.95	≤ 0.025	≤ 0.03	≤ 0.2	≤ 0.25

In previous experiments [2, 11-13] it is revealed that in two-jet gas-shielded welding with a consumable electrode in CO_2 and its mixtures the process stabilizes, and Mn and Si content decreases slightly in the weld metal as compared with samples made with one-jet shielding. This difference in Mn and Si content when welding in similar conditions can be accounted only for different metallurgical processes taking place in a drop of fused electrode metal and on its surface. A lot of scientists emphasized in their works [1, 3, 9, 14 etc.] more favorable conditions for reaction of fused metal with gases and slags in the phase of a drop rather than in the weld pool. As stated in paper [1], a specific surface of fused electrode metal drops is 5-22 times bigger (depending on dimensions of drops) than a specific surface of a weld pool, and a specific velocity of their oxidation is approximately 39 times faster. Hence, it can be concluded that, basically, chemical composition of the weld metal influences composition of electrode metal drops.

Fused (liquid) metal in a drop moves constantly at a high speed. Manganese has the lowest boiling temperature and evaporation heat of all alloying elements in chemical composition of welding wire Sv-08G2S (Table 2). It is the reason for its intense evaporation and oxidation on the surface of a drop when welding, the temperature of which approximates to 3300 K [1, 14].

Table 2

Thermo-dynamical characteristics of elementary substances (in normal conditions) in chemical composition of the welding wire [15]

Characteristic	Substance							
	Mn	Si	Cr	Ni	Fe	Ti	Mo	C graphite
Boiling temperature, K	2235	2623	2945	3005	3134	3560	4885	5100
Evaporation heat, kJ/mole	221	383	342	378.6	340	422.6	590	-
Fusion temperature, K	1517	1688	2130	1726	1812	1933	2890	3820
Density, g/cm^3	7.21	2.33	7.19	8.9	7.87	4.54	10.22	2.25

Increasing consumption of CO_2 in welding, i.e. velocity of flow and number of particles reacting with the surface results in the heightened burn out of silicon and manganese [1, 12, 14].

In two-jet gas shielding the velocity of gas flow is 3.5 times higher than in one-jet shielding, causing, this way, augmentative frequency of drop transfer, drop diminishment and growth of the total active surface of drops carried from the electrode into a weld pool per unit of time. It results in removal of a lot of manganese from the surface of a drop. Therefore, electrode metal with less content of manganese is transferred into the weld pool. Manganese content in a drop decreases with rising CO_2

consumption in welding with any jet shielding. The amount of manganese transferred from the electrode into the weld might make it possible to predict chemical composition and characteristics of the weld metal.

Studying findings of research [1, 11-14], it was revealed that chemical composition of the weld metal, manganese and silicon content in particular, depends on several factors, e.g. temperature of a drop (welding current strength, voltage, technique of gas shielding), time of reaction between a drop and a shielding gas (welding current strength, consumption and gas velocity – the time reduces with their increase), drop dimensions (welding current strength, gas consumption and velocity – a drop gets smaller with their increase), welding wire and base metal composition.

Manganese and silicon content increases in the weld metal [1, 11-14] due to intensifying welding current, whereas it drops with augmentation of voltage. Decreasing manganese and silicon content in the weld metal is registered with rising velocity of shielding gas (CO_2) in open arc welding [1, 12, 14]. Manganese is one of basic elements, since it has a low deoxidizing power and can intensify influence of other elements, resulting, therefore, in change of weld metal properties.

In arc welding with consumable electrode in CO_2 thermal dissociation of carbon dioxide takes place at high temperatures according to an equation:

$$CO2 = CO + O \quad (1)$$

CO_2 dissociates in a high-temperature area of the weld zone [14], approximately 50% gas dissociates when reacting with the surface of a fused metal drop surface (above 3000 °C). Oxygen, which is a product of reaction, can oxidize components of the welding wire:

$$Si + 2O = SiO2\ , \quad (2)$$

$$Mn + O = MnO \quad (3)$$

$$C + O = CO \quad (4)$$

$$Fe + O = FeO \quad (5)$$

Results and Discussion

Experimental studies were conducted to determine an empirical dependence of manganese content in the weld metal on the velocity of shielding gas (CO_2) flow. Mechanized single-pass beading with welding wire Sv-08G2S of a diameter 1.2 mm was carried out in CO_2 with a stationary arc on a plate of steel 45 with two-jet gas shielding. Welding conditions: welding current I = 195...200 A, electrode wire extension L = 12 mm, arc voltage U = 26…27 V, velocity of welding V = 20 cm/min. shielding gas consumption Q = 10, 20, 30 l/min (Table 3). Equipment used: automatic welding unit MECOME WP1500 power source VS–300B and a supplying mechanism PDGO-528M. Outcomes of research are given in Table 3.

Table 3

Experimental results

Consumption, l/min	Gas flow velocity, m/s	Chemical composition of the weld metal, %		
		Fe	Si	Mn
10	1.96	97.55	0.28	1.04
20	3.95	97.73	0.27	0.97
30	5.82	97.91	0.21	0.85

Manganese content, % vs. its initial state in the wire is calculated according to a formula:

$$\Delta C_{(\%)} = \frac{m_{Mn}^{k}}{m_{Mn}} 100\% \quad (6)$$

where m_{Mn}^{k} – mass share of manganese in a drop after passing of a gas jet (Table 3, Mn), %,

m_{Mn} – initial mass share of manganese in a welding wire (Sv-08G2S – 1.8%), %.

Experimental values of manganese content, % in a drop vs. its initial state in the wire (Table 3) were approximated:

$$\Delta C_{Mn} = 62.3 - K*Vg \quad (7)$$

where K = 1.97 – coefficient, %*s/m; Vg – velocity of a shielding gas flow, m/s.

Taking into account theoretical and experimental studies on consumable electrode narrow-gap welding of constructional steels in CO_2, a procedure was developed to predict manganese content in the weld metal depending on the velocity of a shielding gas flow, and manganese content in the electrode wire and base metal.

Predicted manganese content Mn in the weld metal (mass share, %):

$$m_{Mn\_w} = \frac{m_{Mn} \cdot \Delta C_{Mn_w}}{100\%} \quad (8)$$

Manganese content in the weld metal (%) vs. its initial state in the wire and according to the velocity of a shielding gas flow from the nozzle and manganese content in the base metal:

$$\Delta C_{Mn\,w} = 62.3 + Mn_{bm}*10 - K2*Vg \quad (9)$$

where K2 = 1.55 – coefficient, %*s/m; Mn_{bm} – Mn content in the base metal (mass share, %), Vg – shielding gas flow velocity, m/s.

To compare predicted values obtained using this procedure with experimental data, results of experiments conducted before on steels 45, 30HGSA and GL-E36 [11, 12] were used.

Plates of the base metal were joined by mechanized multi-pass keyhole welding with Mn containing welding wire of a diameter 1.2mm in CO_2 with a stationary arc in one-jet and two-jet gas shielding. Welding conditions: welding current I = 195...200 A, electrode wire extension L = 12 mm, arc voltage U = 26…27 V, velocity of welding V = 19...21 cm/min. Shielding gas consumption Q = 10...30 l/min (Table 4).

Table 4

Predicted and experimental values of Mn content in the weld metal

Base material + welding wire	Gas flow velocity, m/s	Prediction Mn, %	Experiment Mn, %	Percentage error, %
30HGSA (Mn=0.96%) + Sv-08G2S (Mn=1.8%). One-jet	0.58	1.27	1.29	1.6
30HGSA (Mn=0.96%) + Sv-08G2S (Mn=1.8%). Two-jet	2.0	1.23	1.26	2.3
30HGSA (Mn=0.96%) + Sv-08G2S (Mn=1.8%). Two-jet	4.95	1.14	1.2	5.0
GL-E36 (Mn=1.25%) + Union K 52 (Mn=1.5%). One-jet	0.5	1.11	1.2	7.5
GL-E36 (Mn=1.25%) + Union K 52 (Mn=1.5%). Two-jet	2.1	1.07	1.05	1.9
Steel 45 (Mn=0.67%) + Sv-08G2S (Mn=1.8%). One-jet	1.45	1.2	1.21	0.8
Steel 45 (Mn=0.67%) + Sv-08G2S (Mn=1.8%). Two-jet	5.06	1.1	1.11	0.9

The results of the research show that the developed method for predicting the manganese content in the weld metal (keyhole welding), based on the escape velocity of the shielding gas and the manganese content in the electrode wire and the base metal, allows calculating the manganese content mainly with an error of no more than 10%.

Conclusion

Two-jet gas shielding intensifies the effect of an active shielding gas jet on the electrode metal drop and surface of the weld pool and facilitates stability of drop transfer. Changing gas-dynamic influence (gas consumption, gas flow velocity), transfer of electrode metal drops, chemical composition of the weld metal, thermal and other processes of consumable electrode welding can be controlled, and required characteristics of the weld metal can be formed.

The developed procedure for prediction of manganese content in the weld metal depending on the velocity of shielding gas flow, manganese content in the electrode wire and base metal allows calculation of manganese content in the weld metal with maximal error 10%.

The work is performed with financial support of the Russian Foundation for Basic Research. №16-38-00194 mol_a. The research is carried out at Tomsk Polytechnic University within the framework of Tomsk Polytechnic University Competitiveness Enhancement Program grant.

References

[1] Novozhilov M. N. The basic principles of the metallurgy of arc welding in the gases. Mashinostroyeniye. 1979. 231 p.
[2] Chinakhov D.A., Zuev A.V., Filimonenko A.G. Gas-dynamic Impact of a Shielding Gas Jet on the Drop Transfer When Welding with a Consumable Electrode. Advanced Materials Research Vol. 1040 (2014) pp. 850-853.
[3] Zhernosekov A.M., Sidorets V.N., Shevchook S.A. Combined pulsed effect of shielding gases and welding current in consumable electrode welding. Welding International, 2014. Vol. 28, No. 12, 962–965.
[4] Brunov O.G., Solodskii S.A, Zelenkovskii A.A. Conditions of arc ignition in welding in shielding gases. Welding International. 2012. V. 26. 9. P. 710-712
[5] Mvola B., Kah P. Effects of shielding gas control: welded joint properties in GMAW process optimization. International Journal of Advanced Manufacturing Technology.
[6] Fedorenko G.A., Ivanova I.V., Sinyakov K.A. Improvement of the technological process of welding in the shielding gases in the wind. Welding production. 2010. 1. P. 6–13.
[7] Gribovsky G., Kravchuk B., Leninvkin V.A. Influence of the double annular flow of the shielding gases upon the process of consumable electrode welding. Welding production. 1996. 4. P. 6–8.
[8] Ramsey G.M., Galloway A.M., Campbell S.W., McPherson N.A., Scanlon T.J. A computational fluid dynamic analysis of the effect of side draughts and nozzle diameter on shielding gas coverage during gas metal arc welding. (2012) Journal of Materials Processing Technology, 212 (8), pp. 1694-1699.
[9] Tarasov N.M. Electrode metal drop detachment by a short-term gas flow (In Russian) // Automation welding. 1986. 7. P. 10–13.
[10] Novikov O.M., Rad'ko E.P., Ivanov E.N., Ivanov N.S. Development of a new technology of shielding gases arc welding on the basis of gas flows pulsations and ionization potentials. Welder-professional. 2006. 6. P. 10–13, 16.
[11] Chinakhov D.A., Chinakhova E.D., Gotovschik Y.M., Grichin S.V. Influence of welding with two-jet gas shielding on the shaping of a welding joint. IOP Conf. Series: Materials Science and Engineering 125 (2016) 012013.
[12] Chinakhov D.A. Dependence of Silicon and Manganese Content in the Weld Metal on the Welding Current and Method of Gas Shielding. Applied Mechanics and Materials. Vol. 756 (2015) pp 92-96.

[13] Chinakhov D.A., Grigorieva E.G., Mayorova E.I. Study of gasdynamic effect upon the weld geometry when consumable electrode welding. IOP Conf. Series: Materials Science and Engineering 127 (2016) 012013.
[14] Zaruba I.I., Kasatkin B.S., Kakhovsky N.I., Potapyevsky A.G. Welding in carbon dioxide. Kiev: Gostekhizdat USSR, 1960. 224 p.
[15] Bigeev A.M. Steel metallurgy. Study guide, 2nd edition, Moscow, Metallurgy, 1988.